HEINEMANN **16 19** GEOGRAPHY

NEW EDITION — FOR EDEXCEL B

Changing Environments

Series Editor

Bob Digby

Authors

Susan Bermingham
Jane Ferretti
Linda King
David Kinninment
Chris Ryan
Celia Tidmarsh
Ruth Totterdell
Graham Yates

D1341464

Heinemann Educational Publishers
Halley Court, Jordan Hill, Oxford, OX2 8EJ
a division of Reed Educational & Professional
Publishing Ltd
Heinemann is a registered trademark of Reed
Educational & Professional Publishing Ltd

OXFORD MELBOURNE AUCKLAND
JOHANNESBURG BLANTYRE GABORONE
IBADAN PORTSMOUTH NH (USA) CHICAGO

First published 2000

ISBN 0 435 352466

04 03 02 01
10 9 8 7 6 5 4 3

Designed and illustrated by Hardlines, Charlbury,
Oxford OX7 3PS

Printed and bound in Spain by Edelvives.

Photographs

P. 6: Figure 1, Popperfoto; P. 7: Figure 3, Hulton Getty; P. 8: Figures 5 and 6, Andrew Johnstone; P. 10: Figure 1.3, Bob Digby; P. 12: Figure 1.7a/b, Bob Digby; P. 14: Figures 2.2, 2.3a/b, Bob Digby; P. 16, Figures 2.5, 2.6, Bob Digby; P. 17: Figure 2.9 a/b, Bob Digby; P. 18: Figure 2.11, Bob Digby; P. 19: Figures 2.13a and 2.14a, Bob Digby; P. 20: Figure 2.15a, Bob Digby; P.21: Figures 2.17, 2.18, 2.20, Bob Digby; P. 22: Figures 2.21 and 2.23, Bob Digby; P. 24: Figures 2.25 and 2.26, Bob Digby; P. 25: Figure 2.30, Bob Digby; P. 28: Figure 3.2, Lake District Herald; P. 30: Figure 3.6, David Kinninment; P. 32: Figure 3.10, David Kinninment; P. 36, Figure 3.14, David Kinninment; P. 37: Figure 3.16, David Kinninment; P. 43: Figure 3.21, David Kinninment; P. 46: Figure 4.2, J S Horton; P. 48: Figure 4.4, J S Horton; P. 54: Figure 4.13, Getty Stone; P. 55: Figure 4.15, Getty Stone; P. 59: Figure 5.2, Australian Picture Library/John Carnemolla; P. 60: Figures 5.6 and 5.7, Bob Digby; P. 61: Figure 5.8, Bob Digby; P. 62: Figures 5.13 and 5.14, Bob Digby; P. 65: Figure 5.20, Allan Cash; P. 66: Figure 5.21, Geography Teacher's Association of Victoria, Murray-Darling Basin Commission and Department of Natural Resources and the Environment, Victoria, Australia, Figure 5.22; Figure 5.22, Ecoscene/Wayne Lawler; P. 70: Figure 5.29, Science Photo Library/Dr Jeremy Burgess; P. 72: Figure 5.33, Murray-Darling Basin Commission; P. 73: Figure 5.34, Geography Teacher's Association of Victoria, Murray-Darling Basin Commission and Department of Natural Resources and the Environment, Victoria, Australia; P. 77: Figure 6.4, Panos Pictures/Bruce Paton; P. 78: Figure 6.5, Tropix; Figure 6.6, Art Directors and Trip/ J Highet; P. 79: Figure 6.7, Panos Pictures; P. 84: Figure 7.2, Art Directors and Trip; P. 85: Figure 7.6, Getty Stone; P. 92: Figure 3, Bob Digby; P. 93: Figure 5, Bob Digby; P. 94: Figure 7, Ecoscene/Ian Harwood; P. 96: Figure 8.2, David Kinninment; P. 99: Figure 8.6, David Kinninment; P. 100: Figures 8.7, 8.8, 8.9, Alan Blackburn; P. 105: Figures 8.20 and 8.21, David Kinninment; P. 107: Figure 8.24, Bob Digby; P. 109: Figure 8.27, David Kinninment; Figures 8.29 and 8.30, Bob Digby; P. 111: Figure 8.31, Airphotos; Figure 8.32, Bob Digby; P. 112: Figure 8.33, Paul Morrison; P. 119: Figure 9.4, Panos Pictures/Bruce Paton; P. 120: Figure 9.6, Trip/M Jelliffee; Figure 9.7, Axiom Photographic Agency/Chris Caldicott; P. 124: Figure 10.2, NHPA; P. 130: Figure 10.9, Jane Ferreti; P. 132: Figure 10.11, Jane Ferreti; P. 135: Figure 11.1A, Robert Harding Picture Library; Figure 11.1B, Getty Stone; P. 137: Figure 11.2, Getty Stone; P. 140: Figures 11.7, 11.8 a/b, Bob Digby; P. 142: Figure 11.9, Bob Digby; P. 146: Figure 12.4, Space Research and Remote Sensing Organization, Bangladesh; P. 147: Figure 12.8, Rex Features; P. 150: Figure 12.15, Trygve Bolstad; P. 159: Figure 1 A-E, Bob Digby; Figure 1 F, Action Aid/David Akhtar; P. 161: Figure 4, Bob Digby; P. 162: Figure 7A/7B, Bob Digby; P. 166, Figure 13.5, Bob Digby; P.s 168-9: Figure 13.8, Bob Digby; P. 171: Figure 13.11, Bob Digby; P. 172: Figure 13.12 Bob Digby; P. 174: Figure 13.16, Bob Digby; P. 177: Figure 13.21, Bob Digby; P. 179: Figure

13.23, Bob Digby; P. 185: Figure 14.4, Ruth Totterdell; P. 186: Figures 14.6 and 14.7, Ruth Totterdell; P. 191: Figure 14.18, Ruth Totterdell; P. 192: Figure 14.19b, Ruth Totterdell; P. 193: Figure 14.23, Ruth Totterdell; P. 195: Figures 14.26 and 14.27, Radhika Chalasani; P. 196: Figure 14.28, Radhika Chalasani; P. 198: Figure 15.3, Elaine Anderson; P. 199: Figures 15.6 and 15.7, Graham Yates; P. 200: Figures 15.8, 15.9 and 15.10, Graham Yates; P. 202: Figure 15.14, Getty Stone; P. 204: Figure 15.16, The National Trust; P. 205: Figure 15.18, John Cleare Mountain Camera; P. 207: Figure 15.22, The National Trust; P. 209: Figure 15.27, The National Trust; P. 211: Figures 15.30 and 15.31, Bob Digby; P. 212: Figure 15.33, Bob Digby; P. 213: Figures 15.34, Life File; P. 221: Figure 16.15, Bob Digby; P. 223: Figure 16.21, GSF Picture Library; P. 225: Figure 16.23, Bob Digby; P. 229: Figure 16.33, Cameracraft Photography; P. 231: Figure 16.36, The Eden Project/Apex; P. 233: Figure 1 A/B/E, Linda King; Figure 1 C/D, National Geographic/Steve McCurry; Figure 1F, Bob Digby; P. 234: Figures 3 and 4, Bob Digby; P. 238: Figures 17.3 and 17.4, 'Manchester, this good old town'; P. 239: Figure 17.11, 'Manchester, this good old town'; P. 240: Figure 17.12, Sarah Jones Photography; Figure 17.13, Jefferson Air Photography; P. 241: Figures 17.14 and 17.16, Sarah Jones Photography; P. 243: Figure 17.19, Jefferson Air Photography; P. 246: Figure 18.2, Magnum/Rai Raghu; Figure 18.4, Magnum/Steve McCurry; P. 248: Figure 18.8, Lonely Planet Images/Chris Mellor; Figure 18.9, Art Directors and Trip; P. 250: Figure 18.12, Images of India; P. 251: Figure 18.15, Still Pictures/Hartmut Schwarzbach; P.256-8: Figure 19.5, Linda King; P. 260: Figure 19.8, Linda King; Figure 19.10, Sarah Jones Photography; P. 261: Figure 19.11, Art Directors and Trip/S Samuels; P. 265: Figure 19.18, Linda King; P. 265: Figure 19.19, S Bermingham; P. 267: Figure 19.25, Trip/Dinodia Picture Agency; P. 268: Figure 19.27a, Lonely Planet Images/David Collins; Figure 19.26 Magnum/Rai Raghu; P. 269: Figure 19.28, National Geographic/Steve McCurry; P. 270: Figure 19.29, Panos Pictures/Paul Smith; P. 271: Figure 19.30, National Geographic/Steve McCurry; P. 272: Figure 19.32, National Geographic/Steve McCurry; P. 272: Figure 19.33, Magnum/Steve McCurry; P. 274: Figure 19.36, Geoslides; P. 275: Figure 19.38, Magnum/Rai Raghu; P. 278: Figure 20.2, Bob Digby; P. 279: Figure 20.3, Bob Digby; P. 281: Figure 20.6, Bob Digby; P. 283: Figure 20.9, Ecoscene/Sally Morgan; P. 284: Figure 20.11, Corbis; P. 286: Figures 20.12 and 20.13, Bob Digby; P. 291: Figure 20.21, Bob Digby; P. 294: Figure 21.2, Bob Digby; P. 295: Figure 21.3, Bob Digby; P. 296: Figure 21.5, Chris Ryan; P. 297: Figures 21.6, 21.7, 21.8, Chris Ryan; P. 299: Figure 21.12, Bob Digby; P. 300: Figure 21.13, Chris Ryan; P. 303: Figure 21.17, Chris Ryan; P. 304: Figure 21.19, Corbis; P. 308: Figure 221., Bob Digby; P. 310: Figure 22.3, Bob Digby; Figure 22.4, Allsport; P. 311: Figure 22.5, Tropix; P. 312: Figure 22.6, Ansett; Figure 22.7, Art Directors and Trip/Eric Smith; P. 313: Figure 22.8, Art Directors and Trip/Eric Smith.

Contents

Acknowledgements

The publishers would like to thank the following for permission to reproduce copyright material.

Maps and extracts

p.6 The European Ltd, 1995; **p.7** Andrew Johnstone, Geography, The Geographical Association 1997; **p.15** Maps reproduced from Ordnance Survey maps with the permission of the Controller of Her Majesty's Stationery Office © Crown copyright; License No; 398020; **p.18** H Toley and K Orrell, Yorkshire / Cambridge University Press; **p.21** D Weyman, Landscape Processes / Allen & Unwin; **p.23** Maps reproduced from Ordnance Survey maps with the permission of the Controller of Her Majesty's Stationery Office © Crown copyright; License No; 398020; **p.28** The Cumberland and Westmorland Herald, 1985; **p.29** T Maps reproduced from Ordnance Survey maps with the permission of the Controller of Her Majesty's Stationery Office © Crown copyright; License No; 398020; **p.29** M The Cumberland and Westmorland Herald, 1986; **p.31** T, **p.38** The Environment Agency; **p.46** Maps reproduced from Ordnance Survey maps with the permission of the Controller of Her Majesty's Stationery Office © Crown copyright; License No; 398020; **p.53** Greg O'Hare, Soils, Vegetation, Ecosystems / Oliver & Boyd 1994; **p.59** B, **p.61** M The Murray-Darling Basin Commission; **p.62**, **p.63** T Geography Teachers' Association of Victoria, Australia; **p.66** JJ Pigram, Salinity and Basin Management in South-Eastern Australia, American Geographical Association; **p.67** Victorian College of Agriculture and Horticulture, Australia; **p.68** T Shepparton News, Victoria, Australia 1989; **p.68** B, **p.69** T, **p.70** Geography Teachers' Association of Victoria, Australia; **p.71** The Murray-Darling Basin Commission; **p.74** T Victorian College of Agriculture and Horticulture, Australia; **p.74** B Geography Teachers' Association of Victoria, Australia; **p.76** B, **p.77** T The International Broadcasting Trust; **p.77** T, **p.80** T Dr Julian Thompson, Wetland Research Unit, University College London; **p.80** B The Independent Broadcasting Trust; **p.81** T Dr Julian Thompson, Wetland Research Unit, University College London; **p.85** M Richard Chorley, Stanley Schumm and David Sugden, Geomorphology / Methuen 1988; **p.86** B, C China Report 1999; **p.86** D World Resources Institute; **p.86** E, F, **p.87** G, I, J Columbia College, NY, USA; **p.87** K Grainne Ryder, A Critique of the Three Gorges Water Control Project Feasibility Study, Probe International; **p.87** L Lu Youmei, Three Gorges Project Development Corporation; **p.90** Maps reproduced from Ordnance Survey maps with the permission of the Controller of Her Majesty's Stationery Office © Crown copyright; License No; 398020; **p.93** BPAmoco; **p.97** B, **p.98** Maps reproduced from Ordnance Survey maps with the permission of the Controller of Her Majesty's Stationery Office © Crown copyright; License No; 398020; **p.104** Blyth Valley Borough Council; **p.109** Northumbrian Coastal Authorities Group; **p.113** T The Independent Newspaper 1995; **p.118** T The New Scientist 1987; **p.118** M, **p.122** T, M Pan African News Agency 1998/1999; **p.124** English Nature; **p.126** The National Trust; **p.130** Maps reproduced from Ordnance Survey maps with the permission of the Controller of Her Majesty's Stationery Office © Crown copyright; License No; 398020; **p.138** Israeli Government Tourist Office; **p.139** T The Independent Newspaper; **p.149** A, B, C Bangladesh Water Development Board; **p.149** Quoted from article published by M A Matin retd. Director Bangladesh Water Development Board; **p.150** T © External Publicity Wing, Ministry of Foreign Affairs, Government of the People's Republic of Bangladesh; **p.154** The New Scientist 1991; **p.161** T Heinemann Atlas 1997 / Rigby Heinemann, Australia; **p.161** M C Bull, P Daniel and M Hopkinson, The Geography of Rural Resources / Oliver & Boyd 1984; **p.164** B © Automotive Association 1999; **p.167** Maps reproduced from Ordnance Survey maps with the permission of the Controller of Her Majesty's Stationery Office © Crown copyright; License No; 398020; **p.168**, **p.169** © David Short, Ashwell Field Studies Centre; **p.172** Halifax Plc; **p.173** B Table of Regional Trends reproduced from the Office for National Statistics with the permission of the Controller of Her Majesty's Stationery Office © Crown copyright; License No; 398020; **p.176** Maps reproduced from Ordnance Survey maps with the permission of the Controller of Her Majesty's Stationery Office © Crown copyright; License No; 398020; **p.177** M Kelly's Directory 1993; **p.179** © Main Estates, Putterill's Estate Agents, Hitchin; **p.180** Village Design Plan for Ashwell; **p.190** T World Population Prospects: The 1998 Revision / United Nations Population Division; **p.190** B Tony Barnett and Piers Blaikie, AIDS in Africa - Its Present and Future Impact / Belhaven 1992; **p.198** B, **p.201** Vietnam News 1999; **p.203** The State of the Countryside Report 1999 / The Countryside Agency; **p.208** B The National Trust; **p.211** Heinemann Atlas 1997 / Heinemann Educational Publishers, Oxford; **p.213** B Bloomfield Lodge, Australia; **p.216** L The Independent Newspaper 1999; **p.216** R, **p.217** T. **p.218** The State of the Countryside Report 1999 / The Countryside Agency; **p.221** B Cornwall County Council; **p.228** T The State of the Countryside Report 1999 / The Countryside Agency; **p.228** B Maps reproduced from Ordnance Survey maps with the permission of the Controller of Her Majesty's Stationery Office © Crown copyright; License No; 398020; **p.230** T West Briton Newspaper, Malcolm Lindsay; **p.238** Whitaker, The History of Manchester; **p.239** T E Gaskell, Mary Barton; **p.239** B Tracy, Port of Manchester; **p.241** T The Rough Guide to Manchester / Rough Guides UK; **p.245** Nelles Maps: Western India / Nelles Verlag, Germany; **p.246** M The Rough Guide to Bombay (Mumbai) / Rough Guides UK; **p.246** R, **p.247** L Lonely Planet Guide, Mumbai / Lonely Planet Publishers; **p.247** R, **p.249** Sharada Dwiredi and Rahul Mehrotra, Bombay: The Cities Within / India Book House PVT Ltd, Bombay; **p.251** T Lonely Planet Guide, Mumbai / Lonely Planet Publishers; **p.251** B M Witherick, Population Geography / Longman 1990; **p.254**, **p.256-258** Maps reproduced from Ordnance Survey maps with the permission of the Controller of Her Majesty's Stationery Office © Crown copyright; License No; 398020; **p.259**, **p.263** A-Z Map Books / Geographer's A-Z Map Co.; **p.264**, **p.265** T OPCS © Crown Copyright; **p.266** B UpMyStreet.com Ltd; **p.267** B D Waugh, Geography: An Integrated Approach / Thomas Nelson & Sons, 1995; **p.269** The Lonely Planet Guide, Mumbai / Lonely Planet Publishers; **p.272** Rahul Mehrotra, One Space, Two Worlds: On Bombay, Harvard Design Magazine 1997; **p.273** T Children's Edition of Agenda 21 / Kingfisher; **p.280** T Joe Cummings, Lonely Planet City Guide - Bangkok / Lonely Planet Publishers 1997; **p.284** B Bangkok Post Newspaper 1998; **p.287** Maps reproduced with the permission of the General Manager, AUSLIG, Dept. of Admin. Services, Canberra, ACT Australia; **p.288** L The Age, 1999; **p.289** T The Public Transport Users Association, Melbourne, Australia; **p.289** M CityLink; **p.290** The Herald Sun Newspaper, Melbourne, Australia; **p.291** The Age, 2000; **p.292** The Age, 2000; **p.295** The New York Times Newspaper, 1999; **p.304** Tom Ridge, State Governor of Pennsylvania; **p.306** T, B The New York Times Newspaper; **p.309** Olympic Co-ordination Authority; **p.314** The Canberra Times Newspaper 1999; **p.315** The Sydney Morning Herald Newspaper.

The publishers have made every effort to trace the copyright holders, but if they have inadvertently overlooked any, they will be pleased to make the necessary arrangements at the first opportunity.

How to use this book

This book is written to help you prepare for the AS examination for Edexcel Geography Specification B. It will also help students who are studying a range of specifications in the UK and overseas.

What is the book about?

The book is about 'changing environments'. It focuses upon issues that concern people and their environments. Geographers study these interactions between people and the physical world in which they live. On one hand, natural processes such as flooding pose challenges for people, while on the other human challenges also exist. Where will the cities of the world find space for their people? How much longer can traffic pour into cities every day before transport systems collapse under the strain? In a world where people demand higher standards of living, there is increasing pressure on space. *Changing Environments* is about some of these changes, and how people are attempting to manage them.

In this book, you will study:

- **River environments**, where you will study rivers as natural systems, landforms that they produce, threats posed by rivers to people (e.g. flooding), how people attempt to manage rivers, and how past mistakes provide further challenges;
- **Coastal environments**, where coastlines are seen as being under threat from competing pressures. The erosive power of the sea threatens some coastlines, while human pressures, such as tourism and increasing populations, are the main concern along others;
- **Rural environments**, in which rural lifestyles are changing. The decline of rural populations in some areas is threatening services and the economic sustainability of the countryside, while in others people in many urban areas are moving away from cities towards more rural lifestyles, and placing pressure on the countryside;
- **Urban environments**, in which urban challenges are investigated. In More Economically Developing Countries (MEDCs), parts of cities are often ageing, and declining in population whilst at the same time other newer parts are demanding more space. In Less Economically Developed Countries (LEDCs), urban populations continue to grow at rapid rates, often leaving urban authorities unable to provide basic services such as housing and transport.

How to use this book

The book is divided into four sections. Each is sub-divided into chapters, which focus on different aspects of issues in different parts of the world. Each chapter is written to enable you to read from beginning to end. You will need to refer to the figures as you read. The following features are threaded into each chapter to help you understand its concepts and content more easily;

a) Theory Boxes, which help to explain geographical processes that are required to understand the issues in each chapter.

b) Technique boxes, to help you to be able to present, interpret and analyse data in each chapter.

c) Student Activities, to help you to interpret data and text, and work with others in understanding different viewpoints on each issue. Some of these are individual, while others are designed for groups.

At the end of each section, a section summary helps you to draw together the concepts and ideas in order to help you reflect upon and revise your learning.

Bob Digby

Introducing river environments

Rivers, people and landscapes

Across the world, several thousands of floods occurred in 1995. Two events are discussed here and show interesting contrasts. One was reported widely across Europe's media, while the second was hardly reported in newspapers, nor appeared on television. One occurred in the heart of Western Europe close to major cities in the Netherlands, while the second took place in a remote area of the High Atlas mountains in Morocco.

The Dutch floods of 1995

In 1953 the Dutch suffered a national disaster. Flooding swept through the country, caused by a 'deadly combination of spring tide and gales'. A low pressure weather system brought heavy rain, while a surge in high tides from the North Sea prevented most water from escaping out to sea. The water breached and destroyed many of the embankments – or dykes – flooding hundreds of square miles of Dutch countryside. The death toll was 1853. A programme of improvements to flood protection was designed to ensure that this would not happen again.

In January 1995, the dykes were breached and the resulting disaster was called the 'flood of the century'. Figure 2 shows a report from *The European*, revealing the impact that the flood had on people.

NETHERLANDS FLOODS ARE THE WORST SINCE 1953

The Netherlands was hit by the worst flooding since 1953. At least one person died and 250 000 people had to leave their homes as emergency services warned that waterlogged flood defences were close to bursting point.

More than 1550km² of agricultural land were submerged and hundreds of homes ruined just 13 months after floods swept the country. Farmers, many of whom refused to leave their holdings, used tractors and trailers to move millions of animals out of danger or sent livestock to market early.

In the largest evacuation this century emergency services cleared the prison at Maastricht and dozens of hospitals, and carried thousands of books to safety.

Emergency workers toiled all night, piling sandbags on to sodden dykes around Tiel, the critical area, lying at the confluence of the Maas and Waal rivers. Police and soldiers patrolled by boat and helicopter to discourage looting while people struggling to move possessions to the safety of upper storeys criticized those offering 'flood cruises' to sightseers as voyeuristic money-grabbers.

Some blamed the damage – put at up to 40 billion guilders ($23.5b) – on environmentalists who campaigned against raising, reinforcing, and extending the dykes which contain the Rhine, Waal, and Maas.

Figure 2 From *The European*, February 1995.

In 1995 this programme proved to have been in vain. Although the Dutch are leaders in coastal protection, much of the Netherlands was flooded in January 1995. One third of the country has been drained from the sea. Dykes were constructed to keep the sea out and to drain water away from land to sea. As land was drained, it shrunk, and much of the country is now below sea level. Transport links and some settlements are built along the dykes, but pressure on space means that much land lies below sea level. The dykes remain at the former level of the rivers draining the land, and are higher than surrounding farmland.

Figure 1 The Dutch floods of 1953.

Effects of the flooding in the Netherlands

1 What were the consequences of the flooding locally, nationally, and internationally?

2 What are the short- and long-term implications for the Dutch?

3 How much should the Dutch treat flooding as:
   a) their problem?
   b) a problem involving other countries?

4 Who should manage investment for Dutch flood protection? Should it be the Dutch themselves, or a wider body such as the EU?

5 Should people in Britain contribute to Dutch coastal and river flood management? Why?

The flood was the result of strong tides, and of river systems behind the dykes. The wettest January on record resulted in threatening water levels in rivers, such as the Rhine and the Meuse. Heavy rain was made worse by premature melting of Alpine snow in Switzerland. This resulted in a two-way attack so that much of France, Belgium, Germany and Holland was subjected to flooding.

Environmental groups have suggested that modifications to channels of the River Rhine, such as straightening meanders and dredging, discharged water quickly to the north of Europe. The channel is now 80km shorter than it was naturally. At an average speed of 2–3km per hour, water from melting snow now reaches the Netherlands 30 hours earlier than it used to. Despite this, *The Observer* stated that the **effects** of flooding were much worse in France, Germany and Britain than the Netherlands, where flooding was the greatest. Dutch preparation for emergencies allowed them to evacuate 20 000 people with little trouble.

Figure 3 Flooding – a European or a national problem?

The Imlil flash-flood of 1995

Imlil is a small village of about 200 people in the High Atlas Mountains of Morocco. It lies at the confluence of four rivers, as shown in Figure 4. Although a traditional village of Berber people, the original inhabitants of Morocco, it is close to a number of tourist treks, and a field study centre is located there. This story comes from Andrew Johnstone, who works there.

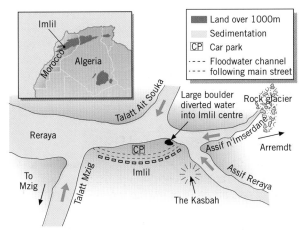

Figure 4 The location of Imlil.

On 17 August 1995, a flash-flood occurred in the upper Reraya valley. Around Imlil itself 70mm of rain fell in $2\frac{1}{2}$ hours. According to one source, the amount of water in the river increased by 27 times, and created a wall of water and rock 6 metres high. Not only was property and land damaged in the valley bottom, but vibration caused by the flood created landslides. In three hours, the river was back to its normal level. Within this time, damage was immense (Figures 5 and 6); 40 parked cars were washed away, and houses, shops and hotels were destroyed. People in the village and surrounding area were evacuated to a mosque further up the hillside. Reports showed that 150 people had died, though one showed that local estimates were greater.

Losses were significant. Maize crops were washed away, as well as fodder crops for winter feed, such as alfalfa. Some animals were lost, including goats, cows and mules. Mules are significant because they work in the fields. In an area where average family income may be £5 per day, a replacement mule can cost £300. Walnut trees are the source of a major cash crop, and their destruction has longer-term significance; they take up to 15 years to mature. Many farmers lost up to 20 per cent of their soil.

Tourists too were affected. August is the height of the tourist season, and hotels and hostels were full. Of the 150 official deaths, between 20 and 60 were tourists, killed while trying to escape the storm and rescue their cars. Further down the valley, campers were killed and swamped in debris (Figure 5). The main hostel was washed away, bringing the tourist season to an early end.

Figure 5 Damage within and around Imlil.

Figure 6 Damage to the fields around Imlil.

What caused the flood?

The storm fell on to steep valley sides where run-off was fast. Although the storm would have caused damage at any time, two other factors aided its destructive power. The previous eight years had been drier than normal, and soil had become compacted and baked, so heavy rain ran off the surface more easily. Secondly, gradual deforestation over many decades had reduced plant cover so that soil was exposed.

Response to the flood

Aid came quickly. Helped by EU funding, the Moroccan government lifted food and water into the area, and the road link was re-established in four days. Local Berber chiefs compiled lists for the authorities so that compensation could be paid. Because no family was insured, compensation was limited; each family expected to receive about £20 for a small family and up to £200 for a family of 10 or more. Local communities rebuilt irrigation channels and cleared soils of boulders.

Which was the greater disaster?

1 Make a large copy of the table below. Complete it, using details from this chapter.

	Netherlands	Imlil
Location		
Duration of flood		
Cause		
No. of people killed		
No. of people evacuated		
Damage to property		
Damage to livelihoods		
Costs of compensation		
Other impacts		

2 Using these, and any other criteria you think appropriate, judge which you feel was the greater disaster. Explain your reasons.

3 Had the two events occurred on the same day, which should have been the greater story to report in the media? Why?

4 Why do you think the Imlil flood received no attention in UK newspapers? Should it have done? Give reasons.

References and further reading

Johnstone, Andrew, 'A flash flooding event in the High Atlas Mountains' *Geography* Vol. 82, No. 354, January 1997

1 Understanding hydrology

The introduction to this section shows how important rivers and their valleys are to human activity. Few rivers react as suddenly to downpours of rain as those in desert regions. However, rivers in the UK do flood, and the management of flooding problems is a major part of the work of government bodies such as the Environment Agency. Part of their job is to look at the management of water in river basins. In order to understand some of these management issues, it is important to know more about how river basins function. The study of water movements within river basins is known as hydrology.

Rivers and rainfall – understanding the hydrological cycle

Study Figure 1.1. It shows measurements of rain which fell over 12 hours, in the 48 hour period which started at midnight on 1 September 1998, at Malham in the Yorkshire Dales National Park in the UK. This rainfall is fairly typical of that which fell across this part of the Pennines on this date. Figure 1.1 also shows the depth of the River Wharfe at Grassington, just a few kilometres to the east, during the same period and just after. This location is part of the upper section of the Wharfe valley, shown in Figure 1.2. The upland area consists of large expanses of moorland, where heather and tussock grasses are the main form of vegetation. A typical view is shown in Figure 1.3. The rainfall shown in Figure 1.1 fell mainly on this moorland, and made its way to the river.

Notice the time lag between the period when most rain fell, and when the river was deepest. This is the result of a number of processes that delay the water in its passage to the river. Together, they help to explain the period that it takes rain water to reach a river. The delay period varies according to:
- the time of year (season) when the rain falls
- the nature of the surface
- the duration of the rainstorm
- the underlying geology (rock type)
- the height of the water table.

Moorland, such as that shown in Figure 1.3, shows a fairly standard response; urban rivers have a much more rapid response to rain, while forested areas and areas where the geology is permeable have a much slower response.

Year and day	Time	Rainfall falling in two-hour periods, in mm	Depth of River Wharfe at Grassington in metres
1 September 1998	midnight		0.103
	2 a.m.	0.6	0.103
	4 a.m.	1.2	0.103
	6 a.m.	3.2	0.103
	8 a.m.	5.4	0.103
	10 a.m.	1.2	0.105
	Noon	0.1	0.103
	2 p.m.	0.4	0.121
	4 p.m.	0	0.473
	6 p.m.	0	0.484
	8 p.m.	0	0.595
	10 p.m.	0	0.531
2 September 1998	midnight	0	0.465
	2 a.m.	0	0.417
	4 a.m.	0	0.377
	6 a.m.	0	0.346
	8 a.m.	0	0.322
	10 a.m.	0	0.303
	Noon	0	0.288
	2 p.m.	0	0.274
	4 p.m.	0	0.26
	6 p.m.	0	0.249
	8 p.m.	0	0.238
	10 p.m.	0	0.228
3 September 1998	midnight	0	0.219

Figure 1.1 Rainfall measurements at Malham starting at midnight on 1 September 1998, and the depth of the River Wharfe at Grassington over 48 hours.

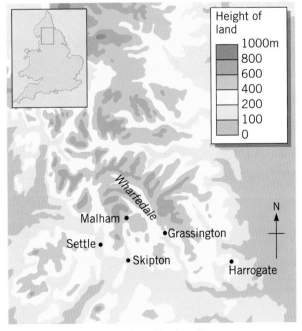

Figure 1.2 Upper section of the Wharfe valley.

1 Construct a graph to show rainfall and river depth, using the data in Figure 1.1.
2 Using Figure 1.8, label the features of the graph you have drawn.
3 How much rain fell in how long a period of time?
4 In what ways is the pattern:
  a) similar to
  b) different from that shown in Figure 1.8? Why is this?
5 How and why might the river's response have been different if the same amount of rain had fallen in:
  a) half the period of time
  b) double the period of time
  c) mid-summer rather than in early spring?
  Draw sketch graphs to illustrate your answer.

Figure 1.3 Moorland landscape in the upper part of the Wharfe valley.

Theory

The hydrological cycle of a drainage basin

The starting point for the study of rivers is the hydrological cycle, also known as the water cycle. The **hydrological cycle** is the circulation and transfer of water in its various forms within and outside a river basin. River basins are also known as **drainage basins**, because they form the area within which rainfall is drained and transferred into river channels. Figure 1.4 shows a theoretical drainage basin. The basin has a border called the **watershed**. From within the watershed, a number of small streams or tributaries flow, joining to form larger streams and a main river course.

The hydrological cycle is shown as a systems diagram in Figure 1.5. Systems consist of three main parts – inputs, flows, and outputs. **Inputs** include everything that is added to the system; **open systems** are those where inputs are added from outside. The hydrological cycle within the drainage basin is an open system, as it gets precipitation from the atmosphere. **Flows** are movements of water within the hydrological cycle. **Outputs** are lost from the system, such as water which is evaporated. In between are stores, where water may remain for short or longer periods of time; surface lakes, soil moisture, and underground water are examples.

1 Make a copy of Figure 1.5. Show the input arrows in one colour, flow arrows in another, and output in a third colour.
2 Annotate your diagram to show how certain conditions could lead to flooding within this system. For instance, what changes in inputs to the system would encourage flooding?
3 How do you think human activity could change flows within the system in such a way as to encourage flooding? What would need to happen to outputs to prevent flooding from happening? Consider urban and rural activities.

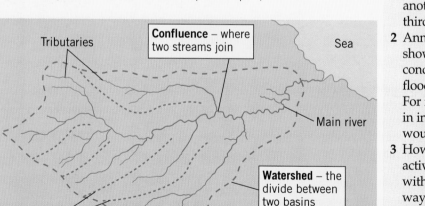

Tributaries

Confluence – where two streams join

Sea

Main river

Watershed – the divide between two basins

Interfluve – the divide between two streams within a basin

- - - - Boundary of drainage basin

Figure 1.4 A drainage basin.

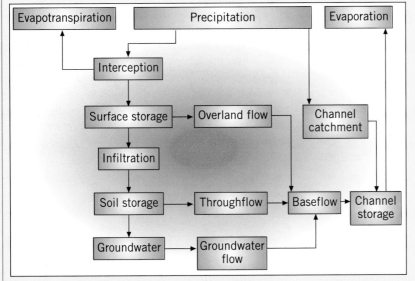

Figure 1.5 A systems diagram of the hydrological cycle.

to absorb water is called its **infiltration capacity**, and is the volume of water that can be absorbed before saturation is reached. If infiltration capacity is exceeded then **overland flow** occurs. The amount of infiltration depends upon certain factors:

- **Antecedent rainfall** – This is the amount of rainfall that has fallen recently. A spell of wet weather may saturate soil and prevent further infiltration.
- **Permeability of the soil** – Sandy soils absorb water easily, and run-off rarely occurs. Clay soils are more closely packed, contain smaller air spaces, and saturate quickly.

When it rains, very little rainfall actually reaches a river channel directly. Figure 1.6 shows a cross-section through a typical river valley. Arrows show the routes taken by water from the valley sides into the channel. Of the rain that falls on the valley sides, a large proportion is held and stored in the leaves and branches of plants, called the **interception zone**. The amount intercepted by plants depends upon:

- the type and density of vegetation – deciduous plants intercept more rain in summer when in leaf than in winter
- rainfall intensity – rain from heavy storms drips from trees more easily than gentle drizzle
- antecedent rainfall.

Some water drips from leaves or flows down stems and reaches the ground surface; some is evaporated straight into the atmosphere. Water that reaches the surface, either directly or by dripping from plants, moves into the soil. This process is called **infiltration**. The ability of soil

- **The intensity of the rainfall** – Rapid downpours are less likely to be absorbed, and water runs off the surface more quickly.

Once water enters the soil, some is taken up by plant roots and transpired into the atmosphere. **Transpiration** is the process by which water vapour is passed through leaves. Evaporation and transpiration are usually referred to together as **evapotranspiration**. Gravity causes some water to move downslope through soil air spaces. This is known as **throughflow**. Some water continues to infiltrate into solid rock below the soil, depending upon how jointed the rock is. Rocks with joints or cracks allow water to move more easily to where the rock is saturated with groundwater. The upper limit of this saturated area is known as the **water table**. Water from this area seeps slowly under gravity towards the river channel as groundwater flow, or **baseflow**, if rock permeability allows it. Of the three, baseflow is by far the slowest process, while surface run-off is the fastest.

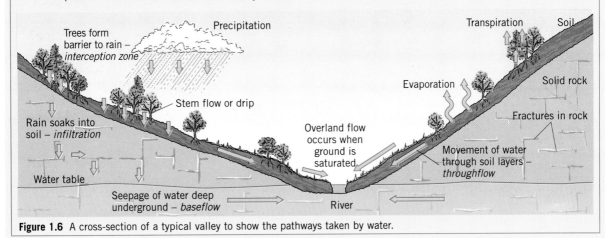

Figure 1.6 A cross-section of a typical valley to show the pathways taken by water.

It is possible to see the different ways in which water reaches a river channel in the landscape. The photographs in Figure 1.7 show different pathways visible in the landscape. Most water seeps very slowly into river channels. Overland flow is the most rapid, with movement by throughflow next, and baseflow by far the slowest. Rates vary enormously, depending upon soil type, slope angle and temperature, but as an example, throughflow has been measured in Derbyshire at a rate of about 23cm per hour. Even a distance of 10 metres could

take nearly two days. Baseflow, on the other hand, has been shown in some areas to take several weeks.

Overall, baseflow makes up the majority of a river's flow throughout the year, accounting for flow when there may have been little or no rain for days or weeks. Throughflow and overland flow therefore contribute storm flow into a river - that is, water which flows in as a result of a downpour of rain. This is explained further in the theory box on storm hydrographs.

Figure 1.7 The different pathways taken by water.

a Overland flow beside the River Wharfe on a day when the soil was saturated.

b Water seeping out of soil over a rock face – evidence of throughflow.

Theory

Understanding storm hydrographs

A storm hydrograph is a graph that shows how a river changes as a result of a period of rain. An example is shown in Figure 1.8 below, where the values are:

- rainfall, measured in millimetres, shown on the scale to the left-hand side of the graph
- depth of the river, shown in metres, shown on the scale to the right-hand side of the graph. In some cases, discharge – or the volume of water in a river – is shown.

Hydrographs show how rivers respond to falling rain. As it rains, a small amount falls directly into the river channel. However, this is a minute proportion of the total; most rain falls on the valley sides. Even at the time of maximum rainfall in Figure 1.8, the river has only slightly increased in depth. Most rain takes time to reach the river – the result of a number of processes described in the theory box on pages 10 and 11.

As the water makes its way downslope, the river rises as its depth increases. This period is shown as the **rising limb** of the graph. Once the river reaches maximum depth, it is possible to calculate the time **lag**, that is, the period of time between maximum rainfall and maximum

depth of the river. From that time, the amount of water reaching the river as a result of the rain storm will begin to decrease, and the river level falls. This is shown as the **recession limb** of the graph, as the river recedes.

From this, it is possible to calculate the proportion of the river's depth which is the result of the rain storm. This is done by taking the original depth of the river before the rain began, and projecting a line across the graph until the point where the river reaches its former depth. Any value above that – shown as the shaded area in Figure 1.8 – is the result of the rain storm, and is known as **storm flow**. The area below it shows the normal or baseflow of the river.

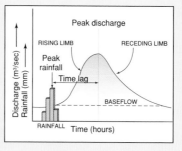

Figure 1.8 A storm hydrograph.

Summary

- Water moves along a number of routes from the time that it falls as precipitation, to its passage along a river channel.
- The movement of water through the landscape can be seen as a system, with inputs, flows and outputs.
- The flow of water in a river is subject to variables such as the surface on which rain falls, how much rain has fallen previously prior to a storm, and rainfall intensity.

Ideas for further study

1 The UK Environment Agency has data for most rivers, and access to rainfall data from many weather stations. You or your teacher can obtain local data by contacting them. However, you need to know precisely what data you want – which river, which points along the river, for which dates.

2 The Environment Agency has a Home Page on the internet which you can use to locate different rivers in different parts of the UK. You can surf for information on different parts of the UK by contacting the address given below.

3 Rivers in urban areas have different responses to rainfall than those in rural areas. Investigate the effect of a sudden storm on an urban area, and consider how and why the response is different.

References and further reading

Web sites

The Environment Agency Home page can be found on
http://www.environment-agency.gov.uk/

Investigating the River Wharfe

Rivers exist because of the movement of water through the landscape which collects together in channels. This chapter will show how rivers are an essential component of river basin landscapes. The effect of weathering on the valley sides, together with slope processes, gradually bring material downslope towards the river channel. There the river removes it, and the material is used in the mechanical erosion of the river bed. By deepening the river bed, more scope is created for the further removal of material on the valley sides. This is shown in Figure 2.1.

Figure 2.2 The valley of Buckden Beck, a small tributary stream of the River Wharfe.

The upper section of the river basin

The upper part of the Wharfe basin lies in the central northern Pennines in North Yorkshire. Many tributary streams flow into the Wharfe, including Buckden Beck which is shown in Figure 2.2. They have similar characteristics so that most streams appear very similar.

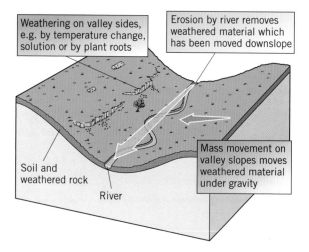

Weathering on valley sides, e.g. by temperature change, solution or by plant roots

Erosion by river removes weathered material which has been moved downslope

Soil and weathered rock

River

Mass movement on valley slopes moves weathered material under gravity

Figure 2.1 The relationship between the erosion caused by rivers and what happens on the valley sides. Weathering acts on the valley sides to weaken and to break down solid rock into fragments. These, over time, move under gravity down the slope to the river, where they are removed. The continual removal of material by rivers is an example of the erosion of the landscape.

Figure 2.3b Small rapids along the course of Buckden Beck. Like the waterfall, these are turbulent waters where the stream spills over different limestone strata (or layers).

Figure 2.3a Waterfall along the course of Buckden Beck.

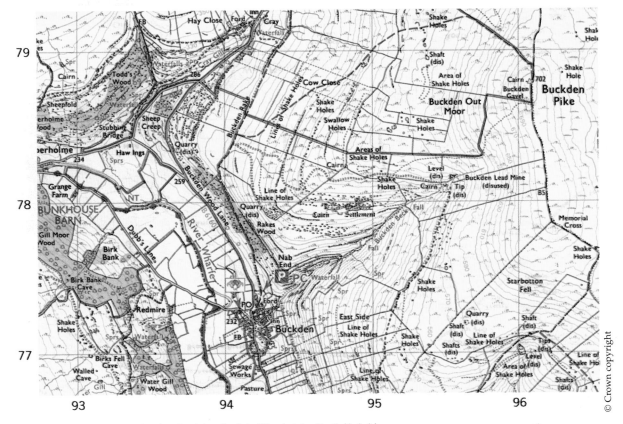

Figure 2.4 OS map extract showing Buckden Beck in Wharfedale, North Yorkshire.

1 Describe the stream of Buckden Beck, using Figures 2.2 and 2.3.
2 Describe the **valley** of Buckden Beck, using Figures 2.2 and 2.4.
3 a) Find the source, or beginning, of Buckden Beck. Now find the point at which it joins the River Wharfe. Give 6-figure grid references for each.
  b) Estimate the height of land at each point, and then find the vertical difference between the two points.
  c) Measure the length of the river course between these two points in kilometres, and then metres.
  d) Find the average gradient for this section of the river.
4 Find other tributary streams in Figure 2.4 that lead to the River Wharfe. What features of Buckden Beck and its valley are common to other streams?

Buckden Beck's general features, such as the steep course, small rapids and waterfalls, and sediment are fairly common to streams in the upper part of a river basin. The gradient of the valley slopes is steep, and the river loses height rapidly over the first few kilometres. The stream is small and generally slow, as energy is lost in overcoming friction offered by the stream bed which is strewn with boulders. However, rain storms may increase stream energy significantly so that it is able to erode rapidly, especially where the limestone geology creates a series of steps leading to small rapids and waterfalls. Stream energy is the result of a complex relationship between a number of factors.

Technique

Investigating rivers and their channels

Figures 2.5 and 2.6 show groups of students taking stream readings along Buckden Beck. They are taking readings to show the following:

- the width of the river. This is not as simple as it sounds, since stream width may vary over a short distance. However, they are measuring width from the right to the left bank.
- the depth of the river. This, too, varies within a short distance. The group decided to measure at five points; the two banks, then one-quarter, half, and three-quarters of the way across. In a wider or more variable channel, this needs to be increased.

With these, the group can calculate the area of the channel. This is obtained by multiplying width by average depth (taken from the five readings). Be careful to convert both readings to the same unit first, e.g. metres multiplied by metres. The result is a channel area expressed in square metres.

Figure 2.5 Taking width and depth measurements along Buckden Beck.

- **velocity**, or speed of the river. This is done by selecting and measuring out a known distance – usually 5 or 10 metres. A float, such as a cork, or table-tennis ball is used to pass over the distance, and timed. Because streams vary in their speed, three velocity readings are usually taken. The speed is found by dividing the distance by the time taken in seconds. The result is a velocity reading in metres per second.

Because the cork measures the surface of the water, which is faster than the stream as a whole, it is usual to correct the reading by 0.8 to produce a reading for the whole cross-section area.

With results for area and velocity, the group are able now to calculate the discharge of the stream. **Discharge** is the amount of water that passes a point in a given space of time. By multiplying cross-sectional area by velocity, a reading is obtained for the number of cubic metres per second. For a stream like Buckden Beck, this is normally a very low value. However, it increases significantly downstream, as more tributaries join the river.

Figure 2.6 Recording velocity measurements along Buckden Beck.

Rivers erode their beds and banks using the sediment in the river bed which is picked up and transported. The gradual reduction in sediment size along a river course is also accompanied by a greater **roundness**, as sharp edges are worn down. Figure 2.7 shows how to measure sediment samples. Size can be measured in different ways, but is usually measured by taking the longest axis. Figure 2.8 shows a scale for measuring roundness.

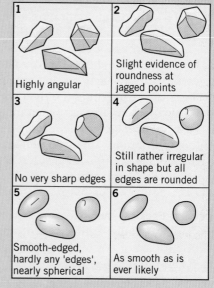

Figure 2.8 A six-point scale for measuring sediment roundness.

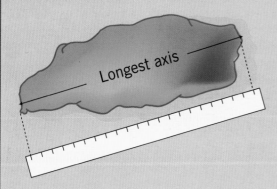

Figure 2.7 Sampling and measuring sediment along Buckden Beck.

The development of a river channel

The photographs in Figure 2.9 are of the River Wharfe in the upper parts of the Pennines, only a few hundred metres from each other. Notice how different they are. A number of questions are worth investigating about rivers: why do river channels vary in size, shape and form? why do they vary in their speed, and in their course? Not only might two or more rivers differ significantly from each other, but the same stream or river may behave differently at different locations along its course.

Figure 2.9b A smoother river channel. Different factors enable the stream to move more rapidly. Its smoothness creates a lack of turbulence and the river therefore flows more quickly. The gradient is steep and friction is minimal, creating faster flow.

Figure 2.9a A river channel in upper Wharfedale. The shape of the channel is wide and shallow, increasing friction and reducing river energy. This is made worse by the roughness, caused by sediment which has been deposited. This, in turn, increases friction and creates turbulence, which slows the river down. Turbulence is deceptive, since it looks as though the river is flowing more quickly than it is.

Theory

Understanding river velocity and channel shape

Chapter 1 has already shown that river depth, width and velocity can be easily measured. The variation in size and shape is significant in assessing competence, that is, how able rivers are to move material along its course, thus eroding its bed and banks; the faster the flow, the more able a river is to erode. Although gradient is important in providing a river with its energy to erode, its channel size and shape is as important.

Study the three channels shown in Figure 2.10. Multiply width by depth. It is clear that each channel has the same area, yet each is very different, and varies from narrow and deep, to broad and shallow. Geographers have found that channel shape very much affects the ability of a river to erode. The single factor which most affects this is the friction offered by the bed of the river and its banks. To move any material, a river must first have surplus energy over that required simply to move.

The unit used to describe this is an index known as the **hydraulic radius**. The hydraulic radius is an index calculated to compare the area of the channel with the amount of friction that the river must overcome in order to have any excess energy. The amount of friction is known as the **wetted perimeter**, that is, the contact of the river between its bed and banks. In Channel A in Figure 2.10, the friction is 36 metres wide, plus 0.5 metres for each of the two banks on either side. The area is therefore 18 square metres, and the wetted perimeter 37 metres.

The formula for calculating the hydraulic radius is as follows:

$$\text{Hydraulic radius} = \frac{\text{Area of river channel in square metres}}{\text{wetted perimeter}}$$

A value of 18 divided by 37 would give a hydraulic radius for this channel of 0.49. The greater the value, the greater the energy available to transport material along the channel and to erode.

1 Study the three river channel shapes in Figure 2.10. By calculating the hydraulic radius, show which one is the most efficient.

2 Explain how the efficiency of each channel is affected by its hydraulic radius.

3 On this basis, is it likely that:
a) rivers are generally slow in their upper course, and increase in velocity lower down, or
b) rivers are generally slow in their upper course, and increase in velocity lower down, or
c) river velocity remains much the same throughout the course?

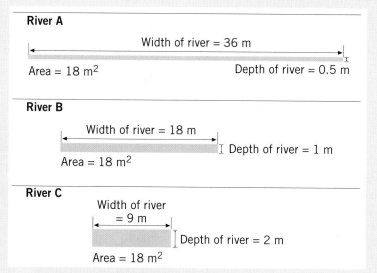

River A

Width of river = 36 m

Area = 18 m²   Depth of river = 0.5 m

River B

Width of river = 18 m

Area = 18 m²   Depth of river = 1 m

River C

Width of river = 9 m

Area = 18 m²   Depth of river = 2 m

Figure 2.10 Three different channel shapes. Each has the same area, but a different shape. This accounts for different energy levels found in each.

Changes in valley shape and form

Figure 2.11 shows the valley of Buckden Beck. The area has varied rock types, or geology, shown in Figure 2.12. Over time these have been attacked and broken down by weathering. The main rock type – the pale-grey limestone shown in Figure 2.11 – forms many features at and below the ground surface. Its structure is resistant, but it is greatly affected by natural vertical cracks – known as joints – that allow water to pass through. Rocks that allow water to penetrate along fractures are known as **permeable**. Other horizontal fractures show the different strata – or layers – of rock and are known as bedding planes. Together these allow water to pass freely in and around the rock.

Figure 2.11 Valley sides along Buckden Beck.

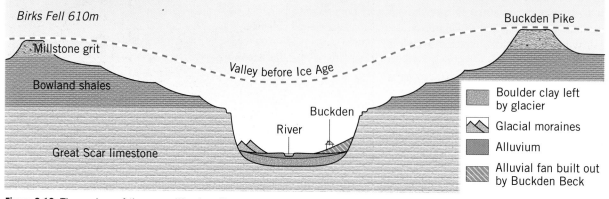

Birks Fell 610m

Buckden Pike

Millstone grit

Bowland shales

Valley before Ice Age

Buckden

River

Great Scar limestone

	Boulder clay left by glacier
	Glacial moraines
	Alluvium
	Alluvial fan built out by Buckden Beck

Figure 2.12 The geology of the upper Wharfe valley.

Weathering of valley sides

Figure 2.13a shows the valley sides of Buckden Beck, covered in fragments of limestone, or **scree**. These are created by **weathering** of the valley sides. Weathering is the mechanical breakdown or chemical decay of rocks in situ, that is, at the point where they are first exposed at or close to the earth's surface. There are different processes of weathering – mechanical, chemical and biotic – shown in Figure 2.13.

Mechanical (or physical) weathering

Mechanical weathering is the action of any process that uses physical force to break up rock fragments. Globally, several processes are mechanical in nature, but the most important in the Pennines is that of freeze-thaw, also known as frost-shattering. The process is shown in Figure 2.13. Limestone cracks – or joints – on the side of valleys such as Buckden Beck, fill with groundwater or rain, which, in winter, may freeze. Freezing at night is common, and during this time the volume of water expands by approximately 10 per cent. In so doing, the ice forces the joints open slightly, so that more water is able to enter. Repeated thawing and freezing gradually opens the joint to further attack, until a fragment of rock is fractured away from the main rock face. The fragment – or scree particle – falls to accumulate at the bottom of the slope.

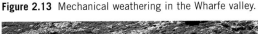

Figure 2.13 Mechanical weathering in the Wharfe valley.

Figure 2.13a Mechanical weathering by freeze-thaw on the valley sides of Buckden Beck. Limestone rock fragments – known as scree – have broken away from the rock face above. Most of these date from colder periods after the last Ice Age, 10 000 years ago, but some are created each year now.

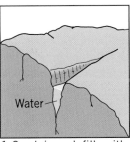 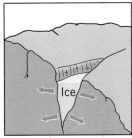 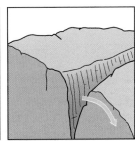

1 Crack in rock fills with water

2 Water freezes and expands by up to 10%, pushing rocks apart

3 Repetition of process forces apart rock fragment which falls

Figure 2.13b The freeze-thaw weathering process. Limestone exposures, or pavements, are exposed to acid rainwater. Joints widen with prolonged exposure to form gaps – or grykes – between limestone blocks or clints.

Chemical weathering

Chemical weathering involves chemical change or decay in order to bring about the destruction of solid rock. Several processes occur, almost all of which involve the direct action of rainwater. Rainwater is not pure water, but mixes with atmospheric gases such as carbon dioxide to form weak acids. The process is shown in Figure 2.14b. Weak acids are able to dissolve alkaline rocks such as limestone (shown in Figure 2.14a) so that joints are widened by solution as water passes through. The widened joints are known as **grykes**, and the blocks of limestone which become separated by them are known as **clints**.

Figure 2.14a Chemical weathering processes in the Wharfe valley.

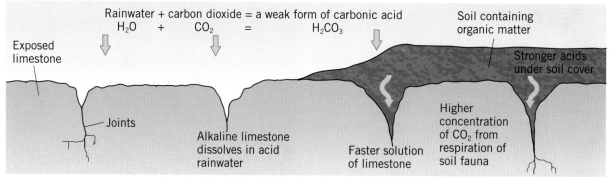

Figure 2.14b The chemical action of rainwater. The diagram shows rainwater – H$_2$O – combining with carbon dioxide (CO$_2$) in the atmosphere to form carbonic acid (H$_2$CO$_3$). This dissolves both alkaline limestones or any other rocks – such as sandstone – whose particles are bound by calcium carbonate cements. This process is accelerated under soil cover, where there are even greater concentrations of carbon dioxide.

Figure 2.15a The action of plant roots in disintegrating solid rock. Plant roots have apparently penetrated rock, fracturing it as they grew.

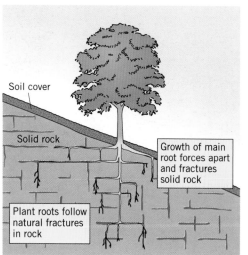

Figure 2.15b The action of plant roots

Biotic weathering

Although rocks may appear solid, small fractures or joints allow plant roots to penetrate as they grow in search of water and nutrients. Microscopic root hairs are able to penetrate even minute cracks in rock faces. As they grow, the cellulose root cells exert a huge force on the rock, widening the joint as they do so, and breaking solid rock into fragments which are held in place until the plant dies. The process is shown in Figures 2.15a and b.

Mass movement on the valley sides

The importance of weathering is that it destroys otherwise solid rock. Having been formed, weathered material is forced downslope under gravity towards the river channel. A number of processes help this movement, which together are known as **mass movement**. Mass movements may be classified, depending upon whether they are rapid or slow, or upon whether material flows or slides down the valley side. Their classification is shown in Figure 2.16. Flows disturb layers, so they are disassembled; slides involve the movement of a mass, so that it remains intact once moved. This chapter will discuss flow movements which are more common along river valleys in the UK.

Rapid flow movements

As the name implies, these are sudden movements of material downslope. Two factors are important: steepness of slope and the nature of the material on the slope. A slope must be able to contain its own weight if it is to remain stable. In the case of the rock fall shown in Figure 2.18, the collapse was due to the steepness of the slope which affected the stability of the rock. More fractured rocks – such as the sandstones and shales shown in the photo – are more likely to collapse, as are those that consist of loose-fitting sediments. Other rock types which are more resistant, such as granite or limestone, are more likely to sustain their own weight, and may even remain vertical.

Type of movement	Flow movement	Slide movement
Rapid	Mud or earth flow, avalanche, rock fall	Landslide, rotational slip (Chapter 8, page 100)
Slow	Soil creep	

Figure 2.16 Classification of different kinds of mass movement found in Pennine valleys.

Figure 2.17 Mudflow on the side of Wharfedale. The mudflow occurred many decades ago and has since grassed over, but its form is clearly visible on the side of the valley.

Figure 2.18 Rock fall on the side of Wharfedale. The steep slope is unable to support the weight of the rock, which has fractured and collapsed.

The mudflow shown in Figure 2.17 is the result of climatic conditions. Steady and prolonged rain adds to the weight of weathered material on the slope. Heavily lubricated by rain, it may become unstable and flow downslope rapidly. Sometimes such flows have disastrous consequences; the Aberfan flow from a coal waste tip in October 1966 was of this nature, and several earth– or mudflows in the upper Andes have led to loss of life in Colombia during the 1990s.

Slow flow movements

These are, as the term suggests, much slower and sometimes barely detectable. Figure 2.19 shows evidence which can be seen on many valley sides in the UK, that an apparently stable slope is far from being so. Terracettes which form on the hillside are the result of constant slow movement of soil particles downslope, usually by rain splash or treading movement. Figure 2.20 shows an example from upper Wharfedale.

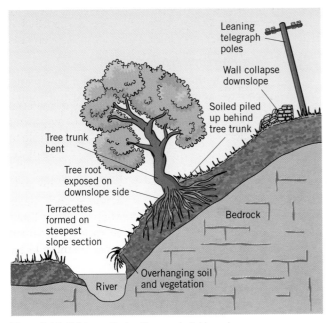

Leaning telegraph poles

Wall collapse downslope

Soiled piled up behind tree trunk

Tree trunk bent

Tree root exposed on downslope side

Terracettes formed on steepest slope section

Bedrock

Overhanging soil and vegetation

River

Figure 2.19 Evidence that soil creep is taking place.

Figure 2.20 Terracettes on the side of Wharfedale.

Figure 2.21 Sediment in the river channel along Buckden Beck.

Erosion by the River Wharfe

Material that moves downslope usually ends up in the river channel (Figure 2.21). This comes either from the sides or banks of the river, or from upstream. Once in the channel, sediment is picked up in times of high flow and carried along as the **load** of the river. The load becomes the tool by which the river is able to erode its bed and banks. The different processes in which the load is carried are shown in Figure 2.22. During periods of heavy rain, greater discharge enables the percentage of load carried as traction to increase sharply. Heavy boulders may be carried along in the flow, and thus increase the rate of erosion.

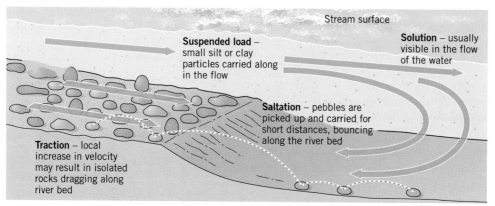

Figure 2.22 How a river carries its load. Load may be carried in one of four ways:
- by **traction**, where the river drags material along its bed that it is unable to pick up; this is known as the **bed load** of the river.
- some material is temporarily picked up, only to be dropped again, which results in a 'skipping' motion along the river bed; this is known as **saltation**.
- in **suspension**, where material is carried in the river current.
- in **solution**, where soluble chemicals are dissolved in water.

Figure 2.23 A pot-hole has been 'drilled' or eroded by corrasion along a limestone river bed, by the pebbles within it.

How a river erodes

Erosion is the process by which a landscape becomes worn away. In transporting material, rivers remove it and carry it to a new location. The process of removal involves the erosion of one landscape, while dep-osition further along the course creates new land-forms. The greater the flow of water, the faster the transfer of material between one location and another.

Rivers erode in the following ways:
- by **corrasion**, also known as **abrasion**, which involves mechanical action. Sand, stones or boulders may drag along a river bed, or may knock into it as saltation bounces them along. Turbulence in the river bed may dislodge stones and

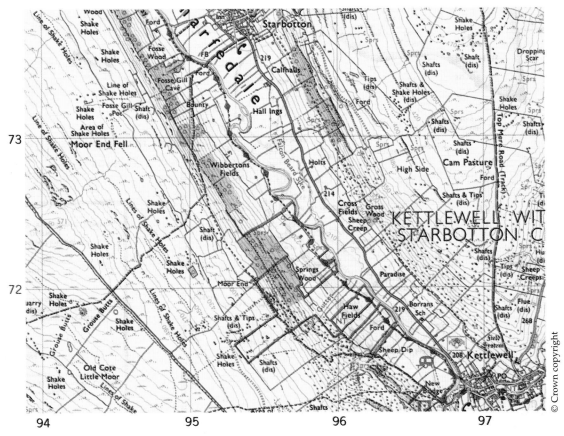

© Crown copyright

Figure 2.24 OS map extract showing the middle section of the River Wharfe between Starbotton and Kettlewell.

pebbles, which 'drill' at the bed as they are caught in the flow, contributing to the erosion of the bed, and of the stones themselves. Figure 2.23 shows a pot-hole caused by erosion.

- by **corrosion**. Rocks, such as limestone are susceptible to chemical attack by acid rainwater. They gradually decay or corrode, in the same way as chemical weathering.
- by **hydraulic action**, where the force of water undercuts banks of loose river deposits.
- by **cavitation**, where pressure from turbulent water flowing at high velocities may have an explosive effect as it is released. This causes shock waves to weaken the sides and bed of the river.

The middle section of the Wharfe valley

Figure 2.24 shows a map of the River Wharfe between Kettlewell and Starbotton. The valley here is very different from that of Buckden Beck, just a few kilometres away. In the upper section of the river valley along Buckden Beck, **vertical** erosion is dominant, and the river loses height quickly as it

1 Compare the river along this stretch of the River Wharfe with Buckden Beck in Chapter 2, using Figures 2.24, 2.26, and 2.27.
2 Describe the valley of this section of the Wharfe.
3 a) Find the point at which Buckden Beck joins the River Wharfe, and then Kettlewell. Give 6-figure grid references for each.
   b) Estimate the height of land at each point, and find the vertical difference between them.
   c) Measure the length of the river course (following the course itself) between these two points in kilometres, and then metres.
   d) Find the average gradient for this section of the river.

moves downstream. In this section, river erosion is more likely to be lateral – to the sides – which broadens the valley.

The clearest difference between the upper and middle section of the Wharfe valley lies in the winding river course, known as a river **meander** (Figure 2.26). On a lower gradient, the energy of the river is directed laterally (to the sides), and faster

Changing river environments

water is forced to the outside of the bend, causing undercutting and eventual collapse. The channel is therefore moved as the river bank erodes. Figure 2.27 shows how the river course now meanders across a flat area known as the **floodplain**, whose edge can be marked by a **river cliff**. Continued erosion creates a narrower neck between neighbouring meanders, leading to an ox-bow lake. The movement of meanders and the formation of an **ox-bow lake** is shown in Figure 2.27.

Figure 2.25 A meander on Oughtershaw Beck, upper Wharfedale. Undercutting has produced a steep bank on the outer edge of the river, while on the opposite side a bank of sediment – known as a point bar deposit or slip-off slope – has accumulated.

Figure 2.26 Meander and floodplain near Kettlewell in upper Wharfedale. The sinuosity value of the river is high. The edge of the floodplain in the far distance is shown as a cliff or sharp break of slope up the valley side.

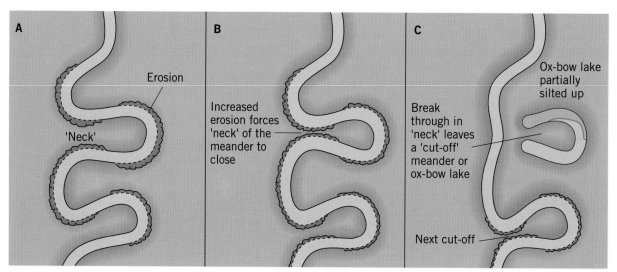

Figure 2.27 The migration of meanders across a meander belt. The migration in Stage A leads to a narrower meander neck in Stage B. The erosion of the neck in Stage C creates a cut-off meander known as an ox-bow lake.

Why do rivers meander?

Figures 2.24 and 2.25 show how different the river channel is along the River Wharfe where the valley floor is flatter. Instead of the straight or gently winding channels seen upstream along Buckden Beck, meandering channels, which sweep broadly over a flat valley floor, are more common. Winding river channels are said to be **sinuous** because they do not take a direct course between two points. The extent to which a river winds is its **sinuosity ratio**. Figure 2.28 shows how a river may travel a much longer distance from A to B than the direct line. The ratio of the distance travelled to the straight line distance is its sinuosity. In this case, the direct distance is 100 metres, and the river channel takes 350 metres to travel; hence the ratio is 1:35.

Straight sections of any river are rare. Water flows in a corkscrew motion known as **helicoidal flow**. Where a river reaches lower ground, the vertical stress on it is reduced and more energy is directed laterally, or to the sides. The greater the volume of water, the greater lateral motion, which erodes the river banks rather than the channel. Meanders are therefore part of the natural flow of a river. This is shown in Figure 2.29.

This process allows water to 'swing' against the bank of a river, where undercutting takes place, followed by bank collapse (Figure 2.30). Note the overhang where further collapse is likely to take place. The corkscrew motion of water means that sediment is actually transferred across the channel to the other side, on the inner part of the bend, where it is deposited. Energy used in erosion results in reduced velocity so that deposition occurs. A bank of sediment develops which is known as a slip-off slope. In this way, sediments are constantly being removed and re-deposited, as shown in Figure 2.31.

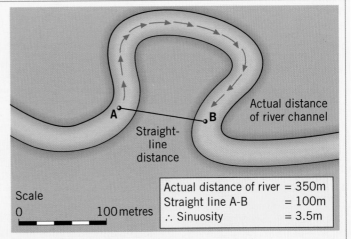

Actual distance of river	= 350m
Straight line A-B	= 100m
∴ Sinuosity	= 3.5m

Scale
0   100 metres

Figure 2.28 Measuring sinuosity along a river channel.

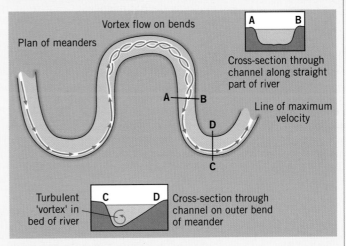

Figure 2.29 Flow of water in a river channel.

Figure 2.30 Erosion on the outer bank of a meander. River energy is directed at the bank, causing collapse. Sediment is transferred by the corkscrew action of water flow across to the inner bank, where sediment is deposited.

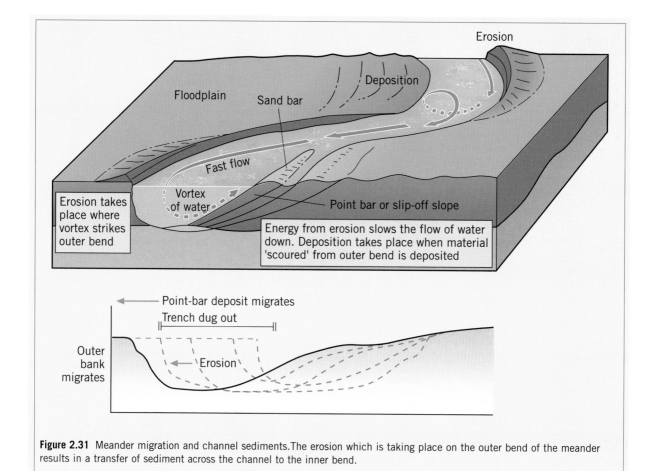

Figure 2.31 Meander migration and channel sediments. The erosion which is taking place on the outer bend of the meander results in a transfer of sediment across the channel to the inner bend.

The flood threat in Wharfedale

Most rivers flood at some stage. The River Wharfe floods several times per year and there are attempts in different locations to manage the flood threat. As yet, however, these are relatively few, because few people are affected. With limited budgets, local councils and the UK Environment Agency have few resources to allocate and usually do so where people are most affected. Particularly in Upper Wharfedale, where village populations are small, there is little land pressure to build further. Buckden and Kettlewell are each sited on land well above the level of the surrounding floodplain.

Chapter 3, a study of Keswick in the Lake District of northern England, shows the issues from flooding in an area where pressures from urban populations are greater. You should compare this with the River Wharfe, and explore how human activity may successfully reduce the flood threat.

Summary

You have learned that:

- Rivers change in their course from upland to lowland sections.
- Rivers may erode more vertically in upper parts of the basin, and more laterally in lower sections.
- In spite of appearances to the contrary and reductions in gradient, they may speed up as they move downstream.
- Rivers alter their velocities according to channel shape and size.
- Rivers develop valleys with distinctive features at different stages of their course.
- Rivers remove and transport weathered material from the valley sides to the river by mass movement.
- Rivers erode and transport material that is later deposited further downstream.
- Rivers may flood in river basins where there is little management or attempt to stop them doing so.

Ideas for further study

1 Compare data for Buckden Beck and the River Wharfe by carrying out studies of streams close to you, or in fieldwork locations. Identify key features of different parts of the river course, and assess how much rock type, weathering process, mass movement, and channel shape and size contribute to the development of a valley landscape.

2 Identify river valleys close to you where management is minimal, and where flooding can be expected during some parts of the year. What causes the flood? Who is affected by it, if anyone? Are there any attempts to manage flood events more in future?

References and further reading

Bishop, V and Prosser, R (1997) *Landform Systems*, Collins Educational

Clowes, A and Comfort, P (1987) *Process and Landform*, Oliver and Boyd

Hilton, K (1985) *Process and Pattern in Physical Geography*, Unwin Hyman

3 Flooding in Keswick

In this chapter, the study of flooding in Keswick in the UK Lake District shows how management of a river becomes a priority, and how successful flood alleviation schemes can work.

Figure 3.1, a news report from the weekly *Lake District-Herald*, reports floods that affected Keswick on Saturday 21 December 1985, the first floods in Keswick for eight years. It tells of families whose homes were flooded to a considerable depth, as the photograph in Figure 3.2 shows. Only parts of the town were affected, shown in Figure 3.3, and while they were not on the scale the 1952 Lynmouth floods, they caused significant damage to property. Look at the shape of the flooded area. To the west, it is narrow and close to the course of the River Greta. Near Fitz Park, it widens and spreads into a broad expanse, flooding several houses.

'Operation Clean-up' after the Christmas flood

Dozens of Keswick families had their Christmas holiday ruined when flood water poured into their homes on Saturday.

People living in High Hill, Crosthwaite Road, and Limepotts Road were worst affected by the water which rose to five feet in places after the River Greta burst its banks.

Mr Sydney Thorley's home in Crosthwaite Gardens was one of the worst affected by the flood.

BUNGALOW HIT

'We had about five inches of water throughout our bungalow and about 18 inches in the garage,' Mr Thorley told the *Herald*. 'The water was coming in so fast that we had to call the police and fire brigade. They managed to commandeer a boat and came right up to the front door to get us.

'It has undoubtedly spoiled our Christmas. Fortunately, we have no relatives or friends coming this year but we were hoping for a quiet Christmas,' said Mr Thorley.

Mr Barry Abbot's bungalow in Limepotts Road was also under several inches of water, and furniture, carpets, and beds were damaged.

DIESEL OIL

'All sorts of things came in with the water, even diesel oil. It all happened very quickly and the water went into every room.'

'We have been busy mopping up and we are all waiting for the insurance men to come.'

Many people living in Crosthwaite Road found their cars damaged by the flood water. In some cases it rose as high as the windscreen and spoiled upholstery.

Ravensfield was not flooded, though residents were evacuated as a precautionary measure because only a wall was holding back the raging River Greta. Some residents were taken to Keswick Cottage Hospital and some to other homes in the county.

Police stacked sandbags against the riverside wall which held and prevented much worse flooding.

The force of the extra water in the River Greta washed away part of a bridge at The Forge under the Keswick by-pass. A bridge which links The Forge with Old Windebrowe, became impassable and was closed by the police.

Keswick Rugby club was under four to five feet of water and players and officials spent the whole of Sunday mopping the premises out.

ROADS CLOSED

The squash club and the town's bowling green were also hit and some roads out of the town, including the Keswick to Borrowdale and Keswick to Thirlmere routes were closed.

However police reported no casualties as a result of the floods.

Police paid tribute to the way people at Keswick aided others in trouble.

'When we were filling sandbags a tremendous number of local people turned out and did their share,' said Superintendent Davidson.

Figure 3.1 From the *Lake District-Herald*, 28 December 1985.

Figure 3.2 Limepotts Road, Keswick during the floods of 21–22 December 1985.

List the effects of flooding mentioned in Figure 3.1. Classify these into tangible (those which can be measured and costed) and intangible (those which cannot be measured but are still important). Which other effects can you think of that the newspaper should have reported?

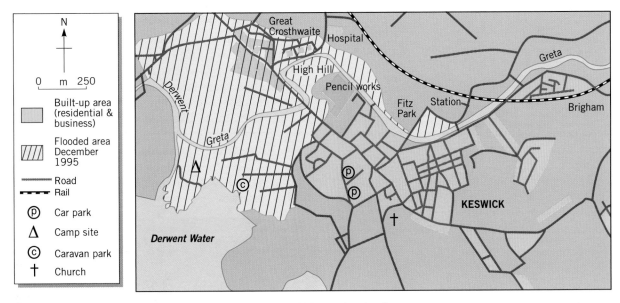

Figure 3.3 The extent of flooding in Keswick 21–22 December 1985. Notice the shape of the flooded area, affecting housing only in the western part of the town.

Keswick flood victims voice their anger at meeting with NWWA

OVER fifty people turned up at a meeting between Keswick flood victims and officials of the North West Water Authority.

The meeting, set up by Workington MP Mr Dale Campbell-Savours, was intended for those affected by the pre-Christmas deluge, but others attended to voice their anger.

The Water Authority are accused of allowing the Thirlmere Resevoir to remain too full, making the river Greta unable to cope with the overflow.

One member of the public who atteneded, Mr Eric Skilton, said he was 'far from satisfied' with the Water Authority's response to the allegations.

'We tried to get the Water Authority to immediately reduce the depth of Thirlmere to avoid another catastrophe but they didn't give much hope of this. We would be more satisfied if the lake was reduced straight away. this is the first thing to do as a safety precaution.

'Until the Authority reduces the level of the lake, the dangers are still here.'

Mr Skilton said Thirlmere's level needed to be reduced for only three months of the year when the lake was at its highest. At present, it was 4 m above danger level, he said.

There was general agreement from the Water Autority that they would pursue the matter with instruments and continue to investigate.

However, the Authority gave no offer of compensation to people whose homes and propery had been damaged.

Figure 3.4 From the *Lake District-Herald*, 1 February 1986. At a meeting with North-West Water Authority (now North West Water, a private water company), angry residents complained about flooding.

Floodings are a natural and essential part of river basin processes and are only a hazard because they affect people. Flooding occurs when a large volume of water enters a river system quickly and cannot be contained within the river channel. By studying floods, we can learn how a river basin works. Through studying a river basin closely, we can learn and predict how likely it is to flood. This information can be used to reduce the damage caused by such events, or perhaps even to modify future events themselves.

Refer to Figures 3.1 – 3.4.
1 a) Calculate the area affected by flooding.
    b) Estimate how many houses were flooded. Describe the location of these houses.
    c) Imagine that your home is flooded. Estimate the cost of the damage, including the cost to replace permanently damaged property and the cost of all the repairs.
2 How did residents of Keswick react to the flood in Figure 3.4? Were they right to feel this way?
3 According to the newspaper articles, who is responsible for managing flood systems in the Keswick area? What are their responsibilities (see Figure 3.4)?
4 According to the newspaper articles, 'somebody was not doing his job'?
    Who was this and what were they not doing?

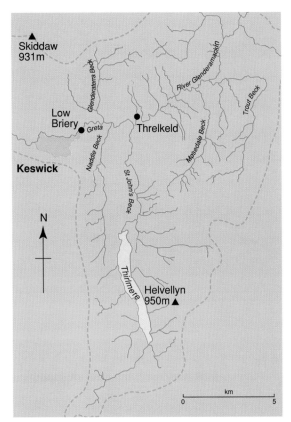

Figure 3.5 The Greta's drainage basin. A number of small streams or tributaries flow from within the watershed, joining to form larger streams and the main river course.

Identifying the causes of flooding

The main town of Keswick was not flooded while Upper and Lower Fitz Park suffered from flooding. Figure 3.6 shows how the main town of Keswick is much higher than Fitz Park. The height naturally protects this part of the town from flooding while Fitz Park is much more likely to flood. However, other factors contributed to the flood, including the regime of the River Greta, the nature of the basin, human structures, and the effect of other rivers. Some are natural, while others are the result of human activity and interference.

Figure 3.6 A view of Keswick from Fitz Park looking south-west across the River Greta. Fitz Park is much lower than this part of Keswick. Locate this view on a map of Keswick.

Factors that affect flooding

Flooding is usually the result of a combination of factors. Figure 3.8 is a Venn diagram which shows factors in a drainage basin, such as slope, human activity, soil properties, and the storms that may affect it. Some of these factors might produce conditions leading to flooding in the river basin, and should be shown in the shaded part of the Venn diagram, where the different factors intersect. Others are less likely to do this, and belong in the clear areas of the Venn diagram.

Read the theory box on *Storm hydrographs* (page 12). Figure 3.9 shows computer models of the storm hydrograph for the rainfall of 21 December 1985, which caused the flood of the River Greta in Keswick.

1 Copy the Venn diagram in Figure 3.8.
   a) Think about each factor in turn. If it will lead to an increased likelihood of flooding, place it in the shaded area. If it will reduce the likelihood of flooding, place it in the clear area.
   b) Complete the diagram by adding each of the factors to it.
      Sometimes several conditions occur together to produce flooding; at other times it only takes one or two.
2 Identify the December 1985 flood in Keswick on Figure 3.7. What does the figure suggest caused it?
3 From Figure 3.7, what conclusions can be drawn about seasonal weather conditions in 1985?
4 Occasional heavy rain storms in summer do not seem to affect discharge much. Why?
5 Even in lengthy rainless periods the River Greta does not dry up. Why might this be? Find out what baseflow is.

The regime of the River Greta and its basin

A river regime shows how the discharge of a river may vary over a year. Figure 3.7 shows how the regime for the River Greta varied through 1985. Discharge in this graph shows the flow of a river for a whole year, taken as daily measurements plotted and joined. Many factors affect the amount of discharge in a river channel. These include intensity and duration of rainfall, gradient or steepness of the slopes in the drainage basin, soil type and depth, **drainage basin density** (the relationship between the number of tributaries and the area of the drainage basin), and the shape and size of the drainage basin. (See also Theory box on pages 10 and 11.)

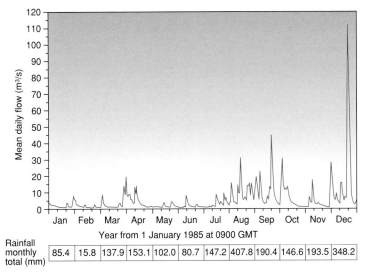

Rainfall monthly total (mm)	85.4	15.8	137.9	153.1	102.0	80.7	147.2	407.8	190.4	146.6	193.5	348.2

Figure 3.7 The regime of the River Greta in 1985. Compare this graph with the rainfall pattern shown beneath the graph.

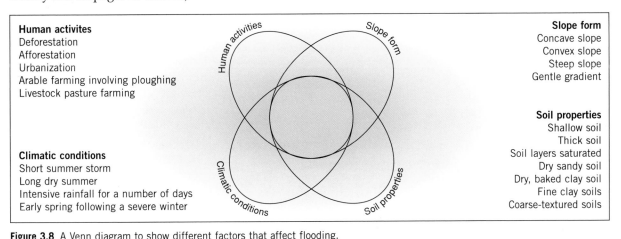

Figure 3.8 A Venn diagram to show different factors that affect flooding.

1 Study Figure 3.9.
   a) Which of the main tributaries contributes most to the flooding of the River Greta? Why?
   b) Which of the main tributaries contributes least to the flood? Why?
   c) Which of the main tributaries peaks latest? Why?
2 In small groups, and using an OS map of the Lake District:
   a) Identify the River Greta and its tributaries.
   b) On your copy, mark in the watershed of the River Greta's drainage basin. Use the contour lines to help you.
   c) Annotate your map to show how the flow of different tributaries affects this drainage basin. Refer to the hydrograph of the Greta's tributaries and to Figure 3.9 to help you.
3 a) Find out about the main rock types, vegetation cover and human influence in the area. Add these labels to your map.
   b) Using a matrix showing these and any other factors, summarise factors that show why the River Greta flooded in 1985.
4 In what ways might the flood in Keswick be linked to the shape of the drainage basin?

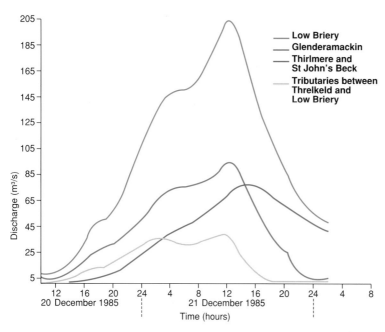

Figure 3.9 Simulated flow hydrographs (based on rainfall and flow data) at Low Briery showing contributions from three main tributaries of the River Greta for the flood of 21 December 1985.

Why did the River Greta flood in the west of Keswick?

The key to understanding why the west of Keswick flooded can be seen in the High Hill area (see Figure 3.3). The flood situation was worsened by the Greta Bridge (Figure 3.10). The bridge is located on a meander bend in the river, which slowed down the flow of the Greta. The arches of the bridge were not large enough to allow the discharge through. They created a back-log of water.

However, as Figure 3.3 has shown, Keswick itself was not flooded, only the area downstream from High Hill and Great Crosthwaite was subject to significant flooding. A major contribution to the cause of flooding here was the confluence of the River Greta with the River Derwent. The increase in discharge created a backflow of water, made worse by the fact that the River Derwent was at bank-full level. This meant that water in the Great Crosthwaite and High Hill areas had not drained away fast enough to prevent serious flooding. Add a note about this on your sketch map.

Figure 3.10 The Greta Bridge, showing how the river is restricted through the bridge. Notice the arches and central support. The support takes up space and prevents much water from getting through quickly in times of flood. However, it is of historical interest, and instead of removing it and constructing a new bridge, the plans are to deepen the river bed and increase water capacity that way.

Solving the flood problem

This exercise is about possible solutions to the flood problem in Keswick, through a 'balloon debate'. There are many different solutions to the flood problem; some may be better than others. A number of people have to argue their case to stay in a hypothetical balloon that is losing height. The balloon stays in the air by losing people. People stay in the balloon through the strength of their arguments.

1 Five speakers, or pairs of speakers, are allocated roles with viewpoints which they must defend.
2 Speakers have two minutes to justify their viewpoint.
3 When all the speakers have presented their view, the audience votes to eject one viewpoint from the balloon.
4 The remaining speakers are cross-questioned by the audience in five-minute rounds.
5 At the end of each round one more viewpoint is ejected.
6 The most persuasive viewpoint wins.

The five roles and viewpoints are:

1 Thirlmere is a flood control device. In fact, Thirlmere is a reservoir built to supply Manchester with water, but people in Keswick believe that one way to control excess run-off should be to store it in lakes, so that no flooding will occur. Local opinion is that Thirlmere could be used to regulate the flow of the River Greta, particularly during peak run-off periods.

2 The engineers' view. Engineering solutions provide the best means of controlling flood waters through Keswick. Engineering schemes are popular and meet with local approval because they are visible and show that something is being done. The engineer's argument is that technology can control flooding.

3 The opposite view to (2) above. Floodwaters can be controlled through engineering, but only so far. Any means of controlling flooding, such as a flood wall, can only be as good as its design and the standards to which it has been built. If its design standards are exceeded by subsequent floodwater levels, damage may be even greater than if there had been no attempt to control floods in the first place.

4 Flood abatement schemes are necessary. By changing land use upstream, such as afforestation, the regime of the river is altered. Increased afforestation would increase interception and delay the time taken for rainwater to reach a river. This view analyses the causes of flooding through river basin processes.

5 The Environment Agency. The Environment Agency is responsible for all river systems in the UK. It argues that Thirlmere is not there to control flooding. It believes that Thirlmere has been successful in supplying water to Manchester for over 50 years, and that if it helps to control flooding at the same time, then this is an added benefit. In addition, it believes that Thirlmere is part of one tributary, St John's Beck, in a river basin that contains many others. Flooding could never be controlled by Thirlmere alone.

Flood alleviation

Flood alleviation is concerned with reducing the effects of flooding on the human environment. There are several possible approaches (Figure 3.11) that range from engineering modifications to river channels and protection of buildings from flood damage, to doing nothing and bearing the cost of any flood which occurs. The four main strategies are:

- structural protection measures through engineering solutions
- river basin management
- modifying the burden of loss
- bearing the loss caused by flooding.

For economic reasons, bearing the loss caused by flooding is more likely to be found in developing countries. It is not likely to meet with support in a developed country – you could consider the reasons for this. The graph in Figure 3.12 shows that in developed countries the loss of life due to flooding is falling with time, but that economic losses are increasing.

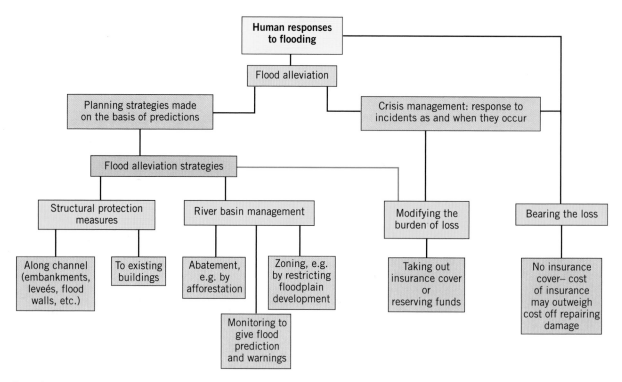

Figure 3.11 Responses to flooding.

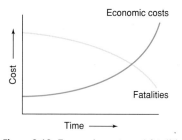

Figure 3.12 Economic costs and fatalities caused by flooding in developed countries. Notice how although fatalities are falling, economic costs are rising sharply. Think about why this might be.

Figure 3.13 shows six approaches to structural defence measures. Figure 3.13a shows the urban area to which different solutions are applied in Figures 3.13b to 3.13g. Study the diagrams carefully as you read the text. Construct a matrix to show the advantages and disadvantages of each approach. Some further reading may help with this. Which approach do you feel is best?

Structural protection measures through engineering solutions

Engineering structures represent the most publicly acceptable response to the flood hazard. 'Flood protection' is a misleading term, as protection is only as good as the design of the structure in question. Since it is the most publicly accepted response, residents may believe flood protection to be without fault. Damage is likely to be greater when a protection scheme is breached. People depend upon it and probably do little else to prepare for the flood. Education programmes need to work alongside structural schemes, to encourage flood-proofing. River management increasingly concentrates on reducing flood losses rather than preventing flooding altogether.

1 Embankments, or levées, and flood walls (Figure 3.13b)

These are designed to restrict flooding to defined limits on the floodplain and to allow controlled flooding of certain areas. They are a relatively cheap form of flood protection and operate to design standards if well maintained. Ideally, flood walls should be located as far away from the river as possible. However, flood walls and embankments can increase flooding both upstream and downstream; upstream because they may constrict flow, or downstream because they may discharge water more quickly into an area less able to absorb it. Figure 3.15 shows flood embankments in Keswick to protect the pencil factory.

Figure 3.13 Six approaches to defending a town against flooding.
Figure (a) shows a town affected by flooding. Figures (b) to (g) show different ways of tackling flooding. Study these carefully as you read the text.

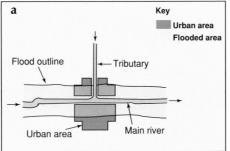

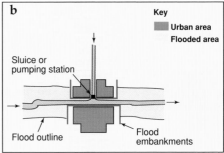

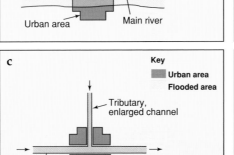

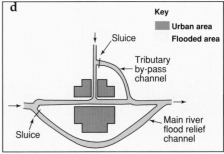

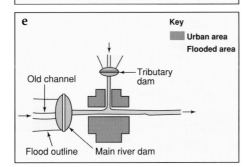

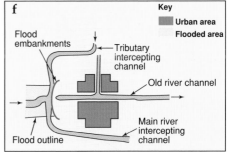

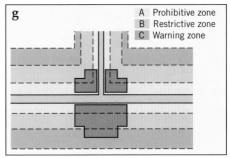

2 Channel improvements (Figure 3.13c)

Channel improvements are designed to confine floodwaters to the river channel in one of two main ways:

- Channel roughness can be reduced by clearing vegetation and other obstacles or by lining the channel with a smooth surface such as concrete. This reduces friction and allows discharge to pass more quickly. Compare the wall in the photograph in Figure 3.14 with the rough river bed. The river bed is strewn with stones and boulders. These increase friction, thus slowing the river down and reducing channel capacity. The wall dividing the river from the road is much smoother and enables water to flow more quickly. Measures like this help to discharge water quickly and reduce the risk of a flood, but may increase the risk downstream.
- The channel can be widened or deepened by dredging. This increases the capacity of the channel.

A disadvantage of these schemes is that maintenance is often ended as the river tries to revert back to its natural state – see the theory box on page 36.

3 Relief channels (Figure 3.13d)

The channel can be shortened by cutting meander loops and steepening the gradient to discharge water away from the area at risk quicker. These schemes are initially very attractive and widely used but they have disadvantages. A faster discharge may increase the likelihood of flooding downstream. In addition, straightening the channel may be self-defeating, as the river goes back to its natural state of dynamic equilibrium (see Theory box). This means that costly maintenance work is frequently needed.

4 Flood storage reservoirs (Figure 3.13e)

These store excess water in the upper reaches of the catchment which is gradually released. Because of the high costs involved, reservoirs have never been built in the UK for the sole purpose of flood alleviation, though some have contributed to flood control. In the UK, only upper river courses offer suitable sites away from lowland areas where most flooding occurs; such reservoirs therefore have minimal effects downstream.

5 Flood interception schemes (Figure 3.13f)

These involve changing the course of the river and work in three different ways:
* the river is re-routed to by-pass settlements under threat
* new channels can be used in addition to the natural channel to store water
* flood embankments help to contain floodwater well away from places under threat.

Figure 3.15 Embankments constructed as part of flood management strategy in Keswick.

Figure 3.14 A comparison to show the roughness of the river bed in Keswick and the smoothness of the wall which divides the river from the road. The wall enables the river to flow more quickly as its surface is smooth and offers less friction.

Theory

Equilibrium in river systems

Rivers can be seen as systems. They reach a state of balance with processes that form them, called **equilibrium**. However, processes vary and so equilibrium is always adjusting itself; this is called **dynamic equilibrium**. Balance within systems is complex, as the flow of energy and materials passing through may vary. If rainfall decreases then a river system has less energy and is less able to carry sediment. Sediment is therefore deposited; over time it builds up and increases the gradient of the river channel, causing the river to speed up as the gradient increases. Sediment is removed again and equilibrium is restored as the river returns to normal. This is known as self-regulation.

Sometimes conditions change so that the river is unable to recover its original equilibrium and a new equilibrium has to be found. When Mount St Helens erupted in 1980 a massive amount of sediment was released into the neighbouring river systems. The original balance could not be recovered because of the size of inputs into the system. A new balance had to be found by the rivers.

Any changes to the river channel create conditions in which the river will attempt to re-create its original channel. Chapter 2 has shown how meanders are created by the natural 'corkscrew' flow of water, which creates bends. Engineering solutions to flooding which attempt to straighten a river channel will therefore be frustrated by the natural energy of the water, which may actually damage work done.

River basin management

River basin management means reducing the harmful effects of a flood, while accepting that a flood may happen. Four methods are well tried: flood abatement, flood-proofing, floodplain zoning, and flood prediction and warning.

Flood abatement

Flood abatement aims to reduce flooding downstream by changing land use upstream. The most frequently used is afforestation, or planting additional trees. The passage of water into river channels is delayed through increased interception. Greater evapotran-spiration through the new vegetation also increases the water losses. This method can be effective but requires large areas of land, and time for forests to become established. For example, channels dug for plantations before tree planting on the slopes of Ingleborough and Whernside in the Yorkshire Dales National Park, actually caused a short-term increase in run-off and flooding in the initial stages. Therefore this method is not suitable for all catchments and needs large areas of land not being used for other purposes to be effective.

Flood-proofing

Flood-proofing involves designing new buildings or altering existing ones to reduce damage that would be caused by flooding. These measures may be temporary or permanent. It includes raising building levels, having steps up to doors, or using sandbags. Flood-proofing tends to be less effective in high-level, fast-flowing and longer-lasting floods. In Figure 3.16 a wall has been built above the river level to protect buildings against a flood.

Figure 3.16 A mix of old and newer buildings in Keswick. Notice how all the buildings are situated well above the level of the river. This is a good example of flood-proofing.

Floodplain zoning (Figure 3.13g)

This management strategy aims to reduce floodplain development. It suggests that floodplains are divided into zones:

Zone 1 *The prohibitive zone* (nearest the river) where no further development is allowed except for essential waterfront facilities.

Zone 2 *The restrictive zone* where only essential development and recreational facilities are permitted. All buildings should be flood proofed.

Zone 3 *The warning zone* (furthest from the river) where inhabitants receive warnings of impending floods and are reminded regularly of the flood hazard.

Flood prediction and warning

River and precipitation levels are carefully monitored by the Environment Agency using data provided by the Meteorological Office. In this way, accurate predictions of flood events can be made in time to reduce damage caused by flooding. Flood warnings are

Figure 3.17 shows the front of a Flood Warning leaflet for Keswick produced by the Environment Agency. Design the rest of the leaflet, including all the information you think residents should have available to them in the event of a flood.

Figure 3.17 Flood warning leaflet for Keswick produced by the Environment Agency.

issued by the Environment Agency, local borough councils and the police. This makes evacuation or removal of possessions possible where necessary. It also enables residents and businesses to put temporary flood-proofing measures into practice.

The measures proposed for Keswick

Consulting engineers Babtie, Shaw and Morton produced a report that considered the different options for Keswick.

- Thirlmere was considered and found to give some protection against floods. It would not be possible to rely upon Thirlmere, however, because it is not designed for this purpose and only controls one tributary in the Greta basin.
- Increasing floodplain storage capacity upstream from Keswick would reduce downstream flood flows. This could be achieved by constructing a flood control weir that would raise water levels upstream during flood conditions, or by constructing an embankment around the edge of the floodplain. This would create a storage reservoir to hold water for longer in the natural floodplain. Farmers whose land would be affected opposed this.
- The cost effectiveness of local flood alleviation schemes in Keswick were considered. The channel between Great Crosthwaite and Greta Bridge, and the bridge itself, have insufficient capacity to maintain floodwaters. Therefore the capacity of both the channel and the bridge have to be improved.

The following five basic schemes have been investigated:

1 The construction of flood walls and embankments at Crosthwaite, the pencil factory and in the vicinity of Greta Bridge.
2 Removal of the weir at Greta Bridge and local re-grading.
3 Re-grading the channel for approximately 400m upstream.
4 Widening of the channel for approximately 350m upstream.
5 Both re-grading and widening of the channel. This design has been based on a 50-year flood flow at 210m^3/sec.

Flood management – a decision-making exercise

You are on work experience with the Environment Agency, working in their Keswick office. Produce a flood management report for Keswick. Refer to all the data supplied in the text so far. Your report should include:

- a statement about the flooding problem at Keswick
- a table showing alternative types of flood alleviation scheme, with the benefits and problems of each type
- an annotated copy of the map in Figure 3.3 showing the types and locations of schemes you advise should be undertaken
- reasons for the schemes that you recommend
- a questionnaire to be completed by residents of at-risk areas in Keswick to assess public opinion of your scheme.

Theory

Flood frequency analysis

It is useful to estimate the probability that a flood of a certain size has of occurring in a given period of time. Flood management depends upon such estimates. To do this, the following steps are taken:

1 The size of the largest flood for every year is arranged in rank order, 1 being the largest, for the time-scale that records are available.

2 The following equation is then used to calculate the interval in time between floods of the same size:

$$T = (n+1)/m$$

where T is the recurrence interval, n is the number of years of observation, and m is the rank order. The value you get tells you the number of years within which a flood of this size can be expected. Note that this calculation is only the probability of flooding, based on what has happened before. Nature is not as predictable.

Flood alleviation schemes are normally built to withstand floods of a 1 in 50-year return period. The Environment Agency now builds alleviation schemes for floods with a 100-year return period. They now consider it likely that a flood will breach any defences yet made.

---

1 Figure 3.18 shows the frequency with which flood levels are likely to be repeated. This is called the flood return period. Estimate from the graph what the peak flows are likely to be for Keswick at:
   a) 25 years
   b) 50 years
   c) 100 years.

2 Compare your answers with the level of the River Greta in the 1985 flood, shown in Figure 3.9. What is the return period for a flood of this level? By which year is it likely that a flood of this type will have been repeated?

3 How should The Environment Agency use such data?

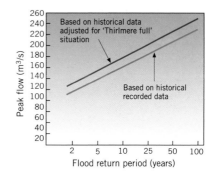

Figure 3.18 Flood frequency distribution at Low Briery gauging station (just upstream of Keswick).

Scheme 1

Scheme 1 was used as a base with which to compare other schemes. It is probably unrealistic as it is likely to result in peak river levels upstream of Greta Bridge up to 450mm higher than those during the flood of 21 December 1985. Flood walls like that in Figure 3.14, and embankments, would be built out of stone with reinforced sheet-piling at certain locations. They would be designed to contain floods to a level predicted to occur only once in 100 years. The flood would be contained within the channel through Keswick with the exception of Upper and Lower Fitz Park areas.

Scheme 2

Scheme 2 concerns Greta Bridge shown in Figure 3.10. At present, the bridge does not allow sufficient water to pass through at peak flow, thus encouraging flooding upstream of the weir. The capacity of the bridge would be improved by removing the weir and lowering the river bed under the arches. This scheme would be insufficient

a) Read the text for Scheme 1. Why do you think the peak flows would be raised upstream of the bridge by the embankments and flood walls?

b) Why would the Upper and Lower Fitz Park areas be left free of embankments and flood walls?

in itself but would enhance benefits from other schemes. Replacing the original bridge with a new single span bridge was also considered, but the cost was likely to be in the region of £0.5 million.

Scheme 3

This involves re-grading the channel and removing the weir. Dredging the channel would allow higher discharge levels to pass through. Between Lower Fitz Park and Greta Bridge, a distance of about 400m, the level of the river bed would be lowered by half a metre and the bank of the channel near the pencil factory trimmed to a gentler slope.

Scheme 4

Under Scheme 4, the channel is widened and the weir removed. Though costly, it is thought to be more reliable than re-grading the channel, as it is less likely to change with time. The initial degree of flood protection is likely to be maintained. The channel would be widened along the pencil factory bank for a distance of about 350m. Embankments and flood walls would be used in conjunction with this scheme.

Scheme 5

This combines schemes 3 and 4. The additional channel capacity created by re-grading and widening the existing wall along Main Street would mean that no increase in height is required. In the long-term, it is likely that maintenance of the re-graded river would be required.

Technique

Assessing the effectiveness of flood alleviation schemes – a cost-benefit analysis

Five schemes for Keswick were considered by the Environment Agency. A means of analysis and evaluation would help to identify the best scheme. One technique frequently used is cost-benefit analysis, in which the costs of a scheme are weighed against the benefits.

Costs and benefits can be divided into primary and secondary categories. Primary costs and benefits are directly concerned with the scheme. Primary costs in a flood alleviation scheme would be purchase of materials or purchase of land. A primary benefit would be saving the cost of cleaning up after a flood. Secondary costs and benefits are related to the scheme but not directly. A secondary cost might be in educating people about the possibility of a flood exceeding the design capacity of the scheme. Secondary benefits might include increased value of property previously prone to flooding.

Public projects of this kind have some costs and benefits which are difficult to measure. Placing a value on the costs and benefits of a scheme can be very difficult. Some factors are easy to cost, such as building materials; others, such as loss of trade caused by flooding, are more difficult. A reduction in human stress caused by flooding is even more difficult to value. Such analysis is bound to be highly subjective, and means that figures vary according to personal opinions of individuals or groups.

A cost benefit might look like this:

Item	Financial cost	Intangible cost	Financial benefit	Intangible benefit

The final decision

In Keswick it was finally decided to:
- increase the capacity of Greta Bridge by removing the weir and deepening the channel shown in Figure 3.18
- increase the capacity of 450m of channel upstream of the bridge
- construct 1300m of earth embankment, and modify and construct flood walls along 300m of river
- improve the wall along different stretches of the river in Keswick.

1 Construct a table to show the primary and secondary costs and benefits involved in the proposals for Keswick. List the primary and secondary costs and benefits of each proposal, and decide whether each is tangible or intangible.
2 Do the benefits of taking action regarding flooding outweigh costs? Or is there a case for the 'Bearing the cost' approach where nothing is done, there is no cost, and flooding is dealt with as and when it happens? Justify your answer.

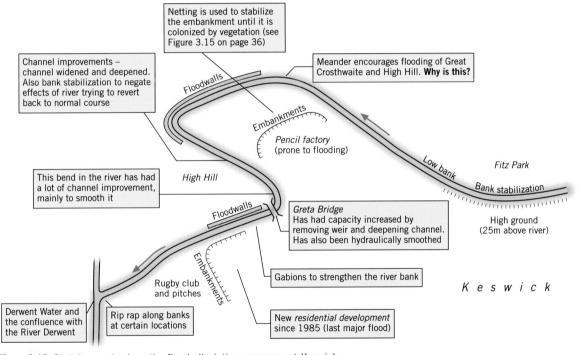

Figure 3.19 Sketch map to show the flood alleviation measures at Keswick.

There were two phases to the scheme. The first included construction of flood defence embankments, walls and sewer outfall improvements. The second included river works and re-grading upstream of High Hill Bridge (deepening and widening the channel to increase channel capacity), retaining walls, deepening under the bridge and constructing a flood bank alongside the pencil works.

You will have your own views about whether or not you feel these were the best solutions. The borough engineers who planned the solutions will always question whether or not they were the best solutions they could have obtained. They won't know until the next large flood occurs. If the solutions work, it is wondered whether people will congratulate the engineers. If the solutions don't work, the newspapers will probably highlight all the negativity.

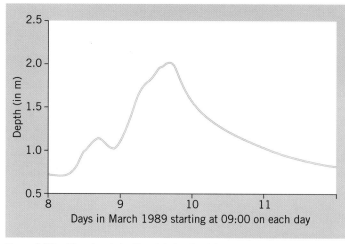

Figure 3.20a River levels in Keswick for 8 – 11 March 1989. The scale in this case is in metres, like data for the River Wharfe in Chapter 2.

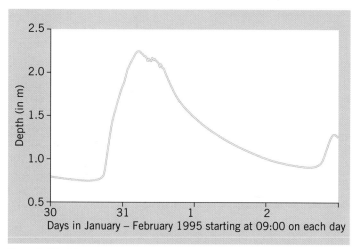

Figure 3.20b River levels in Keswick for 30 January – 2 February 1995.

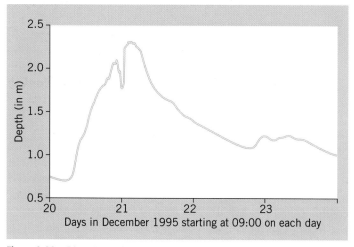

Figure 3.20c River levels in Keswick for 20 – 23 December 1995. This is the flood level originally described earlier in this chapter.

The success of Keswick Flood Alleviation Scheme

The Environment Agency consider the Keswick scheme a success, as it has protected the town from flooding since 1989 when it was completed. By their calculations, flooding would have occurred in Keswick if the scheme had not been implemented. Figure 3.20 supports their case. Figure 3.20 a shows river levels in March 1989, a river level of a 1 in 10-year return period. On the basis of river levels, flooding would have occurred here, especially in 1995 (Figure 3.20b). The 1995 levels are sufficiently high to represent a 1 in 15-year return period. Compared with Figure 3.20c, the river level for December 1995, the river levels are not as great, but without the scheme, they would certainly have been sufficient to cause flooding. That flooding has not occurred since is due to the success of the flood scheme implemented by the Environment Agency.

To test the Keswick scheme to the full requires a flood of up to 1 in 50-year to return. Nonetheless, residents of Keswick are happy with the scheme.

A number of changes have been made to the scheme since its completion. These include:

- further raising of downstream walls below Greta Bridge, to accommodate higher river levels in the River Derwent
- the relocation of flood banks downstream at Greta Grove to allow a housing development. This was not considered to contradict the floodplain policy, as it is located a long way downstream from the town
- the replacement of the floodwall behind Crosthwaite Road, following structural failure and seepage.

The scheme is ongoing. The Environment Agency Flood Warning system for Keswick is widely publicised, and every opportunity is taken to promote measures to reduce the risk from flooding. Regular maintenance is also needed, including inspection of all defences and regular gravel removal from the river bed near the pencil factory. It costs about £5000 to remove and is done every 18 months.

The floodplain policy on development has been maintained although a development has been allowed. The low cost housing Greta Grove development shown in Figures 3.19 and 3.21 was built in an area that is no longer subject to flooding, following completion of the Keswick flood protection scheme.

Figure 3.21 The Greta Grove housing development. Locate this on your map.

Study Figure 3.21. What evidence is there of protection from the flood hazard? What modifications should the Environment Agency insist upon to help alleviate the flood hazard?

Testing the Environment Agency's floodplain policy

The severe flooding of Easter 1998 in England and Wales led to an independent review by Mr Peter Bye. His report, presented to the Environment Agency in October 1998, formed the basis of an Action Plan. The report identifies floodplains as an important component of the river landscape and fundamental to alleviating flood risk. It shows that:
- they are the result of natural processes, and to disturb them would upset the natural equilibrium of a river system
- they are used by the river as a store when river channel capacity is exceeded, until the channel can transport the water again

- they can increase the risk of flooding downstream if they are not used
- they are of great importance to people. Many settlements are located on floodplains and best agricultural land is usually found there.

The Environment Agency has a floodplain policy which ensures that development:
- should not take place where there is an unacceptable risk of flooding, danger of life, damage to property and wasteful expenditure on remedial works
- should not create or make flooding worse elsewhere
- should not prejudice possible works to reduce flood risk
- should not cause unacceptable detriment to the environment
- should ensure that natural floodplains are retained and, where practicable, restored in order to fulfil their natural functions.

New development in Keswick

The Environment Agency were pleased that new development on the Greta floodplain at Greta Grove did not conflict with their floodplain policy. A proposed new development at the pencil factory is not as clear; the owners of the pencil factory have applied for planning permission to expand. As one of the organisations that has to give its approval, the Environment Agency needs to carry out an Environmental Impact Assessment to decide whether or not to grant permission.

The methods of assessing environmental impact shown in the Technique box assume that each factor has an equal weighting in terms of importance. In reality, certain impacts tend to be more important than others. For instance, wildlife habitats would be more important to English Nature than flood alleviation schemes, whereas the opposite may be true to the local council. Rank the factors you have scored in order of importance.

This method focuses upon environmental effects of projects and looks beyond cost-benefit analysis. However, more information is needed for an EIA than for a cost-benefit analysis. Although both are useful techniques for measuring impact, the final decision may often be based on economic or political considerations rather than those of an environmental nature. There are many cases where recommen-dations have been ignored or over-turned.

Technique

Environmental Impact Assessment

It is now a legal requirement that an Environmental Impact Assessment (EIA) is carried out before any major development projects are approved. An EIA is a means of estimating change to the environment that might occur as a result of a project, and weighing up the problems and benefits. It is usually carried out in the following way:

1 An assessment of the existing environment is made.
2 The proposed development is described.
3 The probable environmental impacts of the development are estimated and assessed in relation to the impact upon natural environment, its ecology, habitats, and impact that any change or pollution might have. It is also assessed in relation to the impact upon the human environment, such as recreation, health, aesthetics, well-being of people, and local employment opportunity or reduction.
4 These environmental impacts are scored on a grid like that shown in Figure 3.22.
5 Any modifications that could be undertaken to minimise adverse impacts are considered and their impact reassessed.
6 A decision is made.

Factor	strong negative impact –3	negative impact –2	slight negative impact –1	no impact 0	slight positive impact +1	positive impact +2	strong positive impact +3	Column A Score 1–5	Column B Final weighted value
Environmental factors River quality Appearance of river channel Flooding risk Drainage of land Water quality Ecology Other visual impacts Recreational value Wildlife habitats Employment Housing benefit for people									
								Total	

Figure 3.22 An example of an environmental impact assessment.

The Greta Grove development – an Environmental Impact Assessment

Form small groups and complete the following exercise. Each group should assume the role of one of the following:

• a planning officer for the local council
• an engineering consultant for the Environment Agency
• a member of English Nature
• a member of the local Housing Association, for whom the housing is planned.

In your role:
1 a) Carry out an environmental impact analysis (EIA), using the table in Figure 3.22 and the scoring system shown.
   b) Weight the 12 factors by giving each a score between 1 and 5. For example, if you think that 'water quality' is one of the most important factors, you could give it a score of 5; if you think 'employment' is unimportant, you could give it a score of 1. Write these in column A.
   c) Multiply your weighting by the value you have given in the EIA in **1a**. For example, if you have scored 'water quality' –3 and weighted it 5 in your rank order, then multiply –3 × 5 = –15. Complete column B in the table.
   d) Add all the weighted values together in column B to get a total.
   e) Justify your scoring system.

2 Compare the rank orders of the four different groups. How do they differ? Why should this be so?

3 Write a report, drawing together the attitude of each of the four groups. On the basis of this, would you recommend the Greta Grove scheme?

Summary

You have learned that:

- Floods occur naturally and are hazards only because of human development on the floodplain.
- Floods happen because a lot of water gets into the river quickly. This can be due to human and natural factors. Flood hazards usually occur because of a combination of the two factors.
- Models such as the hydrological cycle, storm hydrographs and river regimes can help us understand the movement of water within river basins.
- People have certain strategies which they use to help alleviate the effects of flooding on the environment.
- It is impossible to completely defend the environment against the effects of flooding; an understanding of flood frequency analysis shows that there will eventually be a flood that is capable of breaching defences.
- Analysis techniques such as Cost Benefit Analysis and Environmental Impact Assessment can be used to choose the most suitable means of flood alleviation.
- Different groups of people will have conflicting opinions on how best to manage drainage basins; there are no simple, absolute answers.

Ideas for further study

In your examination it is useful to have case studies that complement each other to offer different perspectives on the same topic.

1 Research a case study of a flood event in a drainage basin of comparable size to the Greta. It should not one of the large drainage basins of the world. The case study should differ in some way from that of the River Greta. For example, it could be in a different climate zone or be a heavily urbanised basin. Use sketch maps to show the causes and effects of flooding.

2 Examine the flood alleviation techniques in either the same drainage basin or another, more suitable basin. Try to use an example of a basin that has been less successful in its flood alleviation scheme. Use annotated sketch maps to illustrate your study.

References and further reading

Bishop, V and Prosser, R (1997) *Landform Systems*, Collins Educational

Clowes, A and Comfort, P (1987) *Process and Landform*, Oliver and Boyd

Hilton, K (1985) *Process and Pattern in Physical Geography*, Unwin Hyman

Prosser, R (1992) *Natural Systems and Human Responses*, Nelson

There are many articles in *Geography Review* magazine that are relevant to your background reading. The following are a few selected articles:

Higgit, D, 'Where the river runs deep', *Geography Review*, September 1996

Howes, N, 'Equilibrium in physical systems', *Geography Review*, September 1992

Zurawek, R and Migon, P, 'Flooding in the Odra Basin', *Geography Review*, May 1999

4 Freshwater ecosystems

So far, this section has looked at rivers in terms of river water, valley landforms and human manage-ment issues. Rivers are also an essential life force, providing living environments for a range of flora and fauna. This chapter looks at ecosystems associated with rivers and lakes, and how changes in plant communities occur. It shows how types of plants depend on features of the natural environment such as climate, rock type and soils.

Defining an ecosystem

Figures 4.1 and 4.2 show Sweat Mere in Shropshire, a small freshwater lake. This lake has different plant communities on and around it, including:

- open water with floating vegetation such as water lilies
- reeds and marsh plants at the edge where the lake is beginning to silt up
- woodland on the drier land surrounding the lake.

An **ecosystem** is a community of plants and animals together with the environment in which they live. It has two main components:

- **abiotic** or non-living parts of the ecosystem, such as rocks, soils and water
- **biotic** or living parts of the ecosystem, such as plants, animals and micro-organisms.

An ecosystem operates like other systems with inputs, stores, processes and outputs. Ecosystems are found within the biosphere, which is the part of the Earth and its atmosphere that supports life. The **biosphere** includes oceans, land and the lower atmosphere. Life within the biosphere depends on the exchange of materials with the **hydrosphere** (or water environments), the **lithosphere** (or surface environments) and the **atmosphere**.

Figure 4.1 OS map of Sweat Mere, Shropshire.

© Crown copyright

Figure 4.2 Sweat Mere, Shropshire. There are at least three different ecosystems in the photograph. Each community of plants requires different conditions in order to survive.

Ecosystems exist at a range of scales, from a small puddle or rotting log, to a lake, woodland, or even an ocean or forest. They also exist at a much larger, global scale, where they are referred to as biomes. **Biomes** are plant and animal communities characterised on world vegetation maps as, for instance, tropical rainforests, savanna, or tundra.

Boundaries of ecosystems are difficult to identify and define. Plants may not have fixed boundaries, so that the edge of the lake may be indistinct from the beginning of reeds and marsh plants. Inorganic components, such as air, move from one ecosystem to another. An ecosystem is therefore an **open system** because it is affected by factors outside it. The variety of environments within Sweat Mere means there are several different ecosystems within the area shown in Figure 4.3.

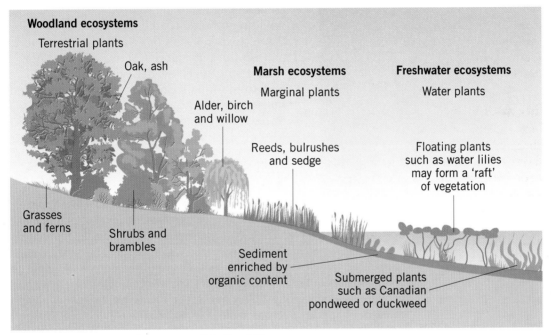

Figure 4.3 Different types of ecosystem within Sweat Mere.

The 'structure' of an ecosystem

To study the **structure** of an ecosystem, both biotic and abiotic elements of the system should be considered. Living components are plants, animals and micro-organisms which are interdependent and survive together. They also depend on the non-living world and must adapt to survive in the surrounding physical conditions. Together, biotic and abiotic elements are arranged in a structure. Three examples are:

- the **layers of vegetation**, which describe the physical structure of an ecosystem
- the **food web**, which shows relationships between different species
- the **biomass and dead organic matter** (DOM).

Figure 4.4 Different communities of plant at Sweat Mere.

Layers of vegetation

A close look back at Figure 4.2 shows that the plants are arranged in layers according to size and height. This occurs in both aquatic and land environments.

Aquatic ecosystems

Within the lake, four distinct communities of plants exist, shown in Figure 4.4.

- **Marginals** are found around the edge of the pond or lake and are able to survive in oxygen-free muds and silt from which they obtain nutrients. These include reeds and bulrushes.
- **Floating plants** depend upon sunlight at the surface of the water for their energy. These include water lilies as well as species that float without anchored rooting systems.
- Plants **suspended** in the water are free-floating and do not rely upon a permanent anchorage, such as algae. These derive their food source both from sunlight close to the water surface and nutrients in the water.
- **Submerged** plants rely upon the bed or banks for an anchorage but depend upon being submerged. Plants such as duckweed and Canadian pondweed are examples.

The terrestrial ecosystem around a pond or lake

On land, the smallest and lowest level of plants make up the **ground layer**. These include most water plants, reeds, sedges and wild flowers that grow close to the water's edge. They are as much part of the terrestrial ecosystem as they are of the aquatic. The **shrub layer** consists of taller woody bushes such as brambles, dog rose and young trees or saplings. The **tree layer** forms the tallest layer and consists of species such as willow or alder. They live close to water where they are adapted to growing in moist soil.

Biomass and dead organic matter (DOM)

Both biomass and dead organic matter are part of the structure of an ecosystem. **Biomass** is the total amount of living plant and animal matter present. It is expressed as the dry weight of tissue in an area, either as tonnes per hectare or kilograms per square metre. Biomass also occurs below the surface, for example as roots, animals and micro-organisms, and above the surface as stems, leaves and animals.

Dead organic matter (DOM) is also an important part of the structure of an ecosystem. Dead and decaying plant matter, consisting of surface litter and soil humus, may actually be greater than living biomass by weight.

Food webs

Wetland areas such as Sweat Mere include the rivers and the surrounding land. They provide a home or **habitat** to many small freshwater creatures and insects. Birds and water fowl use the area to

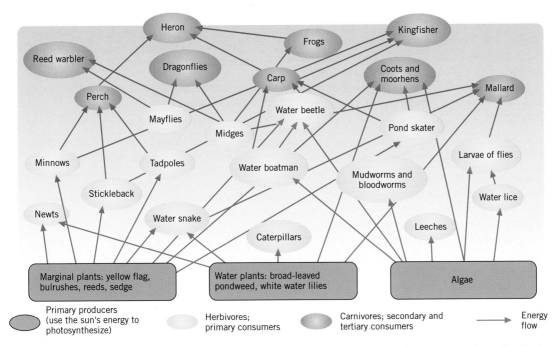

Figure 4.5 Food web of Sweat Mere aquatic ecosystem. Both aquatic and terrestrial environments are shown; the food web shows the dependency of one upon the other.

feed and rodents can often be found living in or close to the water. A large number of plants also live close to the water and, together with algae and protozoa, use the sun's energy to photosynthesize. Together these form the base of the food web, shown in Figure 4.5.

How ecosystems function

Wetland ecosystems, like all ecosystems, function in two ways which are closely linked. These are **energy flow** and the **cycling of nutrients** (sometimes called 'biological cycling'). The Theory boxes explain energy flow and nutrient cycling. Refer to this information through this chapter and throughout Chapter 10, *Investigating coastal ecosystems*.

Theory

Energy flow

Energy is essential for all processes within an ecosystem. Virtually all energy is derived from the sun, although a small amount comes from geothermal energy. Solar radiation, or insolation, reaches the ground in varying amounts depending on latitude, altitude, season and time of day. Green plants use some of this light energy in photosynthesis, where inorganic substances of carbon, oxygen and water are converted into organic compounds, particularly sugars. Oxygen is released during this process.

$$6CO_2 + 6H_2O \rightarrow \text{Energy from sunlight} \rightarrow C_6H_{12}O_6 + 6O_2$$

The rate at which energy is converted into organic matter through photosynthesis is called Gross Primary Productivity (GPP). GPP is measured in kilograms per square metre per year ($kg/m^2/yr$). GPP varies enormously between world ecosystems because rates of photosynthesis differ. Tropical rainforests, hot deserts and tundra regions each have different GPP rates because different amounts of light energy and moisture are available for photosynthesis.

Energy fixed by plants during photosynthesis can be:
- **used** by the plant for life processes, such as respiration
- **stored** as plant material or animal tissue, known as biomass
- **passed through** the ecosystem along food chains or webs.

Plants use energy during respiration and energy is then re-converted back to the atmosphere. Energy fixed by photosynthesis (GPP) minus that lost through respiration is called **Net Primary Productivity** (NPP). Like GPP, it is measured in kilograms per square metre per year.

Energy is stored in plant tissue, but may be passed from plants to animals through a food chain. All animals obtain food either directly or indirectly from plants. Figure 4.5 shows a simplified food web for Sweat Mere, Shropshire:

- **Primary producers** (plants and algae) are consumed by **primary consumers** (herbivores), such as water fowl, snails, insects and some fish.
- These in turn are consumed by **secondary consumers** (carnivores), such as frogs, fish, and birds which feed on insects.
- **Tertiary consumers**, such as otters, and larger birds such as herons or kingfishers, feed on fish.

When plants and animals die they are decomposed by organisms such as bacteria and fungi which use the dead tissue for energy. Each stage in the food chain is known as a **trophic level**. Producers are known as **autotrophs** and consumers as **heterotrophs**. Figure 4.6 shows trophic levels for Sweat Mere, Shropshire.

Energy is passed through the ecosystem, through each trophic level. At each level, energy could be:
- used by the animal for life processes such as moving, eating and respiring
- lost through animal waste
- stored as body tissue, such as muscle, and become part of the biomass.

Whenever energy passes from one trophic level to the next, about 90 per cent is lost and only 10 per cent is passed on. This huge loss is common to all ecosystems and helps to explain why there are rarely more than five trophic levels in any ecosystem. It also explains the inefficient use of world food resources in the production of meat.

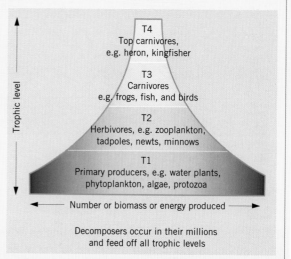

Figure 4.6 Trophic levels for Sweat Mere, Shropshire. In general, smaller numbers of living organisms exist at each successive trophic level; thus there are larger numbers of primary consumers than secondary consumers. The decrease in numbers in successive trophic levels is marked because there are fewer animal species. If decomposers are included, the pyramid shape would alter as decomposers – which technically belong at T5 – are by far the largest in number. The ecological pyramid may also be constructed using biomass, or energy produced by each trophic level.

Figure 4.7 Energy flows in an ecosystem.

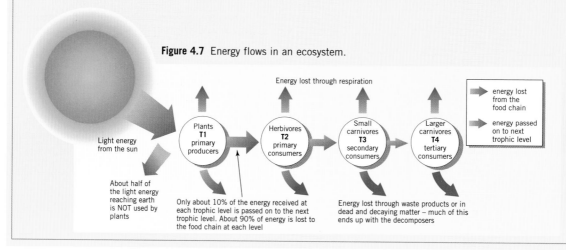

Theory

The cycling of nutrients within ecosystems

Nutrients are chemical elements and compounds needed by plants for growth. Unlike energy, which flows through the ecosystem and is eventually lost, nutrients are cycled within it. Essential plant nutrients include carbon, nitrogen, oxygen, hydrogen, calcium, phosphorus and potassium.

Ecosystems obtain nutrients from the atmosphere, lithosphere and hydrosphere. These are the three main **nutrient pools**. Some nutrients such as oxygen and carbon dioxide, enter and leave the system as gases but others, such as nitrogen, are absorbed by plants in solution from the soil. Nutrients may be introduced or lost either naturally or by human actions. The process of obtaining nutrients and recycling them depends entirely on the flow of energy. Nutrients are used by plants to grow and are converted into plant and animal tissue. When organisms die and decompose, nutrients are released and are used again by living things. Thus there is continuous cycling of nutrients in an ecosystem.

Nutrients may be introduced into an ecosystem by visiting birds and animals, or wind, and may be lost in surface run-off or leaching of soils. Greater changes to the balance can occur as a result of human activities, such as deforestation or the addition of fertilizer which finds its way into water courses.

Figure 4.8 Typical nutrient flows in a freshwater lake.

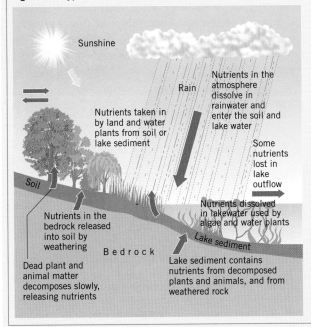

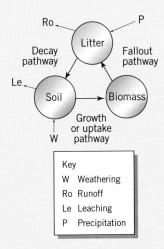

Figure 4.9 Nutrient cycling. Nutrients may be stored within the ecosystem in biomass, soil or litter. They are cycled between the stores by means of growth, fall-out and decay; they may be added or removed by weathering, run-off, leaching and precipitation.

---

1 Make your own copy of Figure 4.9. Describe the processes involved in the cycling of nutrients between:
   a) soil and biomass stores
   b) biomass and litter
   c) litter and soil.

2 How effective is this diagram in showing how nutrients may be added or removed from an ecosystem?

3 What do you understand by the difference between the terms 'structure' and 'functioning' of an ecosystem?

Plant succession in freshwater ecosystems

In any given habitat, vegetation changes over time. Plant species that arrive first will only survive if they are able to adapt to:

- the type of soil (depth, texture, chemistry and nutrient content)
- climatic factors such as temperature, light and moisture
- biotic factors such as grazing and the use of the area by people. Plant species also have to tolerate each other and compete for light and space. This leads to some species dominating others.

Although the photograph of Sweat Mere in Figure 4.2 shows the different types of plant communities on and around it, the communities are related, because they show different stages of development. 'Young' plant communities on the open water are quite different from the older 'mature' communities on the edge of the lake. Now it is mainly a wetland, dominated by aquatic plants but, in time, a series of changes, described in Figure 4.10, would alter this into a land-based ecosystem. Over thousands of years this lake may dry up completely and change into woodland. Such change in vegetation over time is called **plant succession**. Each stage of development is known as a **sere**. Plant succession that occurs in shallow fresh water is known as a **hydrosere**. The sequence at Sweat Mere is shown in Figure 4.10.

Figure 4.10 Plant succession on a shallow lake, a hydrosere.

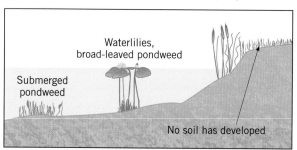

1 Sediment collects at the bottom of the lake and at the edges; the water gradually becomes shallower. Floating plants such as water lilies and broad-leaved pondweed develop on open water. Submerged plants such as duckweed and Canadian pondweed are found in even deeper water.

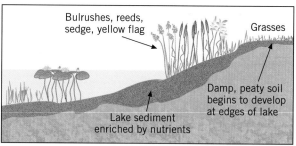

2 Towards the edge of the lake, the water is shallower and marsh plants become established, e.g. bulrushes and reeds which have their roots in the water but flowering parts above it. As dead organic matter accumulates, a damp peaty soil starts to develop (peat is formed from partly decomposed plant matter) enabling sedges and grasses to grow.

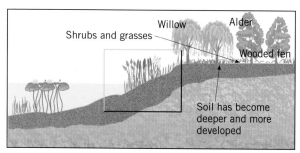

3 The frame shows the area in box 2. The lake becomes shallower and smaller. Soil at the edge dries as sediment and peat collects and becomes deeper. Shrubs such as bramble and willow and alder trees become established, forming wooded fen. The number of plant species is larger.

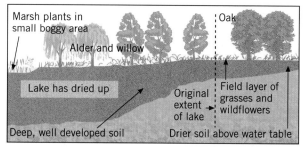

4 Eventually birch and oak trees are established once the level of the soil is above the water table. A field layer of grasses and wildflowers develop and a shrub layer may also be evident. This vegetation needs less acid, drier soil but once established it will become the dominant plant community, known as the **climatic climax community**.

It may take centuries for the process of plant succession to be completed. Eventually the lake will dry up and be replaced by vegetation. Most of the Shropshire and Staffordshire meres, or small lakes, were formed at the end of the last Ice Age, 10 000 years ago. Since then, they have reduced in size as plants have gradually colonized the edges. Other hydroseres develop more quickly, such as the reed beds that colonise the edge of ponds and canals. Loss of canal traffic in the UK since the mid-19th century has allowed marginal plants such as reeds to encroach upon canals which have silted up.

Hydrosere stage							
1	2	3	4	5	6	7	8
Open water; algae, water lilies	Bulrushes	Sedges	Willow, alder	Alder	Alder, birch	Birch	Oak

Plants and habitat							
Habitat description	Reed swamp	Marsh or fen	Open wooded fen	Closed wooded fen	Woodland		
Habitat processes	Accelerated deposition of silt and clay. Floating raft of organic matter forms and thickens		Raft now a mat resting on mineral soil	Black mineral soil revealed in patches; earthworms	Ground level now above water table; oak seedlings	Birch canopy forms; oak saplings	Oak grows through and then over the birch
pH level	–	–	7.3		4.3	3.7	–
Number of species of plant	6	10	14	26	18	14	10

Figure 4.11 Hydrosere at Sweat Mere, Shropshire (Cousens, 1974). This hydrosere is an example of colonization of open fresh water. Sweat Mere is a shallow lake which formed above poorly drained soil in the post-glacial period.

1 Study Figure 4.11 carefully. Draw a graph to show how the number of plant species varies in the eight seral stages.
2 Suggest reasons for the changes in the number of plant species. Do you think the number of animal species would follow the same pattern, or not? Give reasons for your answer.
3 a) Describe and suggest reasons for the change in soil pH in the different seral stages.
  b) How might other soil characteristics such as depth, moisture content and texture change?
4 What natural factors might prevent a lake or area of open water from drying up completely?
5 What sort of human activities might prevent succession from taking place around a lake, and why might this happen?
6 If a lake is close to farmland, excess fertilizer may filter into the water. How might this change the process of succession described in the text? (Refer to the Theory box on Eutrophication on page 55.)

The Florida Everglades, an aquatic ecosystem under threat

The Everglades, shown in Figures 4.12 and 4.13, is a huge area of southern Florida. It is generally thought of simply as a wetland area, but is in reality part of a huge river system draining water from Lake Okeechobee. The River Kissimee brings water from wetter areas of northern Florida into the lake. In the past the lake overflowed seasonally, causing water to move very slowly to the Gulf of Mexico, spreading out across a wide area and covering it with shallow water. It was inhabited by a huge variety of fish, turtles, alligators, birds and many plants, including an unusual plant called saw grass, giving rise to the name 'The River of Grass'. Today the Everglades is no longer free-flowing. The River Kissimee has been controlled, most of the land has been drained, and water is held back by dams and carried along 2250km of canals. The area is used to grow sugar cane, fruit and vegetables. Reservoirs store water so that it can be used for irrigation or to supply water to towns and cities such as Miami, Fort

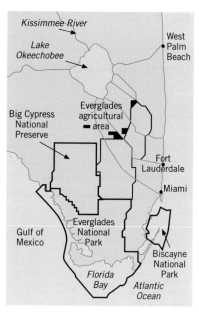

Figure 4.12 Florida and the Everglades National Park.

Figure 4.13 Aerial view of mangrove forest, southern Florida

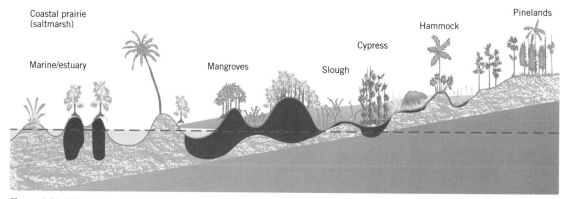

Figure 4.14 Different types of ecosystem within the Everglades.

Figure 4.15
The Everglades.

Lauderdale and Palm Beach. To serve this development, roads, railways and a huge infrastructure have been created. The only part of the area that remains like the original 'river of grass' is the Everglades National Park which covers about 10 per cent of the original wetland. Even here water is controlled by reservoirs and pumps, whereas in the past it flowed slowly and constantly across the land to the sea.

The food web for the Everglades is little different from that of Sweat Mere, except in species. As in Shropshire, primary producers (plants and algae) are consumed by primary consumers (herbivores), such as water fowl, snails, insects and some fish. These in turn are consumed by secondary consumers, such as frogs, fish, and birds which feed on insects. Secondary consumers are eaten by tertiary consumers, such as alligators and larger birds like the wood stork which feed on fish. The variety of environments within the Everglades means there are several different ecosystems within the area as shown in Figure 4.14.

Threats to the Everglades ecosystem

The intensive use of land in southern Florida and the large-scale water management schemes mean that the seasonal pattern of water release into the Everglades has changed. Many animals living in the Everglades are specifically adapted to alternating wet and dry seasons. Water may be released to prevent flooding, or it may be stored to reduce the risk of drought; in either case the natural cycle is disrupted and the ecosystem suffers. For example:

- Alligators build their nests at high-water level; if more water is released their nests are flooded and destroyed.
- Apple snails lay eggs above the water level in the wet season but if excess water is released, the eggs fail to develop. These snails are an important food source for many birds, including the endangered snail kite.
- The wood stork, a large wading bird, has declined in number from 5000 in the 1960s to around 500. These large birds feed on fish trapped in shallow, muddy water. A pair of breeding wood stork need 200 kg of fish which they find concentrated in the shrinking ponds of the Everglades during the dry season. When the natural cycle of water is upset, these birds are unable to find enough fish in the ponds and cannot breed successfully.

The marine ecosystem in the Florida Bay is also threatened. If the bay water becomes too salty because the supply of fresh water falls, shrimp, lobster, crab and fish populations will decline dramatically along the south coast of Florida.

Theory

Eutrophication

Many rivers and wetland areas are altered by excess nitrates and phosphates that run off the surrounding area. Nitrates largely derive from inorganic fertilizers, and phosphates from treated sewage effluent. These nutrients encourage rapid growth of primary producers, particularly algae (or phytoplankton). Initially this may lead to an explosion of animals in the higher trophic levels, such as insects and fish, with a corresponding increase in demand for oxygen.

As the numbers of algae increase, the water becomes turbid and light cannot penetrate, preventing many larger water plants from growing. As their rate of photosynthesis decreases, increased biological oxygen demand cannot be met. Fish and insects start to decrease in number rapidly. Blue-green algae often start to dominate to such an extent that they are clearly visible on the surface of the water. They are less edible to zooplankton than other algae, and some release toxins. There is a rapid decline in all other species and an accumulation of detritus. Unless this sequence is broken, the wetland area will become virtually lifeless, muddy and foul-smelling. The build-up of decomposing material will eventually lead to the area silting up completely. The whole process is known as **eutrophication**.

Eutrophication in the Everglades

The problem of supply

The population of southern Florida is rising rapidly. In the early 1990s migration from elsewhere in the USA accounted for over 900 extra people a day; the state population is now over 6 million. In 1930 Miami had a population of about 22 000 people; now there are nearly 3 million. These people all need a supply of clean water which reduces the amount reaching the Everglades. Increasing standards of living make swimming pools a common feature of most houses.

Agriculture is an important part of Florida's economy, particularly sugar cane, fruit and vegetables. This demands irrigation water, which is diverted away from the Everglades and changes the quantity of water in the wetland.

Tourism is booming; up to 40 million people visit southern Florida every year, many during the dry season. They makes huge demands on water supplies. As more hotels and other facilities are built, the amount of water that can infiltrate and reach natural aquifers is reduced.

The problem of effluent

Agricultural effluent not only makes demands on water but also threatens it. Water released into the Everglades from farmland contains concentrations of nitrates from chemical fertilizers. This affects plants and animals living in the wetlands and can cause eutrophication. Scientists are particularly worried that these chemicals may lead to the decline of saw grass and other important plants. High levels of mercury have also been identified in the food chain.

Alterations to the flow of fresh water

Seasonal variations in flow and demand for water alter the regime of flow into the Everglades. When fresh water runs low, salt water from Florida Bay penetrates the wetlands, upsetting the ecological balance. Freshwater plant tolerance towards salt is low (see Chapter 6).

A complex system of dams and levees protects surrounding settlements from flooding. Excess water is removed from areas where damage may be caused and diverted through the Everglades, flooding alligators' nests, washing away snail eggs and dispersing the concentrations of fish upon which wading birds rely.

Orange Grove Farm
South Florida

Dear Sir

Each year more and more regulations are introduced to stop farmers like me from doing their job. We have regulations controlling how much water we can use to irrigate our crops and we have regulations controlling when we can water our crops. I must declare what fertilizers I use to keep the yields on my crops up and what pesticides I use to control the weeds and bugs which can decimate my fruit. The quality of the water is checked as it comes on to my farm and again when it runs into the drainage ditches nearby. Now we are being told that if there is too much water we must let our fields flood and if there is too little we must hold back on irrigation.

In my opinion all this nonsense about protecting the environment has gone too far. Farmers must be allowed to use technology to produce good-quality crops for the people of Florida and beyond. After all, the economy of this state relies heavily on the fruits that we produce.

Yours faithfully

Tom Jackson

Farmer

Figure 4.16 From the *South Florida News*.

Solutions

Unless the Everglades National Park receives the water it needs, it will not survive. Various attempts to resolve the issues are under way but the fate of the park still hangs in the balance.

- The National Park was created in 1947 to protect the area and recently the boundaries were extended to include more land. It is hoped that this will restore a suitable habitat to many of the endangered species.
- The area has been named an International Biosphere Reserve and a World Heritage Site.
- The National Park and the State of Florida have agreed to work together to enforce water quality regulations.
- The US Corps of Engineers who are responsible for flood control have agreed to consider natural rainfall patterns when making decisions about water management.
- Filtration marshes have been built to clean up the chemical run-off from agriculture.
- The Kissimee River has been altered to allow it to flow more naturally and some areas of farmland are periodically flooded.

The success of these strategies will depend on how far environmentalists can convince planners, farmers and the local people that their own livelihoods are inextricably linked to the survival of this ecosystem.

1 The letter in Figure 4.16 was published in the *South Florida News*. Write a reply to Mr Jackson explaining to him how farming damages the Everglades ecosystem and why the controls about which he complains are necessary.
2 Find out how being designated as a World Heritage Site and an International Biosphere Reserve could help to protect the Everglades.

Summary

You have learned that:

- An ecosystem is a community of plants, animals and micro-organisms together with the non-living environment in which they live.
- Ecosystems function by transferring energy through complex food chains or food webs, and by cycling nutrients.
- The sun is the most important source of energy. Green plants produce food energy (sugars) from the sun through the process of photosynthesis. Energy flows through the ecosystem but at each successive trophic level, energy is lost.
- Nutrients originate in the atmosphere, hydrosphere and lithosphere and are cycled by the ecosystem.
- Ecosystems develop and change over time through plant succession until they reach equilibrium. This natural process can be upset by natural or human changes.
- People are an increasingly important component of ecosystems. Human activities may significantly alter ecosystems. This may be unintentional or for economic gain.
- Ecosystem mangement can prevent or reduce damage but to be successful, the structure and functioning of ecosystems needs to be understood.

Ideas for further study

There are many opportunities for fieldwork in wetland ecosystems. Choose a wetland area in Britain which is under threat, such as the Norfolk Broads, the Somerset Levels or the Fens. Find an OS map of the area and information about it from the library. Look at:

- the character of the ecosystem
- how the ecosystem is used by people
- evidence of damage being caused by the ecosystem
- management strategies and how successful they have been.

References and further reading

O'Hare, G (1988) *Soils, Vegetation and Ecosystems*, Oliver and Boyd
RSPB (1994) *Ecosystems and Human Activity*, Collins Educational

Web sites

The Everglades National Park at
http://www.everglades.national-park.com

5 Managing water supply and quality

This chapter investigates ways in which river basins are managed sustainably. It focuses upon the Murray-Darling river basin in Australia, where management is addressing past mistakes, and preventing future errors from being made.

The geography of the Murray-Darling river basin

The Murray-Darling river basin (Figure 5.1) is in south-east Australia. It consists of two large rivers, the Murray and the Darling, which join near Mildura in Victoria. Over 1 million square kilometres in size, the combined basin is four times larger than the UK – 1500km north to south, and 1000km east to west. To drive from south to north would take two days, or two hours' flying time. It includes Australia's highest point, Mount Kosciusko, within the Great Dividing Range of mountains to the east and south-east. It contains the capital, Canberra, within Australia Capital Territory, as well as parts of the states of New South Wales, Victoria, Queensland and South Australia.

Much of Australia's fresh food comes from the Murray-Darling river basin (Figure 5.2). The basin contains 42 per cent of Australia's farms and produces a wide range of agricultural produce, from livestock (sheep, cattle), fruit and vegetables, rice and cotton, to some of the world's best-acknowledged wine and dairy produce. Yet the basin faces some key environmental issues such as drought, flooding, water supply and water quality, irrigation and the problem of salinity, and land degradation.

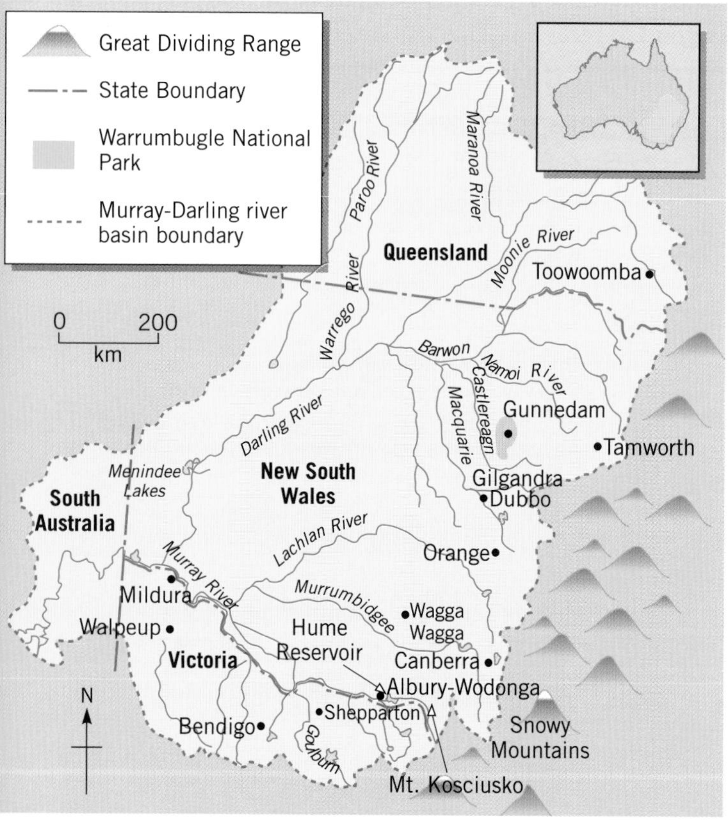

A basin under climatic stress

In the 1990s, food supplies and incomes from farming in Australia were threatened by drought. Droughts in Australia tend to be cyclic. Six major droughts have occurred since 1950. The long drought of the early 1990s was another in a series and was followed by another from mid-1997 until early 1999. The likely cause of these is a change in climate known as El Niño. El Niño produces changes in weather, bringing dry winds in a westerly flow across Australia which are most likely to lead to drought in the east. Study of the climate of the Murray-Darling basin will help to identify causes of drought and its impact. This includes rainfall totals, rainfall variability, evaporation and water budgets.

Rainfall totals

Figure 5.3 shows annual rainfall in Australia averaged over a minimum 40-year period. The Murray-Darling basin lies to the west of the Great Dividing Range of mountains which fringe the south-eastern and eastern parts of the country. Rain-bearing winds from the east deposit precipitation on the mountains, and to the west of the mountains the river basin lies in a rain shadow which has less rainfall than the east coast. The process is explained in Figure 5.4.

Figure 5.1 The Murray-Darling river basin, including its location within Australia.

On the mountains of the Great Dividing Range, totals include all forms of precipitation, such as fog and snow. Snowfall is significant on the Range and contributes considerably to river discharge. Discharge for each major tributary of the Murray-Darling is shown in Figure 5.5. Notice how the greatest flows occur to the south of the basin, along the Murray, reflecting greater snowfall in the mountain ranges.

Figure 5.2 Aerial view of orchards near Mildura, northern Victoria.

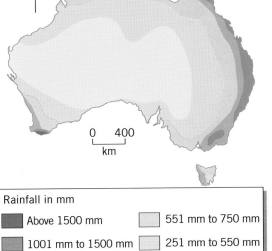

Rainfall in mm

■ Above 1500 mm	□ 551 mm to 750 mm
■ 1001 mm to 1500 mm	□ 251 mm to 550 mm
□ 751 mm to 1000 mm	□ less than 250 mm

Figure 5.3 Average annual rainfall in Australia.

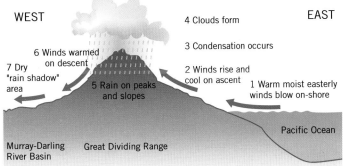

WEST   EAST

4 Clouds form

3 Condensation occurs

6 Winds warmed on descent

7 Dry "rain shadow" area

2 Winds rise and cool on ascent

5 Rain on peaks and slopes

1 Warm moist easterly winds blow on-shore

Pacific Ocean

Murray-Darling River Basin   Great Dividing Range

Figure 5.4 Relief rainfall and the development of a rain shadow to the west of the Great Dividing Range.

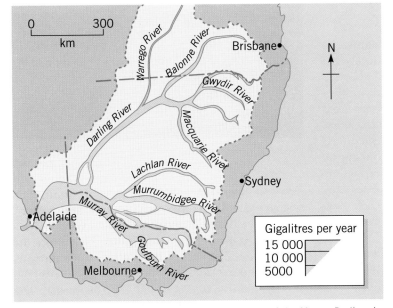

Gigalitres per year

15 000
10 000
5000

River	Min.–max. range (%)
Campaspe	6–210
Castlereagh	7–860
Culgoa	6–766
Darling	0–911
Goulburn	25–219
Gwydir	1–440
Kiewa	31–187
Lachlan	4–550
Loddon	5–256
Macquarie	2–940
Murray	25–214
Murrumbidgee	28–350
Ovens	14–264
Warrego	0–900

Figure 5.5 Variations in river discharge along tributaries of the Murray-Darling river system. Note: 1 gigalitre = 1 billion litres.

1 Study the southern part of the Murray-Darling basin in your atlas. Suggest reasons why snowfall should be higher in these mountains than in those further north.
2 How should this affect the seasonal discharge pattern? Sketch a graph to show discharge patterns during the year.
3 a) Explain how the data in the accompanying table to Figure 5.5 have been calculated and what they mean.
  b) Why, in Figure 5.5, should there be such variations in flow of the tributaries during the year?

Vegetation in the basin

The Murray-Darling basin is variable in appearance because vegetation changes depend on rainfall. There are three broad natural vegetation belts that correspond closely with rainfall totals and distribution. Compare the types of vegetation shown in Figures 5.6, 5.7, 5.8 and 5.9 with Figure 5.3, and also with Figure 5.13.

The slopes of the Great Dividing Range

The slopes of the Great Dividing Range to the east and south are mostly covered in eucalypt forest, or 'gum' trees. Figure 5.6 shows a view of the Snowy Mountains in the Great Dividing Range on the border of New South Wales and Victoria. Here total annual rainfall is higher, and less variable from one year to the next, than in other parts of the basin.

Central and eastern basin

The central and eastern areas are either natural or grazed temperate grasslands. Rainfall totals are lower here and more variable. Figure 5.7 shows a view of the Warrumbungle National Park in New South Wales. The landscape consists of broad areas of tussock grass, interspersed with isolated trees. The dead tree in the foreground of the picture is symptomatic of a grassland in decay. Grazing by cattle has prevented saplings from growing so that, in time, trees will disappear. Tussock grasses may grow up to 2m tall, and have deep root systems to obtain water from greater depth, which is important in drier years.

In the wetter grazing areas of eastern New South Wales mat grasses are found, typical of species encouraged by farmers and grazed by cattle and sheep.

Figure 5.6 Eucalypt forests on the Snowy Mountains, New South Wales–Victoria border.

Figure 5.7 Grassland in the Warrumbungle National Park in New South Wales. Notice the individual tussocks of grass.

Western and central basin

Vegetation in the western and central basin usually consists of different scrub or 'bush' vegetation of drought-resistant species. Figure 5.8 shows the landscape to the east of Broken Hill in New South Wales, where scrub vegetation dominates. This kind of vegetation is found in parts of the basin where rainfall is low and its distribution irregular. In some months, no rain may fall.

Figure 5.8 Scrub and semi-desert vegetation, with mixed eucalypt trees, in central New South Wales. This reflects the vegetation of over half of the basin. Notice that in spite of the low rainfall, several trees are able to survive as a result of their deep roots and drought-resistant leaves.

Year	Annual rainfall total in mm
1976	272
1977	203
1978	406
1979	432
1980	348
1981	380
1982	110
1983	424
1984	282
1985	268
1992	536

Figure 5.9 Rainfall at Walpeup, Victoria, for selected years 1976–92.

European styles of agriculture in the past 100 years or so have had a significant impact upon the natural vegetation. Much of what is left has been affected either by grazing, or by ploughing and crop growth. Figure 5.9 shows parts of the Murray-Darling basin where less than 20 per cent of vegetation that remains is 'natural'. Because of the human impact on the natural ecosystem, there is now pressure to maintain and conserve Australian species of plants and vegetation.

Rainfall variability in the basin

Rainfall totals are sometimes less important than rainfall reliability. Rainfall in the Murray-Darling basin varies significantly, both between and within years, creating occasional drought.

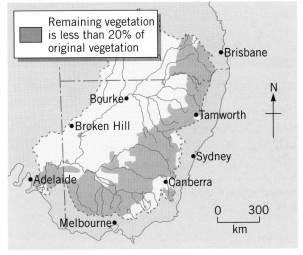

Figure 5.10 Areas of the Murray-Darling basin where less than 20% of the original vegetation is left.

	January	February	March	April	May	June	July	August	September	October	November	December	Total
Average rainfall	18.5	25.5	23	21.4	33.2	30.1	31.9	35.5	32.8	37.3	25.1	23.5	337.8
Highest recorded rainfall	128	191	127	101	98	101	84	89	128	134	88	88	697
Lowest recorded rainfall	0	0	0	0	0	0	2.3	0	1.5	0	0	0	109.9

Figure 5.11 Rainfall variation for Walpeup, Victoria, over several years, in mm

1 a) Use a suitable technique to draw a graph of the data in Figure 5.9.
  b) Using the average rainfall data in Figure 5.11, explain how these figures show that drought in Australia is cyclical.
2 Draw a graph of the data in Figure 5.11 and identify problems for farmers in the Murray-Darling basin.

Drought in the Murray-Darling basin

Figure 5.3 on page 59 shows annual rainfall. However, each year varies significantly from the average, as Figure 5.9 shows. Average annual rainfall at Walpeup over 40 years is 337.8mm. It varies so much that almost any month in the year could receive no rain at all, as shown in Figure 5.11. This unreliability has made farming difficult. Figure 5.12 shows this pattern is true for the whole of Australia. Inland, rainfall is not only less but increasingly unreliable.

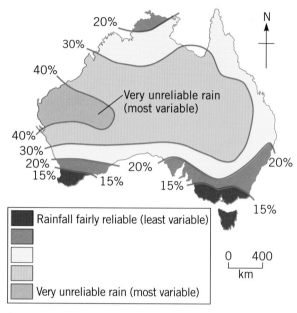

Figure 5.12 Rainfall variability across Australia.

Rainfall variability poses problems for farmers. Lowest rainfall totals coincide with summer when evaporation rates are high. In drought periods, rivers soon reach low levels (Figure 5.13). The Castlereagh River at Gilgandra in New South Wales reached low levels during early 1992 when drought was serious. Yet grasslands offer good grazing for cattle and sheep. Locally, water was, and still is, partly provided by windpumps, drawing water from underground (Figure 5.14). Water is pumped from water-bearing porous rocks – called aquifers – deep underground. This source is known as artesian water. It is sustainable at low densities of population, provided that volumes of water pumped out are balanced by those of rainfall seeping underground. It is not sustainable for large volumes of water and is already under serious threat as a sustainable store. User demand now exceeds rainfall supply by several times. In addition, cattle and sheep grazing around water points may increase pressure on pasture, so that overgrazing is a problem.

1 Study Figure 5.12 carefully. What data are necessary to produce this map? How are the values shown on the map then calculated? How has the map been drawn using these values?
2 Describe the pattern of rainfall in Figure 5.12. When does most and least rain fall? In which season is this?
3 What is the significance of rainfall variability in the Murray-Darling basin in terms of:
   a) its physical appearance?
   b) human activity?

Figure 5.13 The Castlereagh River, a tributary of the Darling, at Gilgandra in New South Wales during the winter 1992 drought. The size of the normal river channel is shown where river sediments have been exposed. In full spate, this and the grassland shown beyond the river would be covered.

Figure 5.14 Windpump in northern Victoria. The water is drawn from underground using wind power and is collected in the corrugated metal storage tank.

Evaporation rates and water budgets

Figure 5.15 shows how rainfall in Australia may be affected by evaporation. Inland temperatures may reach 40°C in mid-summer – between November and February – which reduces the effectiveness of water for plants and for farming. Compare this with the rainfall map, especially the totals per year.

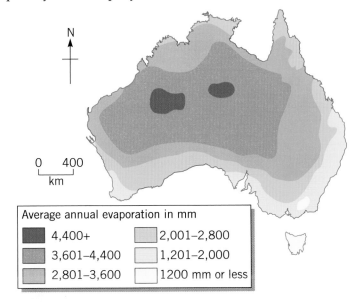

Figure 5.15 Annual evaporation amounts in Australia.

	Average evaporation rates in mm
January	245
February	207
March	167
April	99
May	53
June	33
July	37
August	59
September	87
October	133
November	177
December	239
Total	1536

Figure 5.16 Evaporation data for Walpeup, Victoria. Compare these with the data in Figure 5.11.

Evaporation rates are useful because they help to identify which areas have a surplus of rainfall over evaporation and which do not. Figure 5.16 shows evaporation data for Walpeup, Victoria (referred to earlier in Figures 5.9 and 5.11).

1 Describe annual evaporation in Australia. Why should there be lower rates in some places than others? Which places have the highest amounts? Why should they have such high amounts?
2 Plot a graph to show the data in Figure 5.16. Annotate it to show:
   a) trends during different seasons
   b) reasons for these trends.
3 Using a second colour, draw the average rainfall data as shown in Figure 5.11 on the same graph.
4 Does Walpeup have (a) an annual and (b) a monthly water surplus, water balance or water deficit? Use data to support your answers.
5 Summarise the features of the climate of the Murray-Darling river basin and the issues that different land users there have to face.

Managing water resources along the Murray-Darling rivers

The Murray-Darling river basin has been essential to the settlement and growth of the white Australian population. Without it, the population of south-eastern Australia, in which seven of Australia's eight largest cities are located, could not be sustained without pressure on land or resources elsewhere. Although only one of the seven largest cities – Canberra – is actually to be found within the basin, the remaining six depend upon it for food, water and/or electricity.

Changing river environments

	Total population in census years	
Area	1961	1991
Albury-Wodonga	29 361	66 543
Bathurst	16 938	27 205
Broken Hill	31 267	23 739
Dubbo	14 118	33 859
Griffith	7 696	20 532
Orange	18 994	35 205
Wagga Wagga	22 092	53 447
Bendigo	40 327	57 427
Horsham	9 240	12 552
Mildura	12 279	20 432
Shepparton-Mooroopna	16 085	30 511
Wangaratta	13 784	15 984
Murray Bridge	5 404	15 884
Toowoomba	50 134	81 043
Warwick	9 843	10 393
Canberra-Queanbeyan	67 151	304 264

Figure 5.17 The growth of urban areas in the Murray-Darling basin, 1961–91.

Figure 5.17 shows how urban areas in the basin have undergone rapid population growth in the last 40 years. Some of the towns grew in earlier periods for specific reasons; Broken Hill, in New South Wales, was a major mining centre until recently. However, other towns have grown because the availability of water – for consumption, agriculture and energy – has enabled large population increases to take place. Two factors are important in influencing population growth: irrigation and increasing soil salinity.

---

1 On an outline map of the Murray-Darling basin, locate the urban areas shown, using an atlas.
2 What has been the (a) actual and (b) percentage increase in population for urban areas in the period shown in Figure 5.17?
3 Form groups of three. Using different techniques, draw a map each to show:
   a) the actual increase in population 1961–91
   b) the percentage increase in population 1961–91
   c) the size of cities in 1961 and 1991, using proportional circles of two different colours. Use the information in the Technique box below to help you.
4 Comment on the patterns revealed by each of the maps.

Technique

Proportional circles

Proportional circles represent data where size is important spatially – for example, city populations can be shown in this way. The size of the proportional circle is in direct proportion to the number represented. To draw proportional circles, follow these steps:

1 Decide on the maximum size of circle that can be used to show the largest quantity on the map you are using. For this example, use a 2cm diameter or 1cm radius. The size will always depend on the scale of the map and how many circles you have to draw.
2 Calculate the square root of the largest figure you want to represent. If the value is 10 000, the square root will be 100.
3 The largest radius of 1cm is therefore equivalent to 100 because this is the largest figure you have to show. 1mm is equivalent to 10 units.
4 A second value of 6500 has a square root of 80.62, so the circle would have an 8mm radius. Calculate the sizes of all the circles you will need on this basis.
5 Construct a key as shown in Figure 5.18 beside your map.

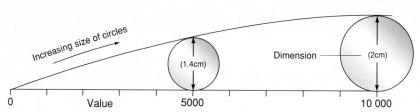

Figure 5.18 Scale for showing proportional circles, using the figures shown in the example.

Irrigation – opening up the bush for agriculture

If Australia was to grow from a nation of 7 million people in 1945, then it had to have a substantial farming base. Until then, most land east of the Great Dividing Range was used for cattle and sheep grazing. Further into the interior few roads existed and survival was not guaranteed for those who set out across the 'bush'. The plan was to farm the interior using the rivers of the Murray-Darling basin.

Figure 5.19 shows the extent of the Murray River valley that is irrigated. Irrigation of parts of the Murray-Darling basin began during the late 19th century, with schemes to store and use the water. Like many major river basins, the majority of the basin is much drier than its source. The highland catchment of the Murray itself is only 2 per cent of the total area of the whole basin but contributes one-quarter of its water run-off. Water can be transported through channels and pipelines to be used in lower parts of the basin. The creation of dams, pipelines, and HEP (hydro-electric power) stations for the new population, was known as the Snowy Mountains Scheme. Figure 5.20 shows one of the HEP stations built as part of the scheme.

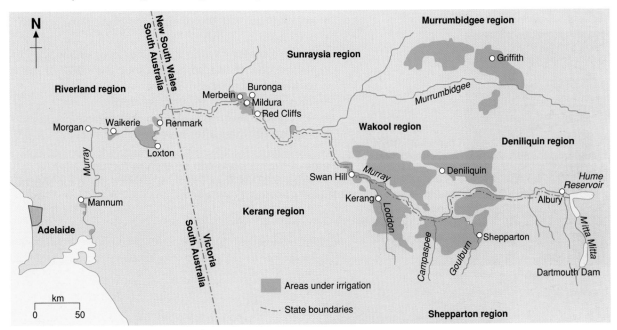

Figure 5.19 Irrigated areas of the Murray river system.

The threat of land degradation

While schemes designed to open up the interior were successful in making water available for farmers and consumers, serious environmental issues have emerged. Settlers applied European farming methods to the land, yet the Australian climate, soil and vegetation were different. Gradually, land has been degraded; it is poorer in quality and less fertile than before. Three serious problems have emerged: soil erosion, soil salinity and poor water quality.

Soil erosion

Grazing cattle and sheep removed woodland cover very gradually. Land was cleared for arable farming much more rapidly. Burning woodland cover was common, in spite of the bushfire risk, and 'ring-barking' trees, where the bark was cut in a circle

Figure 5.20 One of the HEP stations which form part of the Snowy Mountains Scheme.

Figure 5.21 A deep gully on a farm in Victoria.

Figure 5.22 Caked acidic soil in the Murray basin. This photo was taken in New South Wales where excess salt deposits had built up in soil that had been irrigated for several decades.

around the trunk, thus killing the tree. Sudden removal of tree cover exposed soil, either to intense rainstorms, or removal by wind. The effect on the soil is dramatic. Figure 5.21 shows deep gullying caused by run-off on a farm in Victoria.

Soil salinity

Irrigation has produced the problem of salinity in the Murray River basin. Salinity is a measure of the amount of mineral salts in water. Water quality is greatly affected by the saline content of water; too great a concentration of salts reduces plant tolerance for farming and the water may be unsuitable for human consumption. Saline soils are usually present where the following occur:

- isolated bare patches of land and poor plant growth
- trees dead or dying
- pools of shallow water which fail to drain away
- plants show changes in leaf colour, the result of stress
- reduced land productivity
- salt-tolerant plants increase in number
- salt crystals may be visible on the surface
- puddles of water after rain may taste salty.

Some of these effects can be seen in Figure 5.22.

Figure 5.23 shows a graph of salinity along the Murray River. Reading from left to right on the graph, salinity

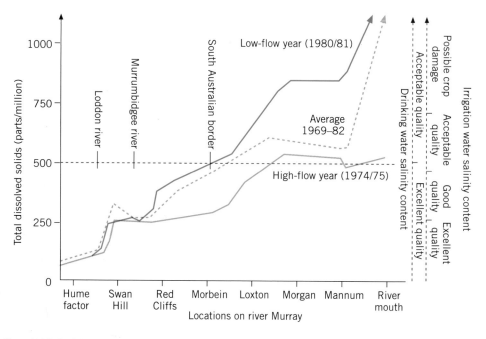

Figure 5.23 Salinity levels along the Murray River.

levels change between Hume reservoir in the upper parts of the basin (shown in Figure 5.1) and the mouth of the river. Notice that three lines are drawn on the graph: one for the average flow between 1969 and 1982, a second for low levels of flow 1980–81, and a third for a year with high levels of flow 1974–75.

1 Explain why salinity is high when river levels are low, and vice versa.
2 What does Figure 5.23 suggest about the tolerance of plants towards salinity? How does this differ from human salt-tolerance? What effects might increased salinity have on plants?
3 Draw a sketch map of the irrigated areas of the Murray valley shown in Figure 5.19. Use a colour scale for different levels of salinity shown along the river between Hume reservoir and the Murray estuary. Which areas appear to be most under threat?

Theory

Salinity – the salt of the earth

Salinity is a particular problem where major irrigation schemes are implemented. All water contains mineral salts; 'salinity' occurs where salt concentration reduces human and plant tolerance. The most common mineral salts are sulphates and chlorides of sodium, calcium and magnesium. The single largest cause of increased salinity is raised water tables that result from irrigation.

When irrigation water is applied to land, it contains mineral salts. Increased salinity is most problematic where volumes of irrigation water (and therefore salts) exceed those required by plants. This process raises the water table – the area of saturated rock or soil – close to the ground surface, bringing salts into contact with plant roots, as shown in Figure 5.24. Concentrations of sulphates and chlorides may even precipitate, or crystallize, from the water. These can be washed out by reducing the water level and 'flushing' them out but in doing so, clay and organic particles are lost and soil structure is destroyed. In addition, plants are unable to tolerate high levels of salt and unless irrigation is carefully managed, soils may become so alkaline that they become unusable.

Other causes include evaporation of water molecules from soil, especially in hot climates, resulting in a greater concentration of mineral salts. Leaching is another process by which mineral salts are removed from soil in solution by rain water or irrigation water as it soaks through the soil. Leached water therefore contains higher concentrations of salts. The salts are returned to rivers or irrigation channels.

As the process of irrigation increases downstream, salinity increases at the same time. This helps to explain the greater concentration of salts along the Murray River in Figure 5.23.

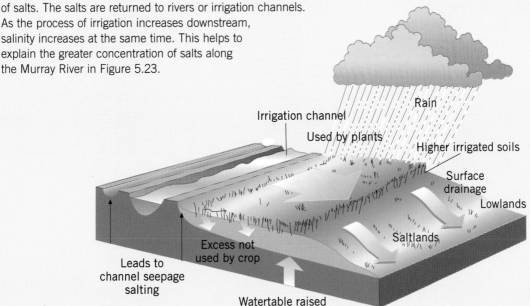

Figure 5.24 Salinity increases where the water table lies close to the surface.

Soil salinity issues in Shepparton, Victoria

Study Figure 5.25, an article from the *Shepparton News*, a local newspaper in a small town of 28 000 people in northern Victoria. The Lonely Planet guide to Australia (1989) describes Shepparton's location as being within 'a prosperous irrigated fruit and vegetable growing area'. However, the news article has a different view.

Increased rates, deteriorating roads, a higher incidence of divorce – even difficulties in burying the dead. They are all effects of the growing salinity problem in the Goulburn Valley.

The recently released draft management plan to deal with the problem is being explained at a number of meetings.

The plan, developed by the Salinity Pilot Program Advisory Council (SPPAC), makes a number of daunting predictions, apart from the obvious effect salinity will have on the environment.

It is expected there will be a 4.3 per cent rise in Shepparton City rates if the plan goes ahead. The plan is estimated to cost $880 million.

Money aside, SPPAC Chairman John Dainton brought home the reality of the problem with some chilling accounts.

'In a number of cemetery locations in the region, it is already difficult to bury the dead with high water tables making it necessary to pump graves prior to the coffins being lowered to their final resting place,' he said.

'We have all experienced what salinity, high water tables and poor drainage do to our road network.

'Rodney Shire estimates it is spending $1.2 million annually on road edging and foundations.'

Mr Dainton said that if the problem was not addressed roads would continue to self-destruct at an unprecedented rate.

The snowballing effects of these losses on processors, suppliers and the service sector in the region would be devastating.

SPPAC estimated this reduction in wages and salaries paid would manifest itself in job losses – 3500 in this city by the year 2020.

Perhaps the most frightening effect would be on the entire social, environmental, and economic fabric of the region.

'The breakdown of family structures which includes child abuse, divorce and suicide can be expected when a community becomes stressed,' Mr Dainton said.

'We can also expect illness rates to increase, a migration of our talented youth from the area, and an increase in the need for counselling at all levels.

'If nothing is done to arrest the alarming rise in water tables and the ensuing salinisation processes, the problems now faced by many land holders will multiply rapidly as the true extent of the problems unfolds,' Mr Dainton said.

The SPPAC report also envisages that river salinities will rise dramatically as a result of the dryland saltload doubling over 100 years from 196 000 tonnes to 460 000 tonnes a year.

Figure 5.25 From the *Shepparton News*, 4 September 1989.

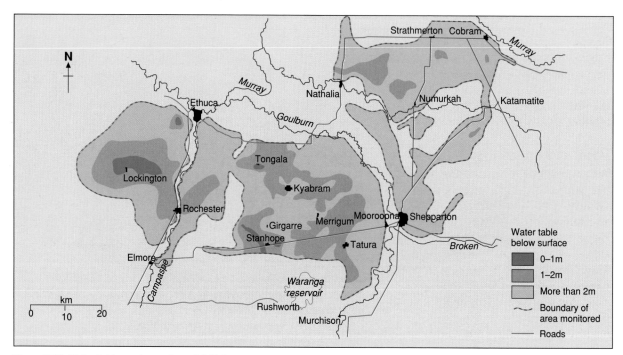

Figure 5.26 Water table contours, August 1982.

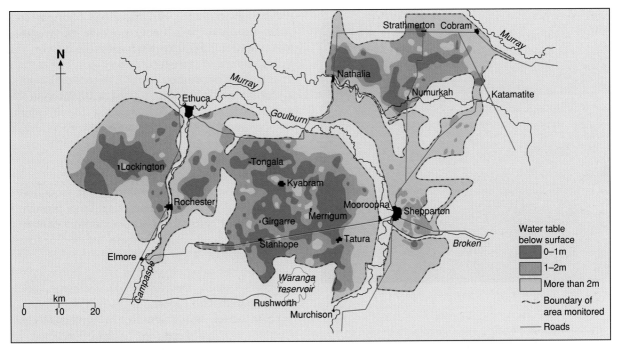

Figure 5.27 Water table contours, August 1990.

Figures 5.26 and 5.27 show the extent of water-logging of soils in the Shepparton region in 1982 and 1990. By 1990, over 50 000ha of the irrigated area had a water table less than 2m below the surface, or one-quarter of all irrigated land shown. Unless management practice is changed, this will increase to 90 per cent by 2020. Long-term productivity is estimated to fall by 30 per cent as salinity increases. In a region where orchards and irrigated pasture are the most significant part of the economic base, this is serious.

Poor water quality

Water quality has been a serious issue in the Murray-Darling basin. Other than salinity, another major cause is eutrophication, described on page 55. Algae are small aquatic plants that exist either as single cells or as filaments. They thrive in water systems at locations where water flow is static or slow. Algae are harmless and, like most plants, produce oxygen on which a healthy freshwater ecosystem depends. However, they are subject to stress and are reliable indicators of environmental

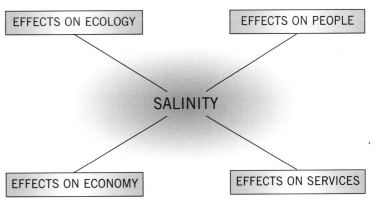

Figure 5.28 The effects of salinity – a futures wheel.

1 Make a copy of Figure 5.28 in the middle of a large sheet of blank paper. Draw pointers on your futures wheel to show the effects of salinity on 'People', 'Services', the 'Economy', 'Ecology', and any other headings you can think of.
2 Read the article in Figure 5.25 and study the maps in Figures 5.26 and 5.27. Show on your futures wheel the predicted effects of salinity. Link up the different points to show how you think these effects might happen.
3 Are the most significant futures that you have predicted:
   a) beneficial or problematic?
   b) social, economic or environmental?

Figure 5.29 Water affected by blue-green algae.

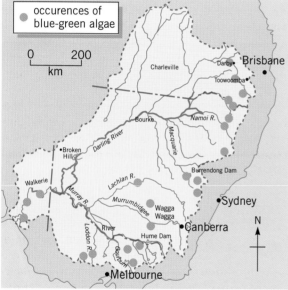

Figure 5.30 Blue-green algae occurrences in the Murray-Darling Basin in the 1990s.

quality. Often their presence alerts farmers and water suppliers to situations where poor water quality and salinity threaten river ecology.

Algae vary in size, type and colour, but certain types may be dangerous. In 1991, 'scum' or 'blooms' of blue-green algae formed along a 1000km stretch of the Darling. Figure 5.29 shows a 'bloom' along the shores of a dam. Three factors encouraged its emergence:

- low levels of water during a dry period intensified the concentration of algae
- nutrient was caused by pollution from local sewage and run-off from fertilized farmland accelerated growth of the algae and of water weed. Placed under stress of a confined environment, algae produce toxins. Some are short-lived neuro-toxins, or nerve poisons, but others, such as liver-toxins, have been associated with cancer.
- poor water management, in which oxygen deficiency prevents a healthy balance of aquatic plants or fauna.

The map in Figure 5.30 shows the extent of the issue in the Murray-Darling basin.

Summarizing the issues

1 Summarize the key issues that face the Murray-Darling river basin.
2 To what extent are the issues:
   a) natural and part of the physical background?
   b) the result of human interference?
3 Why does management of a river basin require a 'holistic' approach? Identify ways in which water supply, water quality, the salinity problem and land management are linked together.

Designing a management plan for the Murray-Darling river basin

Although climate causes of many of the environmental issues in the Murray-Darling basin, human responses have created the problems in the basin today. As a result of using European farming methods in a significantly different climate, land across the eastern half of the basin has been degraded and became poorer in quality. Promises by past governments to open up the interior with limitless water were, in the long term, false. Decisions made then, although understandable at the time, were responsible for the problems experienced today.

It is difficult to separate the problems of water supply and quality, soil erosion, salinity, and land degradation from each other. Land degradation includes issues such as decline in natural vegetation and the invasion by weeds and noxious plants. Water quality covers pollution problems of surface water and groundwater resources; in particular, algal and nutrient issues. Water and soil salinity are the biggest degradation issues in the basin.

There are now moves to repair losses and amend some past mistakes. The Murray-Darling Basin Commission was established to take responsibility for water and basin management issues. Also, Landcare, a community-based movement, uses both government and local farming groups to promote greater quality of land management across Australia.

Managing water supply and quality

Each Australian state traditionally regarded water within its boundaries as its own. New South Wales therefore regarded the Castlereagh River as its own, since all of it is contained inside its borders.

Where different states are involved, questions of water entitlements arise. Adelaide, in South Australia, relies upon the Murray River for over half its water supply. Greater demands during dry periods mean that little water may reach South Australia. Figure 5.31 shows the extent of the issue. Note how in average conditions, the largest proportion of water along the Murray River is surplus and flows to South Australia. Compare this with flow during dry conditions.

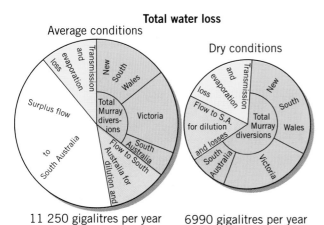

Total water loss

11 250 gigalitres per year   6990 gigalitres per year

Figure 5.31 Water use along the Murray River.

1 Identify the contrasts in water usage along the Murray River between normal and dry conditions.
2 Whom would this affect? Why?

Water trading between farmers

Water trading was introduced in the early 1980s to address some concerns. Now any increase required for human consumption or agriculture has to be met by trading quotas. Each user or property owner is allocated a quota that limits individual water use.

There are water property rights in all states and territories. Traditionally, landowners have rights to water. If a landowner wanted more water, the only way to achieve this was by purchasing land with a water entitlement from an existing irrigator and combining the two. Now water entitlements can be bought or sold independently of land. The Commission believes that trading allows water to be moved to sites where it can be used for higher-value uses. This, and the fact that irrigators will be able to benefit by selling water they do not need, should lead to greater efficiency in its use. Irrigators who wish to expand will be able to buy water to develop their activities. Alternatively, an irrigator who may wish to sell water that is not needed, or take land out of irrigation, can sell all, or part, of a water entitlement. This is more sustainable use of water and ensures that irrigation takes place where it is most needed.

Interstate water trading

Water trading takes place within river valleys, so the effects are local. However, permanent interstate trade is now being introduced, with one region on the Victoria–South Australia border on trial. Water is taken from one state to another in ways that do not result in increased levels of salinity, reductions in environmental flows or land degradation. This region has been chosen because of its location close to three states, and because its agricultural activities (citrus, stone and other tree fruits, vines for fruit and wine, and numerous vegetables) are dependent on high water requirements.

Step 1 A private seller decides to sell a water entitlement.

Step 2 A potential buyer is sought, either through personal contacts or through an intermediary, such as an estate agent.

Step 3 A contract is agreed between the buyer and the seller.

Step 4 The buyer and seller simultaneously lodge applications to transfer water with the licensing authority in the state of destination.

Step 5 Copies of both applications are forwarded by the state of destination licensing authority to the state of origin.

Step 6 The licensing authorities of both the state of destination and the state of origin judge the applications, based on water entitlements within the state and environmental issues such as salinity in the state to which the water is going.

Step 7 The Murray-Darling Basin Commission (MDBC) advises each state whether or not the water can be delivered.

Step 8 On the basis of the states' requirements and the MDBC advice, the application is either approved or rejected. The sellers have their water entitlement reduced by the amount sold and the buyers have theirs increased.

Figure 5.32 The process of buying and selling water entitlements.

The Commission is investigating ways for farmers in South Australia to purchase water easily from Victoria or New South Wales. In this way, more water should be guaranteed to reach South Australia in drought years. Transfers downstream will result in the water remaining in the river longer, which should have a positive environmental impact.

Water is monitored between states for quality, to ensure that no environmental degradation occurs as a result of interstate water transfers. If water is sold from a site that is not suitable for irrigation, water trading should result in environmental improvement.

In its 1999 campaign, the Australian Conservation Foundation has noted that 'it is significant that in many parts of the Murray-Darling basin, the irrigation community has shown a preparedness to accommodate environmental concerns by giving up watering the environment.' It is concerned that interstate agreements are not yet secure: 'Unfortunately, Queensland's commitment to environmental flows in the Upper Darling system is very weak; so weak, that Queensland is actually planning a dramatic increase in the size of the irrigation industry there.'

Irrigation management is helping the salinity issue. Mass irrigation and application of water only increases the problem. Over 90 per cent of irrigation water from the basin can be lost in this way, through seepage, evaporation or increased salinity. Many farmers now try different irrigation techniques, such as:

- **furrow irrigation** (Figure 5.33) which concentrates water along furrows where it gets to plant roots more easily and which is 50 per cent efficient
- **trickle irrigation** which pipes water directly to the area around the roots of a tree, and which is up to 90 per cent efficient
- **lasers**, used to level ground to reduce drainage of water into natural or artificially created hollows where waterlogging might occur.

Form small groups.
1 Using the evidence, decide how you would put the case either *for* or *against* the decision of the Queensland government to increase the size of its irrigation industry.
2 Compare your views with those of other groups. Are your decisions similar?

Figure 5.33 Furrow irrigation of a cotton crop. Strategies such as this take more water to within reach of plant roots and therefore reduce waste.

Managing land degradation

A combination of soil erosion and increasing salinity have led to gradual land degradation. Because of falling incomes and drought, farmers have become willing to be involved in schemes that will improve their land. The scale of the problem means that farms are unlikely to improve unless all do so simultaneously. Unless all farms in a stream catchment are involved, any programme which seeks to improve land is unlikely to last long. Local community groups have developed from Landcare and other groups such as Saltwatch.

The main aim in controlling land degradation is to lower the water table and develop planting schemes for native tree species. Two areas are important within a basin:

- **Areas of recharge** on the upper slopes of a basin, from which water drains downslope by a combination of processes (see Chapter 1). They are called recharge areas because these are the areas where water enters the water table and 'recharges' it. Clearance of these areas over the last 100 years or so has led to increasing drainage of water towards lower ground.

Figure 5.34 A contrast between controlled grazing and over-grazed land.

- **Areas of discharge** are the lower land areas into which more water now drains from the areas of recharge.

The solutions lie in managing areas of recharge, because the cause of salinity starts here. Strategies include reducing the amount of water moving downslop, for example by:

- planting more deep-rooted, native trees to draw water from deeper levels;
- planting different species of pasture, such as lucerne, which have deeper roots and draw up more moisture;
- fencing and controlled grazing, limiting numbers of cattle or sheep to those which land capacity is able to maintain. This reduces over-grazing and allows soil to develop under plant cover. A clear contrast is shown in Figure 5.34.

---

1 Identify strategies that have been adopted for lowland areas of discharge shown in Figure 5.35. Make a large copy of the diagram and label it with reasons why different strategies have been adopted and how these improve land quality.
2 Show how the land in Figure 5.35 might appear in 20 years' time if all the intended strategies succeed.
3 How might the information in Figure 5.36 be used to promote re-planting of native tree species?
4 Prepare a document for farmers to show how re-planting of native species and the sale of water quotas together can improve land quality.

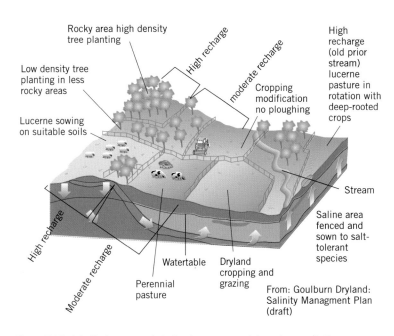

Rocky area high density tree planting

High recharge

moderate recharge

High recharge (old prior stream) lucerne pasture in rotation with deep-rooted crops

Low density tree planting in less rocky areas

Cropping modification no ploughing

Lucerne sowing on suitable soils

Stream

Saline area fenced and sown to salt-tolerant species

High recharge

Moderate recharge

Watertable

Dryland cropping and grazing

Perennial pasture

From: Goulburn Dryland: Salinity Managment Plan (draft)

Figure 5.35 A holistic approach to land management to reduce salinity.

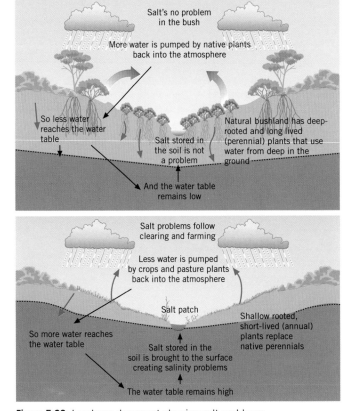

Salt's no problem in the bush

More water is pumped by native plants back into the atmosphere

So less water reaches the water table

Natural bushland has deep-rooted and long lived (perennial) plants that use water from deep in the ground

Salt stored in the soil is not a problem

And the water table remains low

Salt problems follow clearing and farming

Less water is pumped by crops and pasture plants back into the atmosphere

Salt patch

Shallow rooted, short-lived (annual) plants replace native perennials

So more water reaches the water table

Salt stored in the soil is brought to the surface creating salinity problems

The water table remains high

Figure 5.36 Landcare document showing salt problems after clearing and farming.

More farmers now have a greater understanding of how their land might be managed. Landcare advisers and local community groups encourage farmers to become aware of land quality, in order to recognize how land improvement might bring about changing fortunes. The greatest change lies in how many farmers recognize the necessity for re-planting of native species. Most farmers now speak of 'recharge' and 'discharge' areas of their farms. Figure 5.36 shows how this awareness has come about.

Conclusion

This chapter shows that water is not just a necessity for human life and survival. It is an instrument of power, used by governments and individuals. The study of the Murray-Darling has shown how water is now being allocated between states in Australia. On a wider scale, international water agreements have been in force since the early 20th century, for example between the USA and Mexico regarding the Colorado River. Water as a resource is valuable and the tensions it creates could prove to be a real conflict in the long term.

Summary

You have learned that:

- Climate plays a major part in the nature of a river basin; its rainfall totals, distribution and reliability are all major influences in what happens in a water catchment.
- Natural vegetation and ecosystems are fundamental to a river basin. Changes in one part of an ecosystem may lead to changes in other parts, and be reflected across the river basin as a whole.
- Environmental problems often arise from unintended human actions. Land clearance affects the hydrology of a river basin and causes soil erosion, land degradation and salinity.
- The potential of dry lands is increased greatly by the application of river water for irrigation. However, land quality may deteriorate as salts increase in concentration.
- Sustainable agriculture is the only long-term solution to marginal lands in river basins, where a fine balance exists between nature and human activity. This is a necessary part of river basin management strategies, which require the involvement of all players to be successful.

Ideas for further study

1 *For the sake of a country's economic future, it is more important to maximise what can be produced from the land than it is to protect the natural ecosystem.*

   Discuss this, referring to the Murray-Darling river basin.

2 Compare issues in the Murray-Darling river basin with those in another large river basin, such as the Colorado in the USA. What are the environmental impacts of water management there? What attempts have been made to manage water supply, water quality and land degradation?

References and further reading

Pilger, John (1992) *A Secret Country*, Vintage Publishers
BBC *Australia 2000* series (1999), programme 4 *The Spreading Desert* (although made for 13–15 year olds, the programme deals with land salinity in western Australia)

Web sites

Australian Conservation Foundation at http://www.acfonline.org.au
Murray-Darling Basin Commission at http://www.mdbc.gov.au/

6 The Komadugu-Yobe river basin

This chapter examines the Komadugu-Yobe river basin in north-east Nigeria and how it is managed. Figure 6.1 shows the main rivers that feed the Komadugu-Yobe River, flowing from the Bauchi Plateau 500m above sea level, into Lake Chad. Since the mid-1970s, the Nigerian government has built several large dams and barrages to manage seasonal flows in these rivers. This has caused enormous changes to the hydrology of the area and, as in most cases of dam construction, has brought benefits and problems.

The basin characteristics
The climate of north-eastern Nigeria
Compared with much of Nigeria, the northern part of the country is much drier and has a more habitable climate than the hot and humid southern rainforest on the coastal belt. Although temperatures can be very high in summer, the heat is drier; temperatures of 45°C are not uncommon. This is typical of a dry savanna climate.

Kano is the main city in this area. The climate graph for Kano, Figure 6.2, shows a savanna climate with pronounced wet and dry seasons. Rainfall occurs almost entirely between April and October,

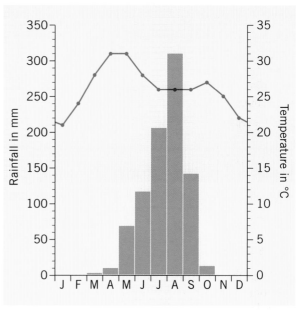

Figure 6.2 Climate graph for Kano, Nigeria.

with rainfall highest in July and August. During the long dry season there is very little rain and, in some months, none at all. Total annual rainfall in the region decreases from south-west to north-east, and the length of the dry season increases.

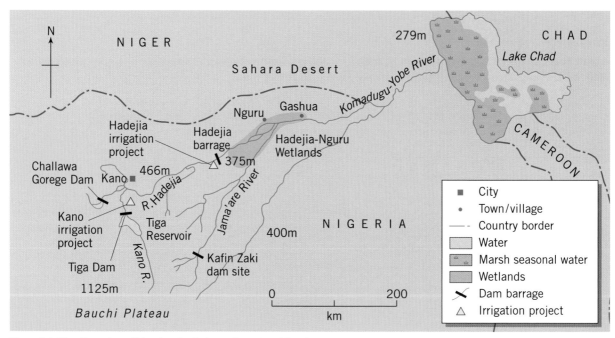

Figure 6.1 The Komadugu-Yobe river basin in north-eastern Nigeria.

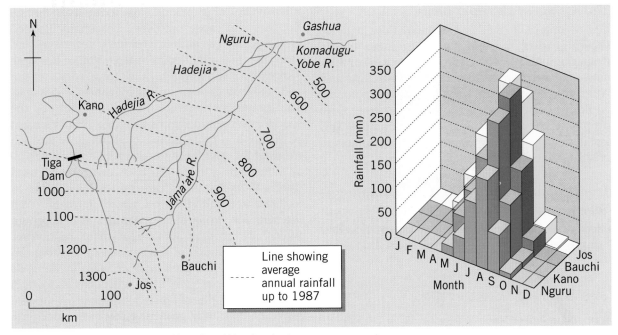

Figure 6.3 Rainfall variation in the Komadugu-Yobe river basin.

Since the mid-1960s, average annual rainfall in this part of west Africa has declined. In northern Nigeria lack of rain means that the Sahara Desert is creeping southwards. Persistent drought has reduced plant cover, killing grasses and shrubs and destroying root networks that bind soil together. Once exposed, dry season winds easily pick up and remove soil cover, leaving the area unable to support plant life – hence the extension of the desert.

Theory

The savanna climate

Areas that experience a savanna type climate lie between the tropics ($23\frac{1}{2}°$N and $23\frac{1}{2}°$S) in South America, Africa, South-east Asia and Australia. Northern Nigeria lies between 10°N and 13°N. Temperatures throughout the year are high, usually between 20°C and 30°C, reflecting their tropical location. The annual range of temperature is moderate, 10°C in Kano but as low as 4°C in other areas with this climate type.

The savanna climate can be seen as a transition zone between equatorial climates (associated with tropical rainforest areas) and hot desert climates. On the ground, vegetation shows this transition with tropical rainforests giving way to dry grassland. Total annual rainfall in savanna areas in west Africa varies from about 1200mm in the south to 500mm in the north. The most important feature is the very pronounced wet and dry seasons with virtually all the annual rainfall occurring in 4–6 months. Highest temperatures occur just before the start of the wet season when there is an increase in temperature, but once the rains have started temperatures fall slightly.

Figure 6.4 Savanna landscape in northern Nigeria. With the grassland here there are low trees and shrubs. Only grasses, deep-rooted trees and shrubs, or drought-resistant species are able to survive the dry season.

The regime of the river basin

The regime of a river is its annual flow pattern, showing when discharge rises or falls. In the Komadugu-Yobe river basin there is a clear link between river flow and rainfall. Between June and August, intense tropical rains usually bring floods, while between November and March, river flow drops considerably, causing local tributaries and streams to dry up. Farming in this tropical area is greatly influenced by rainfall distribution.

Traditional farming in the Komadugu-Yobe basin

Each year, between June and August, flooding brings renewed fertile silt deposits that help intensive farming on land close to the river, though flooding can cause problems. Seasonally flooded land is known as *fadama*, and fruit, vegetables and rice are the main cultivated crops. Between November and March, the long, dry season restricts crop growth unless irrigation is possible. Traditional methods of irrigation, such as the shadoof (shown in Figure 6.5), or a simple system of ditches and gates, enable crops to grow through much of the year, once floodwaters have receded.

Land further from the river can only be used during the wet season. These rain-fed fields, known as *tudu*, are cultivated by poorer farmers, often women, for crops such as millet and sorghum which are able to survive in drier conditions. However, for much of the dry season nothing can

Figure 6.5 A traditional shadoof irrigation system. A channel has been dug from the river bank so that water can be lifted using devices such as the shadoof. This allows land to be used during the dry season after floods subside.

be grown at all; wealth depends entirely upon what can be grown in the rainy season. Figure 6.6 shows a boy gathering cowpeas, which are planted amongst the main crops of millet and sorghum. Planting a range of crops is important when the rainfall is unreliable. Cowpea plants have two advantages: they are a good source of protein, and they fix nitrogen in the soil, thus fertilizing it.

The Hadejia-Nguru wetlands

The Hadejia-Nguru wetlands lie at the confluence of the Hadejia and Jama'are Rivers. The wetlands are a major resource for agriculture; over 2 million people live there, including farmers and traders. Some of the area, for example Lake Nguru, is permanently flooded and is important for fishing. Most land is seasonally flooded *fadama* and is intensively farmed. Fulani cattle herders also travel through this river basin, seeking water and pasture for their cattle. The wetlands provide important grazing for this semi-nomadic group as well as crop fields for the settled farmers. They are a unique habitat for many migratory birds and water fowl, which has helped the development of some tourism in the region.

Figure 6.6 Growing cowpeas on the rain-fed fields or *tudu*. These fields are too far from the river to be flooded, so they rely on rainfall during the June–September period. They cannot be used in the dry season.

Most people live in small villages that are relatively isolated. Their irrigation water supply is often controlled by small-scale dykes, gates and channels that they have built themselves. In the dry season small petrol pumps help to lift the water onto the land. Drinking water is sometimes drawn directly from the rivers but most villages rely on hand-dug wells reaching groundwater supplies that are recharged during the annual floods. Trees growing where water supply is more reliable are an important source of fuelwood, and most villages have a clay pit which they use to make bricks.

River management

Damming the river basin

Seasonal flow of water in the river basin and pressures of a growing population have led to a series of development projects to dam the waters for irrigation. High oil prices in the 1970s enabled Nigeria to invest in several large devel-opment schemes, including the Tiga Dam which was built on the Kano River in 1974. A second, the Challawa Gorge dam, was built in 1992 followed by the Hadejia River barrage. Another large dam, Kafin Zaki on the Jama'are River (Figure 6.1), is planned; construction was started and then halted. It may never be completed.

Water stored in dams is partly used to supply Kano, a large and rapidly-growing city of over 1 million inhabitants. Regular supplies of water for both domestic and industrial use are essential. Its main purpose, however, is for the Kano Irrigation Project, in which canals (Figure 6.8) and dykes carry water to 13 000 hectares of land for irrigation. Small motorized pumps, available at subsidized prices, and many tubewells which reach groundwater supplies, are used to irrigate the land. A barrage along the Hadejia River also provides water for a further 12 500 hectares of land.

Impacts of the dams on local people

A World Bank survey in 1994 showed that the Kano Irrigation Project has been successful for farmers who are able to irrigate their land. It has led to increased yields of crops, such as rice, and to a greater variety of fruit and vegetables, such as onions, tomatoes, garlic and peppers. During the dry season wheat is planted and, if fertilized, gives good yields. The Nigerian government has banned imported wheat so there are good returns on this crop. Better transport allows farmers to send crops to Kano and to large urban centres in southern Nigeria.

Wealthier farmers who own their land and are able to buy motorized pumps, fertilizers and quality seeds, have benefited most. Poorer farmers, especially those who rent land, are not able to afford pumps and fertilizers and lose out if landowners want to expand. Initially, more jobs were created for landless farm labourers, especially at harvest, but in future it is likely that machines will replace manual labour.

Not everyone who lives beside the river benefits. Some villages were flooded and people were re-housed; many lost their land and are having to learn new skills such as fishing. Even land quite close to the dams may not be included in the irrigation project. Without canals to carry water, farmers must continue to farm in their traditional way, even though storage of water in the dam has changed the pattern of flooding and may make this more difficult. Some farmers, especially poorer ones, are worse off than they were before.

Fulani cattle herders have not benefited either; it is difficult for them to find grazing areas because more land has been claimed for cultivation. Access to the river itself is also a problem and the water has blocked traditional livestock routes.

Figure 6.7 A canal carrying water from Tiga Dam. A system of canals carries water from the Tiga Dam to Kano for domestic and industrial use. It also goes to fields in the Kano irrigation project, and farmers can control the flow of water onto their fields using sluice gates. Crops can now be grown all year round.

Fishermen have also suffered. Fish numbers and species have declined during the last decade. The construction of dams has disrupted fish movements and more people are trying to make a living from fishing, especially those who have lost land or can no longer rely on floods to water their crops.

Changes to the environment

With irregular rainfall and river flow, annual flooding has always varied, but Figure 6.8 shows the enormous difference since dams were built upstream of the wetlands. It shows how flooded land has been reduced in the Nguru wetlands since dams were used to store water in the wet season. Land that no longer floods cannot be used to grow rice or other crops which rely on abundant supplies of water. The extent of flooding on the Nguru wetlands fell by almost half after the construction of the Challawa Gorge dam in 1992.

The reduction in the extent of flooding has also affected small-scale irrigation. Dykes and gates built by local people no longer work because the floods are not high enough. Rice production is declining and farmers are having to grow other crops.

The hydrology of the river basin has altered significantly. Discharge on the Hadejia River is now controlled by sluice gates on the dams and the annual flood is much reduced. Discharges have fallen from a peak of over 400 cubic metres per second to under 50 cubic metres per second. Figure 6.9 shows discharge on the Hadejia River before and after the construction of the Tiga Dam.

Changes in the extent of the wetlands are also affecting the huge numbers of migratory birds that use this as an important feeding and nesting area. They may also affect the growing numbers of tourists who come to see the birds and wildlife. Groundwater levels have fallen dramatically since 1974, partly because of periods of low rainfall but mainly because of the dam construction, shown in Figure 6.10. This is a particular problem where farmers rely on groundwater supplies for agriculture and for all their own needs.

Much of the fertile silt brought down by the floodwaters is trapped behind the dams. Many dams of this scale have an estimated lifespan of less than 100 years because of the problems of sedimentation. Even more important, these valuable, fertile deposits are no longer reaching farmland downstream.

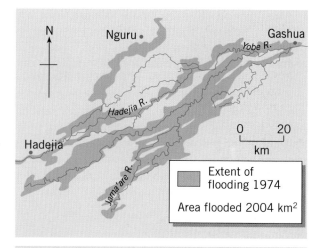

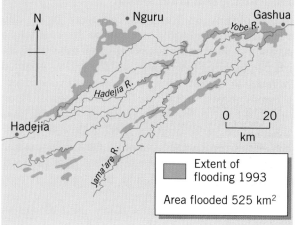

Figure 6.8 The extent of flooding in the Nguru wetlands in 1974 and 1993.

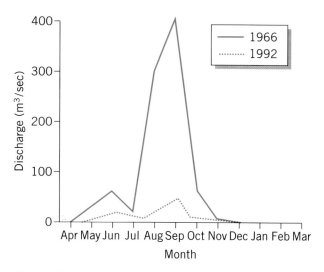

Figure 6.9 Discharge on the Hadejia River before and after dam construction.

1 Describe the extent of flooding in the Hadejia-Nguru wetlands in 1974 and 1993, shown on Figure 6.8. Use named places and look at the key to estimate distances and area.

2 Suggest reasons for the changes in the extent of flooding, considering natural changes as well as river management strategies.

3 a Describe changes in discharge on the Hadejia River, shown in Figure 6.9, before and after dam construction.

   b Describe the changes in groundwater levels 1964–84, as shown in Figure 6.10.

4 Suggest reasons for the changes in discharge.

5 What patterns in groundwater levels since 1984 would you expect a new monitoring programme to show? Why?

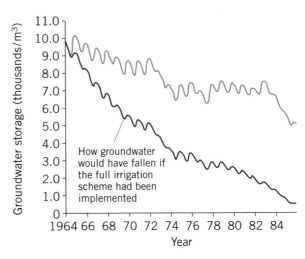

Figure 6.10 Changes in groundwater levels 1964–84. No data are available since 1984; no monitoring takes place.

Option	Reduction in extent of flooding
1 Expand irrigation on the Kano River Irrigation Project to 27 000ha, and operate both the Tiga and Challawa Gorge dams for irrigation water	11%
2 Construct the Kafin Zaki dam and irrigate 84 000ha along the Jama'are Valley	30%
3 Fully implement all three dams and irrigation schemes	50%
4 Release large wet-season flows from the Tiga, Challawa Gorge and Kafin Zaki dams into rivers	20%

Figure 6.11 The impact of dam construction and irrigation schemes along the Hadejia River.

Options for the future

One of the main problems of the Kano Irrigation Project and the dams is the lack of an overall plan for the river basin. Each project has been carried out individually and not as part of an overall management scheme. Hydrological modelling has shown that if all the schemes currently planned for the basin are completed, including the Kafin Zaki dam, then the hydrology of the area will be fundamentally altered with serious consequences. The extent of the flood will be severely reduced and the rate of groundwater replenishment will drop, so that people will not be able to maintain their current way of life. The important and densely populated Nguru wetlands will probably be most seriously affected.

A number of options exist for the Komadugu-Yobe river basin in the near future. Current projects could be maintained, while other new possibilities exist.

The possibilities for the basin's future are summarised in Figure 6.11. One option is to expand irrigation on the Kano River Irrigation Project to 27 000ha, and operate both the Tiga and Challawa

Gorge dams for irrigation water. This alone would reduce the flow of the Hadejia river by 46 per cent. Another option, the completion of the Kafin Zaki dam on the Jama'are River, would provide irrigation for a further 84 000ha of land. The dam would stretch 17km and inundate 235km^2 of land at an average depth of only 3m. The cost would have been US$250 million in 1993. The dam would hold back yet more of the floodwaters that feed the Nguru wetlands and the Komadugu and Yobe Rivers.

1 a) In pairs, identify the effects of damming the Komadugu-Yobe river basin, including the Nguru wetlands. Write each effect out on a card.

   b) Classify each of the cards on a large copy of the table below, according to whether these are social, economic or environmental effects, and whether they represent benefits or problems.

Effect	Benefit	Problem
Social effects		
Economic effects		
Environmental effects		

2 With another pair, compare the effects.

3 Has the management scheme been a success or a failure? Justify your argument.

4 Discuss the advantages and disadvantages of completing the Kafin Zaki dam on the Jama'are River, where construction was started but never finished. Write a report of 750 words, recommending whether the dam should be completed or not and giving clear reasons for your choice.

Sustainable development

Reference is made throughout this book to projects that may or not be sustainable. A common definition of sustainable development is 'one that does not compromise quality of life for future generations by current practice'. Judging whether an existing or proposed development is sustainable, or not, may be done on social, economic or environmental criteria:

Social   deciding whether it brings social benefits for all people, rather than just a few, and will continue to do so in the future

Economic   deciding whether it brings an improvement in the economy, locally as well as nationally, and that all people will be better off now and in the future

Environmental deciding that there are no environmental impacts which create short- or long-term environmental damage.

Most indicators of sustainability are not just social, economic or environmental, but may be evaluated on more than one basis. Some examples are shown in Figure 6.12.

Factors that act against a sustainable future include those that:

- are environmentally destructive, e.g. those that pollute, remove a resource permanently, or damage it for a period of time; examples include any form of air or water pollution, soil damage or erosion
- use a resource at a rate which cannot be renewed before current stocks run out
- are short-term and use high energy-consuming materials, whose futures are finite. Not only does the raw material itself decline in stock but may use other exhaustible resources in the process. An example is the manufacture of steel plate which requires the use of fossil fuels such as coal in its manufacture, without making provision for recycling or re-using any of the products.

Social	Affordable land, housing and rents, within reach of everybody
Social and environmental	Mixed land uses in a city that avoid long commutes, such as work and housing. Development and maintenance of features that enhance quality of life, such as parks.
Economic and environmental	Re-cycling activities that sustain stocks of raw or manufactured goods for longer and which reduce energy consumption
	Energy-saving strategies that reduce consumption of fossil fuels, and environmental pollution
Economic	Developing an infrastructure to ensure that varied economic activities can take place both now and in the future, e.g. local services such as medical, roads, schools

Figure 6.12 Sample indicators of sustainability.

The expansion of large-scale irrigation within the basin without acknowledging and making allowances for the needs of downstream water users will have severe implications on the wetlands. The loss in wetland productivity is not offset by any increase resulting from the irrigation schemes, since limited water resources of the river basin are shown to be incapable of meeting demands made by irrigation users.

The Nguru Wetlands Conservation Project

Option 4 in Figure 6.11 shows a regulated flooding regime which would allow irrigation on the existing areas of irrigation on large schemes, assured and reliable flooding in the wetlands, and regulated flows for use in small irrigation schemes. Part of the purpose of such a project is to conserve the value of wetlands, both economically and environmentally. The Nguru Wetlands Conservation Project is a field project established to raise awareness of the problems of the wetlands. It aims to encourage sustainable development in the area for the benefit of both people and wildlife.

One of the most important tasks is to ensure that water is distributed fairly, and this means persuading the authorities to release water from the dams upstream to maintain the wetland, rather than keep it for irrigation elsewhere.

A range of activities is designed to ensure that whatever is carried out can be done so without damaging the environment for the future, and without risking future economic potential. The priority is to persuade authorities to release water from the dams upstream. In addition, they propose to:

- survey the extent of the flood and of changes in vegetation and wildlife
- teach local people about wetland management resources
- monitor use of water from village wells, in order to check levels of the water table
- investigate competition for land between animal herders and crop farmers
- distribute fuel-efficient wood stoves, designed to reduce wood consumption
- help people to resolve conflicts.

Look carefully at the proposals of the Nguru Wetlands Conservation Project.
1 Do you consider this project to be more or less sustainable than the rest? Give reasons for your answer.
2 Why do you think the priority of the group is to persuade the authorities to release water from the dams upstream? Why may this be difficult to achieve?
3 Select three other conservation project activities that you think are important. Give reasons.
4 Suggest one other approach which you think would be useful. How might it be applied?

Summary

You have learned that:
- Rivers in different parts of the world have different regimes.
- Seasonal floods cause problems but also bring benefits to people who live on the floodplain. People adapt their way of life to make use of the river. This may be particularly important in an LEDC.
- River management interferes with the hydrology of a river basin and may have far-reaching social, economic and environmental effects, some of which may be unforeseen.
- Some people benefit from large-scale river management and irrigation schemes but there may be disastrous consequences for others.
- River management is a continuing challenge and needs to be seen in an holistic way, involving a full understanding of the processes involved. A helpful framework is to judge a proposed change against whether or not it is sustainable.

Ideas for further study

Working in groups, find out about a river management scheme in a large river basin in another LEDC, for example the Nile (Egypt), the Indus (Pakistan), the Volta (Ghana), the Amazon (Brazil), or the Ganges-Brahmaputra (India).

1 Draw a map to show the location and extent of the river basin and its main tributaries, its climate and the river's regime.
2 Describe problems that exist in the river basin, for example flooding, demand for irrigation or pollution.
3 Discuss management strategies that exist or are planned.
4 Compare management of your chosen river with the Komadugu-Yobe in Nigeria. Are there similar problems or management strategies? In your opinion, which river basin has been managed more efficiently?

References and further reading

Channel 4, Programme 2 *Dammed Water* in 'Geographical Eye over Africa', series first broadcast Spring 1995
Dammed Water (1995) (photo pack and information booklet) available from Action Aid, Chard, Somerset

Web sites

Kano climate details can be obtained at
    http://www.onlineweather.com/Africa/climate/Kano.html
Wetlands Research Unit, Home Page at
    http://www.geog.ucl.ac.uk/~jthompso/hnwcp.html
or http://www.geog.ucl.ac.uk/~jthompso/hadjam.html

Super dams, super problems?

The final chapter in this section focuses on water management. It investigates a new breed of 'superdams', that is, dams that are larger than any yet built in the world, and which will be completed within the first 20 years of the 21st century. This chapter will explore plans for, and possible impacts of, one of the world's major water projects, the Three Gorges Project along the Yangtze River in China.

Why is water so important?

In 1999 the global population passed 6 billion. The rapid growth in global population resulted in huge demands for water. Not only do more people require more water, but they have increasing expectations. China has one quarter of the global population, and is intent upon a programme of rapid economic growth. This will not be possible without water to generate HEP, for consumption in its 40 cities, and to manage flood threats.

The Three Gorges Project in China

Figure 7.1 shows the location of the proposed Three Gorges Project along China's Yangtze river, to be built upstream from Yichang. Find the same area in your atlas. This is one of the world's largest rivers; it is 6 300km long, the distance between the UK and the USA. Figure 7.2 shows the new Xiling Bridge across the Xiling gorge. This bridge is half a mile wide and spans the river 1 000km inland.

Figure 7.2 The new Xiling Bridge across the Xiling gorge along the course of the Yangtze. Even here, 1000km inland from the sea, the river is half a mile wide.

The proposal to construct a huge dam along this section of the river has been alive for over 60 years. It was finally approved by the Chinese government in 1992. Work on the dam began in 1994 and is expected to finish in 2009. The proposed dam will be 185m high, with a lake behind it, 600km long and 1.1km wide. Estimates of its cost in late 1999 range between US$17 billion and US$40 billion. When completed, it will be the largest HEP project in the world and will provide one-seventh of the total electricity supply produced by China in 1990.

Why the dam is considered necessary

Figure 7.3 shows the seasonal flow of the Yangtze downstream of the dam – estimated after the completion of the dam – as well as actual levels for the present. Notice that it has distinct seasonal variation. The monsoon climate of south-east Asia brings most rain within the space of a few summer months. Water supply for economic use, electricity production and human consumption has to be available all year.

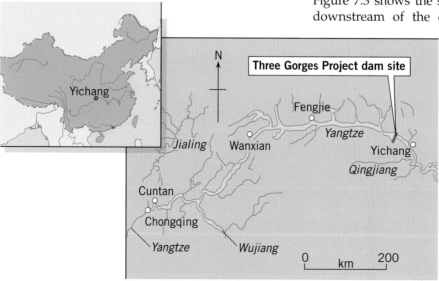

Figure 7.1 The location of the proposed Three Gorges Project.

The heavy rain in each monsoon season creates a huge increase in river flow during June–August. Added to this is the seasonal snow melt from the Himalayas. In the past, huge floods have killed thousands of people. In August 1999, flooding threatened human life, housing and livelihoods. Storage of the summer increase would make water available throughout the year.

> 1 Annotate a copy of Figure 7.3, identifying on it seasonal changes in water levels and reasons for these changes.
> 2 Why should water levels be lowest in the dam during the period of highest flow (Figure 7.4)?
> 3 What benefits and problems can you see in the altered flow of the Yangtze after the completion of the dam (Figure 7.3)?

There are constraints in the construction of the dam.

- The Yangtze floodplain is densely populated. On the one hand, people are threatened by floods but on the other, their displacement would take them away from their homes and livelihood.
- The Yangtze transports huge volumes of sediment along its course. Sediment has already substantially reduced the volume of water behind the Hoover Dam on the Colorado River, as shown in Figure 7.5. Ways have to be found to prevent this along the Yangtze.
- The Yangtze is the main artery for navigation in this region of China. The design of the dam would have to allow for ships to be raised by a series of locks alongside it.
- Tourism is being developed in China. Among the areas that would be flooded are the gorges themselves, considered to be spectacular scenery (Figure 7.6).

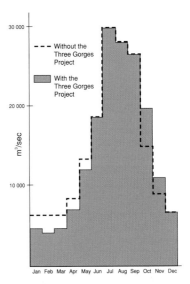

Figure 7.3 The seasonal flow of the Yangtze downstream of the dam, estimated after the completion of the dam, as well as actual levels for the present.

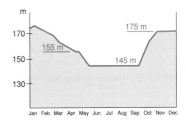

Figure 7.4 Planned water levels within the Three Gorges Dam.

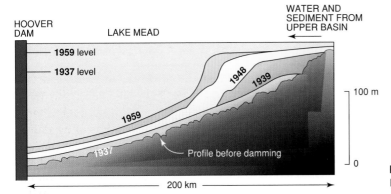

Figure 7.5 Sedimentation behind the Hoover Dam, along the Colorado River.

Figure 7.6 The gorge scenery along the Yangtze River. These are among the attractions that will be flooded once water collects behind the dam.

Different viewpoints on the Three Gorges Project

The Three Gorges Project has its supporters and opponents, so a variety of views need to be considered before assessing its impact. In its favour, the Project would generate economic development. Against it, criticism focuses on its social and environmental impacts. Chinese officials estimate that the river will flood two cities, 11 counties, 140 towns, 326 townships and 1351 villages. About 23 800 hectares of cultivated land will be submerged. According to estimates, more than 1.1 million people would have to be resettled, an expense which accounts for one third of the project's cost. Consider the range of views in Figure 7.7.

Figure 7.7 Viewpoints on the Three Gorges Project.

A   The water conservancy and hydro-power project on the Three Gorges of the Yangtze River, will greatly promote the development of our economy, for present and future generations.

President Jiang, People's Republic of China, 7 November 1997.

B   The dam will generate 9% of China's current capacity. The 26 hydro-electric turbine units will out-perform 15 nuclear power stations and have the annual equivalent of burning 50 million tons of coal without atmospheric pollutants; i.e. 100 million tons of carbon dioxide, 10 000 tons of carbon monoxide, 2 million tons of sulphur dioxide, 370 000 tons of nitrogen oxides and huge amounts of fly ash.

China Report (1999), stating the Chinese government view.

C   The Yangtze, third longest river in the world, has brought its share of mayhem to its banks. In Shashi, residents have had to raise its embankment 10m over the last 450 years. In 1931, Wuhan's customs house was flooded and 140 000 died. The Long River (Yangtze) continued to bring death and disaster. It took another 2 700 lives as recently as 1996.

China Report (1999), stating the Chinese government view.

D   China's power output must rise by 8% annually to keep pace with a 6% annual increase in GNP. That means the nation's total 1990 power capacity of 130 000 GW must grow to 580 000 GW by 2015.

World Resources Institute, 1994-95.

E   The dam will flood over 12 000 cultural and archaeological sites, some as old as the Palaeolithic Age. The government recognises the need to rescue 20 000 fossils in the area. However, it has not estimated the time to complete this. It is impossible for archaeologists to recover all of these in ten or even twenty years.

Seth Goldstein, Columbia College, USA.

F   When the project is complete, over a million people will have been relocated. These people are often unfairly compensated and poorly treated during relocation. Many will lose their livelihood. Because the Yangtze region is a rich one for farming, they will not be able to relocate to an area with such rich farmland.

Seth Goldstein, Columbia College, USA.

G   According to Earthscan's *Damming the Three Gorges* (1994), resettlement funds are not paid to those who are moved, but to authorities to use for job creation. The money 'provides local government with revenue, and its leaders with considerable powers' - that is, it makes corruption more likely, putting money into the hands of individuals to use as they see fit.

H The scale of the project is troubling considering the previous record of the Chinese government. In 1975, the Shimantan and Banqiao dams collapsed, claiming 200 000 lives. Although the government has kept most information confidential, in 1981 it admitted that the number of collapses had risen to 3 200. These were small in comparison to the proposed Three Gorges dam ...

Seth Goldstein, Columbia College, USA.

I The area is known for its beauty. The breathtaking gorges will cease to exist because of flooding ... The project endangers 112 rare species of aquatic creatures specific to the river. The Chinese alligator, the Yangtze River dolphin and the Chinese paddlefish are among creatures affected by changes in the river.

Seth Goldstein, Columbia College, USA.

J The reservoir is located on a major geological fault; it will hold so much water that it could trigger an earthquake that would devastate nearby populations and damage the dam itself. Landslides common to the area could occur, causing a tidal wave to breach the dam and unleash terrible floods on the valley.

Grainne Ryder, Exposing the Secrets of the Three Gorges Dam.

K The Three Gorges Project is designed to be protected against an earthquake of intensity 7. The possibility of reservoir-induced earthquakes is not ruled out, though the magnitude should not be high. Analysis shows that the maximum earthquake in the region should not exceed 6.

Lu Youmei, Three Gorges Project – A Progress Report.

1 Assess each viewpoint in Figure 7.7. Say which represents an attitude and which represents data. Are any 'data' presented as fact, or are they no more than viewpoints put across in a convincing way?

	Benefit of the dam	Costs of the dam
Social		
Economic		
Environmental		

2 Make a large copy of the table above. With each viewpoint:
   a) classify it as social, economic or environmental
   b) say whether it represents a benefit or cost.
3 Should the Three Gorges Project go ahead as a result of this analysis? Explain your views.

So who do we believe?

The extracts show different attitudes because the values of each writer are different. When people interpret the likely effects of the dam, they focus upon attitudes and upon data. 'Attitudes' represent different views about people (or society), the economy, or the environment. 'Data' may be presented as factual, but may be no more than one person's view. Technocratic views, which focus upon technology in solving a problem, may be presented as facts. Often held by engineers, they promote technology and the capabilities of people in building a dam. They are often convincing. However, their work involves risk assessment, an intelligent form of guess work. Risk assessment measures whether or not a disaster is likely to happen. One person's risk assessment may not match another's, depending upon how it is done.

Financing the Three Gorges Project

A review of investors and those who created and are to build the dam, shows that companies from the UK, France, Sweden, Switzerland, Germany, and Canada have joined Chinese companies in bidding for US$1 billion worth of engineering work until 1999. Yet the Chinese government has had difficulty raising money for Three Gorges. It recognises that *'absent from participating in the Three Gorges Project are US companies, whose ability to compete is hampered by the US Export-Import Bank's refusal to give US companies loans to participate in the project, due to concerns about the environment and the resettlement of over a million people.'*

The role of the World Bank and the IMF

Usually, the World Bank and the International Monetary Fund (IMF) are involved in financing projects in LEDCs. These two institutions draw together money from banks in MEDCs to invest and earn interest overseas. Both are primarily controlled by the USA, which has nearly one vote out of five in controlling what happens to its money; 40% of votes come from the USA, Japan, Germany, France and the UK. Some people question whether China's interests are best served by such an arrangement.

The World Bank has declined to have anything to do with the Three Gorges Project, because it was considered a major political issue. The problem of resettlement has scared it as an investor. It funded a feasibility study of the Three Gorges Dam in 1988, designed to assess costs and benefits of the project. One official reported that as a result of this, everyone at the Bank was 'freaked out over resettlement'. This is not surprising; China has been the largest recipient of World Bank project funds. One World Bank report has showed that over one-third of past World Bank projects in China result in compulsory resettlement. The UK Guardian newspaper reported that over 30 million Chinese citizens have been evicted from their homes in the last four decades, 'forced to move by road, railway and reservoir projects'. Over 150 000 people, evicted over 30 years ago to make way for a large dam in China, are still living in temporary shelters.

In total, 32 projects financed by the World Bank before the building of the Three Gorges Dam required the removal of nearly half a million people. In China's previous largest resettlement project, the Sanmenxia Dam on the Yellow River in the 1960s, 280 000 people were removed from the site to arid land up to 500 miles away. Eventually, 150 000 were allowed to return to their original land on condition that they build their own houses. The World Bank admits that these people are still moving in search of adequate livelihoods.

The Chinese government approved the project in 1992 through the National People's Congress (NPC), in spite of opposition from some of its members. Finance has so far been provided from a number of sources, including Japanese security funds, and loans from Canada, Germany and Switzerland. The reason is fairly clear; the Chinese government itself has stated that *'As China changes financing regulations to further attract foreign enterprises, its market will become increasingly attractive.'* A population of 1.2 billion people is a tempting market for many companies and governments.

> Write an essay of 1 000 words saying whether or not you agree with the following statement from the Chinese government about the Three Gorges Dam: "The potential benefits from financial growth and commercial development greatly outweigh any potential drawbacks".

Summary

You have learned that:

- Countries such as China see significant advantages in building super dams.
- The Three Gorges Project will have impacts on people, economy and the environment.
- Assessment of the costs and benefits of projects like this depend upon people's values and priorities.

Ideas for further study

Tthe issue of super dams is much greater and challenges ideas about how space is used and by whom. Other schemes include:

- the Sardar Sarovar Dam along the Narmada River in western India
- the Yarmuk River schemes in the Middle East
- the Mekong River scheme, affecting people in Thailand, Laos, Cambodia and Viet Nam.

Research these using search engines on the Internet. The GTAV Home Page of the Geography Teachers Association of Victoria may help you. Its reference is given below. Select one scheme and identify the social, economic and environmental issues involved.

References and further reading

GTAV Internet Home Page – address http://server.netspace.net.au/~gtav/gtav.html
Lean, Geoffrey, 1994 *Promises to soothe turbulent waters* in The Independent on Sunday, 18 December 1994

Changing river environments: summary

Enquiry questions	Key ideas and concepts	Guidance and possible examples
Introducing fluvial systems 1.1a How do fluvial systems operate within the global hydrological cycle? 1.1b What impact does the hydrological cycle have on fluvial systems?	• The global hydrological cycle – role of atmosphere, oceans, biosphere in a closed system. Relationship to drainage basin (open system) and links to coastal system. • The hydrological cycle – key processes, and factors affecting their variability. • Impact on soil moisture budgets, storm hydrographs and river regimes.	• *Chapter 1 (Hydrological cycle)* • *Chapter 2 (River Wharfe)*
Process and change in river environments, landforms and ecosystems 1.2a What physical factors and processes influence channel characteristics and valley landforms? 1.2b What ecosystems exist in a river environment?	• A number of processes (e.g. weathering and mass movement) interact to create variations in cross and long profiles. Factors such as geology are important locally. • Hydraulic geometry of river channels. Changes downstream of river channel variables such as width, depth, velocity, discharge and river channel efficiency. • River channel load. Competence and capacity. Factors influencing sediment budgets. Linkage to river processes of erosion, transport and deposition. • Features associated with changing river channels, both meandering and braiding, to include within the channel and on the valley floor. • The basic structure and functioning of hydroseres, to include energy flows and nutrient cycling. • The importance of wetlands in a river environment.	• *Chapter 2 (River Wharfe)* • *Chapter 2 (River Wharfe)* • *Chapter 3 (Keswick)* • *Chapter 4 (Freshwater ecosystems – Sweat Mere and Florida Everglades)*
Environment – people interactions 1.3 How can changes in river landforms and river environments have an impact on people's daily lives? 1.4 How have human activities, some of which may be conflicting, influenced river environments? What are some of the consequences that can occur as a result?	Impacts of • changing river channels • changing sedimentation levels • changing discharge - flooding and low flow conditions: • Recurrence level and risk.	• *Chapter 3 (Flood management strategies in Keswick)* • *Chapter 4 (Komadugu -Yobe River Basin, north-eastern Nigeria)*
	• An **overview** of activities within a river channel and its catchment. Potential conflicts between activities. • ***Choose two*** of the following to illustrate enquiry questions in 1.4: – channelisation (re-sectioning and realignment) – dams and reservoir construction – urbanisation – water quality issues – over extraction for domestic, agricultural and industrial use	• *Chapter 3 (Keswick) – channelisation (re-sectioning and realignment)* • *Chapter 4 (Komadugu – Yobe River Basin, north-eastern Nigeria) – dams and reservoir construction,* • *Chapter 6 (Murray-Darling River Basin) – dams and reservoir construction, water quality issues* • *Chapter 7 (Three Gorges Dam) – dams and reservoir construction*
Management and the future 1.5 How does the management of river systems pose a continuing challenge for people?	• Human pressures on drainage basins create the need for management schemes. • Successful management requires an understanding of landform systems and processes and can be evaluated in terms of costs and benefits. • Bad management can lead to disastrous consequences. Practice has evolved over time, with increasing moves towards holistic management of the catchment and sustainable strategies.	• *Chapter 3 (Keswick) – channelisation (re-sectioning and realignment)* • *Chapter 4 (Komadugu – Yobe River Basin, north-eastern Nigeria) – dams and reservoir construction)* • *Chapter 5 (Florida Everglades)* • *Chapter 6 (Murray-Darling River Basin) – dams and reservoir construction, water quality issues* • *Chapter 7 (Three Gorges Dam) – dams and reservoir construction*

Introducing coastal environments

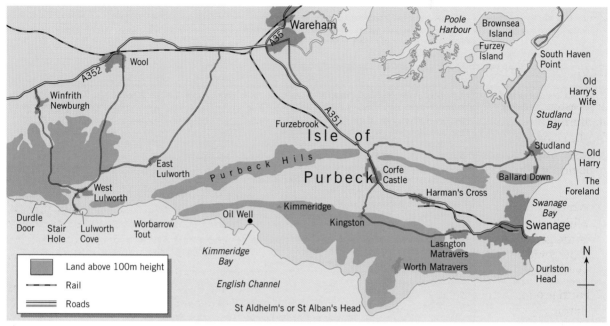

Figure 1 Map showing part of the Isle of Purbeck and its location within the UK.

This section is about coastal environments. It shows how a range of coastal features make coastlines fascinating for study, and how human activity has as strong an impact on coasts as it does upon rivers. It shows how coastlines need to be studied as 'whole areas', rather than just their landforms or ecosystems.

The Isle of Purbeck

The Isle of Purbeck on the Dorset coast is popular for Geography fieldwork because its coastal and geological features are outstanding, and because it has a concentration of human activities, each having a different impact on the environment.

Figure 1 shows the areas within this chapter. The Isle of Purbeck is not an island

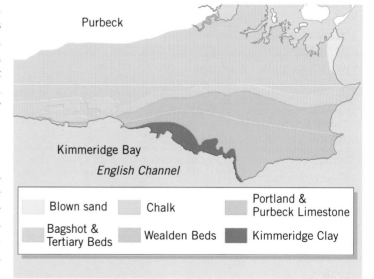

Figure 2 Geology map of the Isle of Purbeck

in the traditional sense, but it is isolated. It lies between Poole Harbour to the north and Swanage Bay to the south and east. To the west, the Dorset coast extends towards Devon. A combination of geological structure, physical process and human activity have led to an environment which is under pressure, for which several competing bodies are responsible.

The geological background

The Isle of Purbeck area has a mixed geology, with rock types of Jurassic age and younger (Figure 2). The rocks lie in 'bands' which run parallel to the southern coast. They run into the east coast at right angles, giving it a distinctive shape. Notice how some resistant rocks appear to extend further into the sea as headlands while others have formed bays.

Geology and coastline formation

Rock type has not only influenced the broad shape of the coastline. At a smaller scale, it has also influenced landforms. The area between Lulworth Cove and Durdle Door is known for its spectacular scenery and geology, including some of the finest examples of geological folding and coastal form in the UK.

---

1 Which types of rock appear to be resistant? Which are less resistant? Draw a sketch map of the geology of this area, and annotate it to show which are the more or less resistant rock types. Use evidence in your annotations.

2 Form groups of two or three and obtain a few simple geology textbooks, or rock identification kits or handbooks. Research into each rock type shown in Figure 2.
   - What does it consist of?
   - How hard or resistant is it?
   - How was it formed?
   - What strengths or weaknesses does it have?
   - Of what use is it to people?

3 From the information, identify a rank order of rock types, from most to least resistant. Does this rank order confirm what you identified in question 1 above, or not?

Figure 3 shows three features of the coastline close to Lulworth Cove. Durdle Door in photograph A is a natural arch cut into Purbeck limestone. The process of erosion has occurred at a major joint, widening it by wave action. The strength of the rock has enabled a roof to survive, so that a natural arch is created.

Lulworth Cove in photograph B shows a link between coastal form and geology. On the right, a headland of Purbeck limestone closes the entrance to the bay. Here the sea has had difficulty in breaching its resistance; once breached, however, erosion of less resistant clays and shales behind has

been rapid, and the bay broadens out. Behind that, on the left, chalk cliffs mark a different rock type. Their cliffs are steep but above the water line it is possible to see points where the grassy slope leading to the water edge has 'slipped', exposing white chalk beneath. These cliffs are unstable; in 1976 a fieldwork group of six students was killed during a sudden landslip after a period of wet weather.

Photograph C shows Stair Hole, west of Lulworth Cove, where sustained wave attack has breached a resistant bed of Purbeck limestone. Once breached, erosion of the weaker, surrounding rock is accelerated. This is exactly the process that created Lulworth Cove.

Coastal processes

As with river valleys, coastline form is the result of a combination of processes. Figure 4 shows how coastlines are a result of four processes, and their combination along any stretch of coast. Like valley slopes, the effect of **weathering** on a cliff face, together with slope processes of **mass movement**, gradually brings material downslope towards the sea. Together, these are known as **sub-aerial processes**. There wave action removes it and uses it in the mechanical **erosion** of the cliff. By undercutting the cliff, more material is threatened by collapse, only to be further eroded and broken down into beach material; this either accumulates or is removed by the sea. Wave currents may remove beach material and take it along the shore in a process of **longshore drift**. The balance between these four processes is fundamental to an understanding of human impact on coastlines.

---

4 Using the text in this section:
   a) draw an annotated sketch of Lulworth Cove, using photograph B in Figure 3. Including features such as headland, bay, beach, cliff line, etc.
   b) label your sketch with the three different rock types, using Figure 2 to help you.
   c) construct a sequence diagram of three or four stages, using photograph C in Figure 3 to help. Show how the cove has developed its shape from the time the sea first breached it to the present.

5 How might the coastline appear in future? Construct two more diagrams that show future possible changes and add them to your sequence.

Figure 3 Landforms showing erosion on the Isle of Purbeck coastline. Each of these landforms shows the importance of geology in determining landform appearance. Photograph A shows Durdle Door, a natural arch supported by Purbeck limestone. Photograph B shows the curved shape of Lulworth Cove, where resistant headlands surround a bay of weaker strata. Photograph C shows where wave attack has breached the limestone and the rate of erosion has accelerated in weaker clays lying behind the limestone (see Figure 2).

Geology as a resource base

Geology has also been responsible for the economic base of the Isle of Purbeck. Traditionally, Purbeck limestone was used as building stone. However, removing the stone adds pressure on the

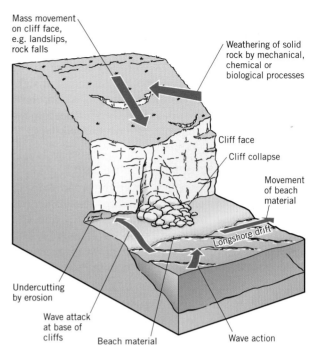

Figure 4 The relationship between weathering, mass movement, wave erosion and beach processes along coastlines.

landscape, especially now that most stone is quarried, rather than mined underground.

The oil industry puts further pressure on the area. In the 1960s, Purbeck was the only oil-producing area in the UK, before North Sea oil. On the cliff above Kimmeridge Bay, a 'nodding donkey' has pumped small amounts of oil from beneath the cliffs for over 40 years. The oil is found in oil shales within the Kimmeridge series of strata. Its impact is small and mainly in the seven journeys per week that take the oil by road to Furzebrook rail terminal. Further north, on Furzey Island, BP has a major oil production site from Wytch Farm oilfield, opened in the early 1990s. It employs over 300 people, including those at the production site, where eight 'nodding donkeys' pump oil from 60 wells. Here the gathering station draws oil by pipeline across Poole Harbour to the rail terminal at Furzebrook. Current production of 90 000 barrels per day is being increased to 100 000.

Coastal ecosystems

Two major coastal ecosystems are prominent in the Isle of Purbeck: sand dune ecosystems, and estuary and salt marsh ecosystems. Some of the most difficult environments are found on exposures of sand, in which dunes first develop, and which later support plant communities able to withstand strong on-shore winds and salt water. Their fragility is made more difficult by pressures from

Figure 5 Photograph showing Kimmeridge Clay and the oil-bearing shales lying beneath it. Although this is a highly oil-rich rock, it is geologically weak. Cliffs formed in Kimmeridge Clay erode quickly and are unstable.

tourism. Studland Bay is a major focus of this section in chapter 10.

Based upon tidal changes which expose areas of silt and mud at low tide, followed by flooding by salt water at high tide, salt marshes are also sensitive ecosystems. Not only are plant species adapted to cope with salinity and immersion beneath water, coastal storms also add to the hostile environment, in which survival is difficult. Fauna are highly specific, from mud-inhabiting creatures such as crabs or worms, and shellfish such as oysters, to wading birds. The pressure caused by coastal development is often greatest at estuarine sites, whether for industry (port and harbour development, oil terminals), service industries (sewage disposal) or leisure (yachting marinas).

Figure 6 shows the range of sensitive sites around Poole Harbour and Wytch Farm, on the northern shore of the Isle of Purbeck.

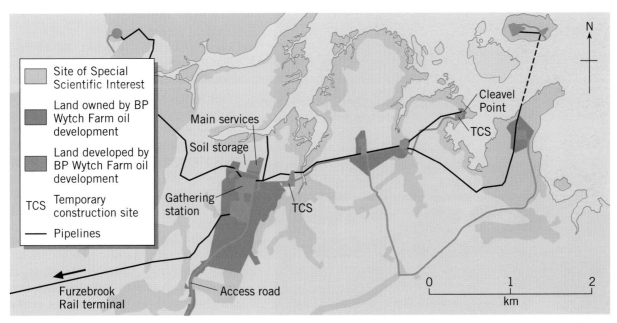

Figure 6 Sensitive sites around the BP Wytch Farm development.

1 Identify Furzey Island and the three sites involved in the Wytch Farm oilfield on Figure 6.
  a) Why is oil production a sensitive issue in this area?
  b) Production involved consultancy with 14 organisations, including MAFF (Ministry of Agriculture, Fisheries and Food), Wessex Water, Poole Harbour Commissioners, the Nature Conservancy Council, the RSPB (Royal Society for the Protection of Birds), and the Council for the Protection of Rural England. In small groups, decide how and why these organisations might have been involved in the consultation about oil production.

2 Draw a sketch map of Figure 1. On it, identify and label different pressures which occur along this coast.

3 Decide which pressures are social, economic, or environmental. Use different colours to highlight your annotations accordingly.

Pressures from tourism

Tourism is a major industry in the Isle of Purbeck, with day and long-stay visitors, and holiday-home-owners. Swanage is a traditional 'resort', but many people are also attracted to a range of scenic and cultural attractions. Each brings its own pressure. Lulworth alone attracts over half a million visitors each year. The road leading from Swanage to Studland Beach (Figure 7) and then by ferry to Poole frequently becomes choked with traffic, especially at summer weekends.

Historic sites

Corfe Castle is a major attraction, owned by the National Trust. Other historic sites include villages such as Worth Matravers. Lulworth and other locations have been designated Heritage Coast sites.

Figure 7 Tourist pressures; car parking at Studland Bay in summer.

Nature attractions

The area is rich in bird life, saltwater fauna and flora, heathlands – among others – and both coastal and inland ecosystems. Several sites are Sites of Special Scientific Interest (SSSI), and the Isle of Purbeck as a whole is an Area of Outstanding Natural Beauty (AONB).

Managing the coastal environment

In such an environmentally sensitive area, management is required to minimise impacts of different pressures. Lulworth alone has numerous designations such as SSSI, AONB and Heritage Coast, which help to conserve its unique natural heritage. Each involves different organisations in managing different aspects of its development and change. In fact, a range of organisations are responsible for management of the Isle of Purbeck. They are categorised into private companies, government bodies and voluntary organisations.

Private companies

These, by definition, are people whose business in the Isle of Purbeck is focused on the generation of wealth. They vary from small, independent businesses, to large transnational companies such as BP. Their interests vary and it is difficult to manage these as a whole. They tend to promote private rather than public interests.

Government bodies

Government bodies consist of:

- County Councils, elected to control planning, roads, and services such as education, libraries, and tourist information. Where areas are designated Heritage Coast or AONB, county councils co-ordinate work such as footpath maintenance, and different interests such as landowners across whose land these lie. They also allow or refuse planning permission which might alter communities, or work against their interests.
- Borough Councils, which are smaller units within the county, responsible for some housing, and environmental concerns such as managing the coastline. They have budgets delegated by County Councils and receive some money from central government.
- Parish Councils, who have little responsibility except for local parks.

National government controls some finance, such as coastal management funding from MAFF. In the Isle of Purbeck, it also controls direct funding into defence and land controlled by the Armed Forces. The Isle of Purbeck contains large areas which have limited public access because of Army Training around Lulworth Camp.

Much work is done by government agencies, such as the Environment Agency which controls river and freshwater quality, beach and bathing water quality, and pollution. English Heritage manages historic sites. Each agency is given a budget by national government. National government is also responsible for implementing national laws, e.g., rights of access to property.

Voluntary organisations

These generally represent single issues or interests. They usually campaign on behalf of members in return for membership fees or donations. The Ramblers Association represents those who campaign for the Right to Roam, the RSPB represents ornithologists, the National Trust purchases and maintains historic property, while the Council for the Protection of Rural England campaigns for and seeks preservation of a non-urban lifestyle.

References for further study

Isle of Purbeck web page, address
http://www.purbeckweb.co.uk/
Kimmeridge Bay web page created in partnership with Bournemouth University
http://csweb.bournemouth.ac.uk/consci/
coastlink/purbeck.htm

1 On your map of pressures on the coastline, identify and annotate who might have responsibility for managing different pressures.

2 What difficulties might there be in bringing different organisations together?

3 In pairs, identify the groups involved in the following issues in managing the coastline;

Issue 1 A proposal to extend the BP oil extraction at Wytch Farm

Issue 2 A proposal to designate the whole of the Purbeck Coast a National Park

Issue 3 A proposal to build a new yachting marina in Swanage.

For each issue, decide:
a) what interests exist for different groups
b) how some of these might conflict.

4 In groups of four, discuss the case for having a central body responsible for all aspects of social, economic and environmental issues in the Isle of Purbeck.

8 Managing coastlines

Why do coastlines need to be managed?

A coastline is the boundary between land and sea. This definition is not as simple as it appears, because the boundary is different at low tide and high tide, or during storms and in times of calmer weather. It is therefore more usual to refer to a **coastal zone**, a wider part of the coastline that may be affected by a variety of processes. These include erosion, transport and deposition, and are affected by both human and natural influences. The coastal zone is marked by a variety of landforms, including beaches, cliffs and sand dunes. Each adjusts to a range of processes, from intense wave energy and storm-force winds to periods of calm. The area of coast may also be greatly affected by processes that occur further out to sea. The coastal zone and the extended area out to sea is known as the **littoral zone**.

Coastal zones are the focus for much human activity and development; pressures from housing, commerce, leisure and tourism all play their part in influencing the character of a coastline. This chapter shows how the coastline in Northumberland is under pressure, and that there is a need to manage it carefully.

The Northumberland coastline from St Mary's Lighthouse to Druridge Bay

This study is designed to show the many different processes and issues that affect a small stretch of coastline. Figure 8.1 shows the location of the south Northumberland coast, between St Mary's Lighthouse at Whitley Bay, 10km from Newcastle-upon-Tyne, to the northern end of Druridge Bay. In total, this stretch is about 26km long.

This study explores issues of erosion at Collywell Bay and Sandy Bay, and between Collywell Bay and St Mary's Lighthouse. These areas suffer from cliff erosion but each has been dealt with differently. The study also looks at Druridge Bay, an area of deposition in which a number of different management issues are emerging.

The South Northumberland coast is strongly influenced by the geology and geomorphology of the area. It is typified by long sandy bays separated by stretches of cliffs. Compared to other British coastlines, a relatively small proportion of the coastline is protected by sea defences. It has a relatively low population density and consists of important natural habitats, including several Sites of Special Scientific Interest (SSSI). Industry has also had an important impact upon the coast here. Centuries of coal mining have led to much land subsidence and dumping of spoil. More recently, coal has been mined using open-cast methods. Some areas of subsidence or open-cast mines have been reclaimed and are now quasi-natural and nature reserves. Sand extraction from beaches and dunes has also had an impact upon coastal form, or morphology.

Collywell Bay

Figure 8.2 was taken on the clifftop near Collywell Bay, along the southern part of this stretch of coastline. The area is located within the

Figure 8.1 The south Northumberland coast.

Figure 8.2 Warning sign on cliff top at Collywell Bay.

village of Seaton Sluice, shown in Figure 8.3 and 8.4. This section is designated a SSSI because of the geology. A typical erosion feature – a stack, named Charley's Garden – is located just north of Collywell Sands, shown on Figure 8.4. Differences in geology within this small bay help to account for its shape and form.

What part has geology played on this coastline?

Study Figure 8.5 which shows the geology of Collywell Bay. Consider the role that geology played in the development of the coastline. The geology is complex, broadly consisting of sandstones, shales, mudstones and coal seams at right angles to the coast. Sandstone is the most resistant of these, but Figure 8.5 shows that there is more than one kind. The Crag Point sandstone has created the headlands of Crag Point and Rocky Island. Clearly, Charley's Garden sandstone has eroded more than Crag Point sandstones, and is weaker. However, it produced a small 'stack' known as Charley's Garden (Figures 8.3 and 8.4). Given that no other rock type of those remaining produced a similar feature, we can deduce that sandstones are more resistant than other rocks.

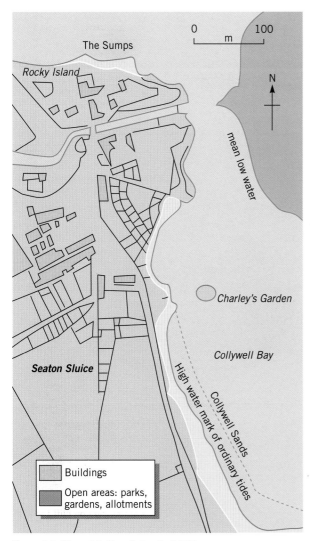

Figure 8.3 Map of Collywell Bay in 1887.

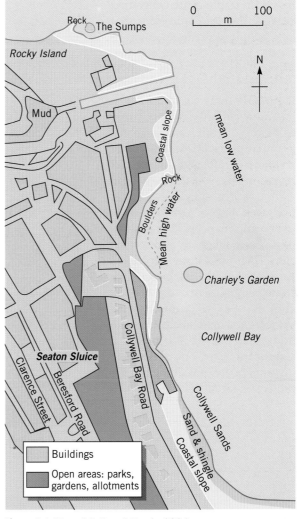

Figure 8.4 Map of Collywell Bay in 1994.

The sandstones are thickly bedded, well jointed and have a 'blocky' appearance. The debris created by erosion at the cliff foot and on the shoreline consists of large blocks of sandstone, as these fall away from the cliff. The rocks were all formed between 220 million and 190 million years ago, during the Permian period.

Why is there a problem of erosion at Collywell Bay?

Figure 8.6 shows housing along the cliff top at Collywell Bay. Find this area on Figure 8.3 and 8.4. How rapid do you think the rate of cliff erosion is here? Why is erosion a problem in Collywell Bay? Geographers are concerned because these are questions that link people with their environments. Figures 8.3 and 8.4, two maps of the same area in 1887 and 1994 respectively, can help to provide some answers.

Collywell Bay does not suffer as badly from erosion, for example, as the Holderness Coast on North Humberside, where cliff recession exceeds 2m per year. Although it is slower, erosion at Collywell Bay is still a serious issue, and expensive measures have been taken to protect the cliffs. Pressure from geologists has led to the area of exposed rock strata being designated as a SSSI, and the Bay is popular with students for field study. Pressure from residents has also played a significant part in attracting investment and providing further protection measures.

1 On Figure 8.4, locate where the photograph (Figure 8.2) was taken. What does this photograph suggest about the geology?
2 Draw a sketch map to show Collywell Bay. Annotate it with features of the coast, such as headlands and the bay itself, but also pay attention to smaller-scale features shown on Figure 8.4. Include information about the Geology on your map.
3 What do Crag Point to the south and Rocky Island to the north suggest about the resistance of rock types there?
4 Using evidence, create a rank order of all rock types shown, from most to least resistance.

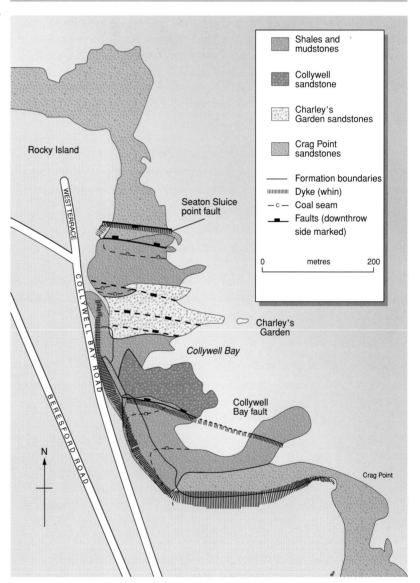

Figure 8.5 The geology of the Collywell Bay area. All rocks here are of approximately the same age, from the Permian period.

Figure 8.6 Collywell Bay, seen from the south.

1 a) Identify changes that have taken place within Collywell Bay between 1887 and 1994. How rapid has erosion been on the cliff top and on the cliff base?
  b) Measure the amount of coastline eroded between the dates of the two maps in Figures 8.3 and 8.4. Calculate the rate of erosion in metres per year for this stretch of coast.
2 Draw the area of Collywell Bay between Rocky Island and Crag Point to show changes to the bay between 1887 and 1994.
3 Add annotations to show how the area has been developed between 1887 and 1994. Do these developments suggest that erosion is a rapid problem?

Theory

Erosion processes and cliff formation

Cliff processes can be sub-divided into two categories: those that operate at the foot of the cliff, and those that operate on the face of the cliff.

Cliff-foot processes

The main agent of erosion at the foot of a cliff is wave energy. Waves are a direct result of wind which creates friction on water surfaces; the greater the wind strength, the greater the size of the wave. Waves may be generated a long distance from where they eventually reach a shoreline; the distance travelled is known as the fetch. The greater the fetch, the greater the wave size. The greatest wave sizes are therefore likely to be found on coasts that face major oceans, such as the east coast of Australia, or the Atlantic coast of Ireland.

Figure 8.6 shows how the cliff foot at Collywell Bay has been undercut. The headland over-hangs the base which has been undercut, whereas in the foreground and at other points around the foot of the cliffs, boulders show where cliff collapse has created rock debris. As waves approach the shore, friction with the shoreline slows them down. The lower layers of water are slowed down, and upper layers over-ride them, creating a build-up of water which eventually spills over as a wave. At the foot of a cliff, this may have several effects.

- *Hydraulic pressure* Solid rock faces such as sea cliffs are often fractured by joints or bedding planes between rock strata; these contain air spaces. The force of the waves compress air within the fractures, which explodes as each wave recedes and air is released. The repetition of this process causes further fracture of the rock, which then falls apart from the cliff.

- *Abrasion or corrasion* As waves approach, wave energy enables material to be picked up and thrown against the foot of the cliff. The mechanical action of material against the cliff foot is known as abrasion. This can be particularly effective against cliffs with little resistance to erosion such as those at Sandy Bay. It leads to undercutting of the base of the cliff and subsequent cliff collapse.
- *Corrosion or solution* This is the chemical effect of salts or of acids held within sea water. Weak acids dissolve or corrode alkali rocks containing calcium carbonate, such as limestone or chalk. Salt crystallisation occurs on all rocks facing the sea, as salt rehydrates on the surface. The stresses created by crystal growth and chemical effects of salts weaken many rocks, particularly those which are already weak and fragmented.

Cliff-face processes

Cliff-face processes involve the action of weathering and mass movement of rocks on the face of the cliff. These are collectively known as sub-aerial processes, and consist mainly of weathering and mass movement.

Weathering

Weathering is the disintegration or decay of rocks at the point where they are exposed at the earth's surface. More details are included in Chapter 2. Processes include mechanical processes such as freeze-thaw, and chemical processes, such as solution and corrosion. Together, these act on the cliff face and weaken rocks already exposed to the sea.

Figure 8.7 Wave attack at the foot of a cliff at Collywell Bay. Notice how waves attacks the cliff foot head-on at this point. This particular attack is against weak muds and clays which erode easily.

Mass movement

Mass movement is the downslope movement of rock material under gravity. It is significant in the way that it causes cliff recession. The simplest example is cliff collapse (Figure 8.7). Other processes include land-slips or landslides, where weakened cliff material slips intact along a line of weakness, often following heavy rain or a storm. During this time the weight of the mass increases as a result of the rainwater (Figure 8.8). Slumping is similar but involves disintegration of the whole structure, unlike slips which remain intact as they move. Figure 8.9 shows a slump that appears to have been caused by a spring, which added water and weight to a part of a cliff face, causing the cliff to fail. More details are included in Chapter 2.

Figure 8.8 Slip lines along a cliff, showing mass movement along the cliff face. In this case there are lines of weakness at right angles to the surface. Stress lines are set up caused by structural failure of the rock. These lines or fractures become lubricated with water and fail easily with additional weight during wet periods. As a result, the cliff edges slip, still intact.

Figure 8.9 Slumping along a cliff line. This material has become saturated with water, probably from an underground spring. The weight of saturated clay has caused the cliff to fail and the movement has brought the material down rapidly, mixing as it moves so that it is no longer in tact.

How does erosion affect people's attitudes?

Living close to an area where erosion is taking place does affect people's attitudes. You have already seen how housing developments have taken place at Collywell Bay in the past 100 years, in spite of erosion. The evidence seems to suggest so far that erosion is slow, but that remedial action and warnings are needed. Do people see erosion as a threat? If they do, what is it like to live where they perceive a threat? Do local councillors believe that opinions of people who live in Collywell Bay are significant?

How can Collywell Bay be protected?

Look at Figure 8.10 which shows one aspect of coastal management, that of coastal defence. The map shows annual spending on coastal defence in Britain by section of coast. The amounts vary considerably in different parts of the country. This money is spent by local councils, who are responsible for coastal defence work. Once borough council engineers have identified work areas where they believe work is necessary, funding is sought from council budgets and from the Ministry of Agriculture, Fisheries and Food (MAFF). Decisions are made by elected councillors acting on the advice of engineers. Like many studies in Geography, you will find that people who decide whether or not to take action are affected by their own perceptions and preferences. Study the theory box for more detail on decision-making and coastal protection.

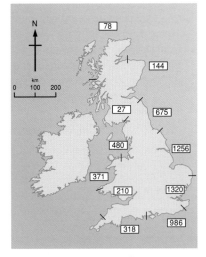

Figure 8.10 Annual cost of coastal defence work in Britain.

1 Using your atlas, identify areas of high, medium and low population density in the UK. Compare this map with Figure 8.10. What is the relationship between the two maps? Why should there be this relationship?

2 Why should the amount spent on coastal defence in Figure 8.10 be uneven around Britain?

Theory

Who is responsible for managing Britain's coastal zone?

Coastal management is a large responsibility, covering a wide range of interests and organisations. It includes coastal protection, flooding, protection of both human and natural environments, and managing problems that recreational activities and visitor pressure may bring. No single government agency is responsible for it, and in the past there has been a piecemeal approach, involving different organisations. The coast has often been studied from a small-scale perspective and protection has been designed for a particular area with scant regard for the impact that it may have on another. For example, coastal groynes developed to retain and control to sand movement in one resort may mean that beaches further along the coast become starved of sediment supplies. If nothing is done in the latter areas, then their own beach will narrow and their cliffs become more exposed to erosion. The construction of sea walls can have similar effects, such as those built along the south Hampshire and Dorset coast between Bournemouth and Hurst Castle, and along the Holderness coast between Flamborough Head and the Humber Estuary.

Figure 8.11 shows the number of organisations responsible for managing Britain's coastline. Local borough councils traditionally had most of the responsibility, identifying areas in need of protection, costing the work, and submitting plans to the Ministry of Agriculture, Forestry and Fisheries (MAFF). These plans were independent of those submitted by other areas, even in the same county or local authority.

- County and Borough/District Councils
- The Environment Agency
- English Nature (SSSI and areas of outstanding natural beauty
- National Nature Reserves
- Countryside Commissions (Heritage coasts)
- National Park Authorities
- Miistry of Agriculture, Fisheries and Food (MAFF)

Figure 8.11 Organisations Responsible for Managing Britain's coastlines.

More recently, there has been greater co-operation between councils and local interest groups. In 1995 Shoreline Management Plans (SMP) were introduced by MAFF, to provide the basis for sustainable coastal defence policies for coastlines, and to set objectives for their future management. The administrative areas are defined by sediment cells, which are areas in which coastal processes are relatively contained. Their size is often too large for workable projects and they are divided up into sub-cells. MAFF divided the coast of England and Wales into eleven sediment cells, shown in Figure 8.12.

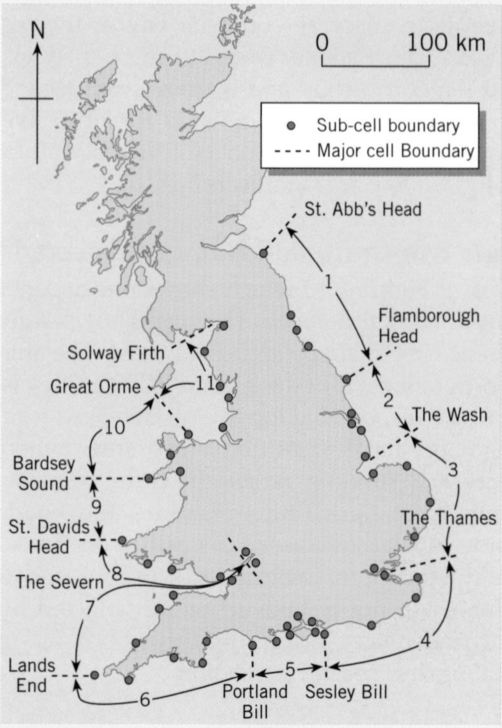

Sediment cells are based upon natural processes rather than administrative boundaries, making coastal management much more co-ordinated and effective. However, the cells cross council boundaries and co-operation between the different groups is therefore essential. The key issues in the preparation of SMP are:

- coastal processes
- coastal defences
- land use and the human or built-up environment
- the natural environment.

Sub-cells are further divided into management units, even smaller lengths of coast for more effective management.
Each management unit can decide independently, on the following options:

- do nothing (other than measure and monitor)
- hold the existing defence line
- advance the existing defence line
- retreat the existing defence line.

Figure 8.12 The major sediment cells of England and Wales.

Coastal management in Northumberland

The following authorities are responsible for coastal management in south Northumberland:

- Alnwick District Council
- Blyth Valley Borough Council
- Castle Morpeth Borough Council
- North Tyneside Council
- Wansbeck District Council
- Environment Agency
- Northumberland.

Together, these organisations work with Berwick upon Tweed Borough Council and Scottish Borders Council as well as special interest groups from the Northumbrian Coastal Authorities Group in designing plans that are sustainable for this coastline.

Defending coastlines against erosion is different from coastal protection, which protects coasts from flooding. Like all work related to flooding, this is managed by the Environment Agency, who are also answerable to MAFF. There are two broad options for defending areas of coasts such as Collywell Bay: 'hard' or engineering solutions, or 'soft' approaches.

Hard approaches to coastal defence

'Hard' solutions involve engineering work. Figure 8.6 (page 99) shows the broad sweep of Collywell Bay and the sea wall constructed at the foot of the cliff to protect it from further erosion. Although effective in many ways, there are some problems

with this method of defence. Undercutting of a sea wall itself can take place. Although wave energy no longer attacks the cliff foot, it is only deflected, and the energy is often used in scouring the bed at the foot of the wall. This eventually leads to collapse (Figure 8.13). A further effect is that wave energy is then directed at the beach, leading to beach erosion and removal.

Stone revetments protect the beach through the use of stone boulders which absorb most of the wave energy by increasing the surface area exposed to wave attack (Figure 8.14). However, these are regarded as ugly and are therefore unpopular. Another alternative is to drain those parts of the

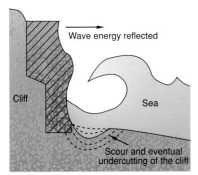

Figure 8.13 The effect of wave attack on a sea wall.

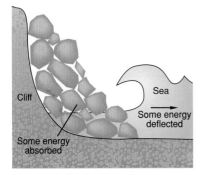

Figure 8.14 Section through a revetment. A revetment is a structure consisting of large angular boulders, placed to protect a cliff. The surface area of the boulders and air spaces absorb wave energy and reduce the potential rate of erosion.

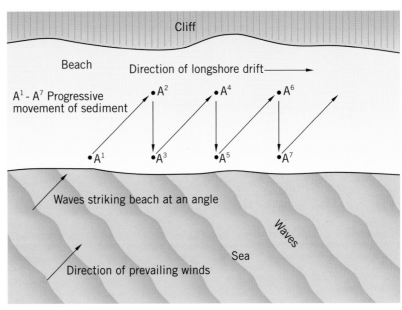

Figure 8.15 Longshore drift.

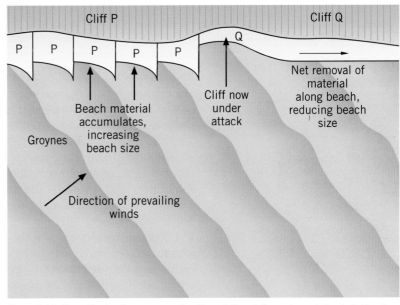

Figure 8.16 The effects of groyne construction at P and Q can be unintended. Beach accumulation at P has resulted in beach starvation at Q. As a result, wave energy is now increased closer to the cliff, increasing the threat of erosion.

cliff which are likely to become saturated, thus preventing slumping. This is done by boring into the cliff and pumping out water at the base.

Beaches are fundamental to cliff defence. The energy of approaching waves is absorbed, or dissipated, by friction. The best source of friction is a wide beach, since much wave energy is then absorbed before a wave reaches the cliff foot. You may have noticed in comparing the two maps of Collywell Bay in 1887 and 1994 that the beach has diminished. The effect of the sea wall, therefore, has been to remove the very feature which would have protected the cliff foot in the first place!

Other ways of designing cliff-foot defences have been attempted. Groynes attempt to retain beach sediment. These are usually wooden structures, built at right angles to the beach. They are designed to trap sediment which would otherwise be removed by longshore drift. However, by preventing removal of beach material from one point, other locations may suffer. Figures 8.15 and 8.16 show how the construction of groynes may protect the cliff at point P, while threatening further erosion at point Q. Point Q becomes starved of sediment, thus reducing beach size and increasing wave energy at the cliff foot.

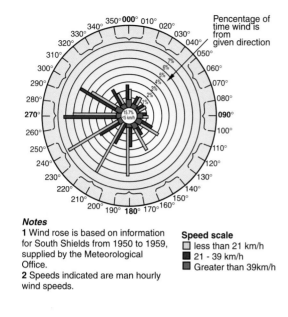

Notes
1 Wind rose is based on information for South Shields from 1950 to 1959, supplied by the Meteorological Office.
2 Speeds indicated are man hourly wind speeds.

Speed scale
☐ less than 21 km/h
■ 21 - 39 km/h
■ Greater than 39km/h

Figure 8.17 Wind speeds and direction along the southern Northumberland coast.

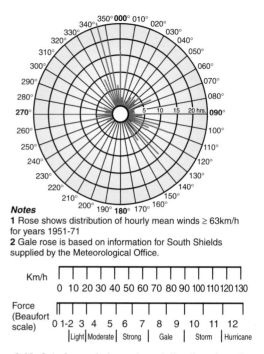

Notes
1 Rose shows distribution of hourly mean winds ≥ 63km/h for years 1951-71
2 Gale rose is based on information for South Shields supplied by the Meteorological Office.

Km/h 0 10 20 30 40 50 60 70 80 90 100 110120 130

Force (Beaufort scale) 0 1-2 3 4 5 6 7 8 9 10 11 12
|Light|Moderate| Strong | Gale | Storm |Hurricane

Figure 8.18 Gale-force wind speeds and direction along the southern Northumberland coast.

Soft approaches to coastal defences

The cost of 'hard' engineering solutions is high. Attempts to reduce central government expenditure have led MAFF to re-assess its preferences for coastal defence. 'Soft' options for coastal defence adapt to and supplement natural processes. This means that some cliff-lines might be left to retreat naturally, or at least be protected by cheaper or more sustainable options. 'Sustainable' in this case means something that protects the coast and maintains its protection, preferably through natural processes such as off-shore breaks and beach nourishment.

Off-shore breaks involve the creation of an artificial reef at sea, which reduces wave energy before it reaches the shore zone. It can comprise a range of materials, from dredged sediment to old car tyres which are dumped and anchored strategically.

In Portobello on the Firth of Forth, east of Edinburgh, during the 19th century, extraction of sand from the beach provided raw material for a sand and brick works. This activity depleted the beach so much that by 1926 it had almost disappeared. Storm waves were severe and caused much damage, as waves reached even as far as the houses on the promenade. In 1970, a beach nourishment project was introduced to import sand from Fisherrow Sands. Reducing the gradient of Portobello beach, together with groyne construction and the use of coarser, heavier sand, has succeeded in keeping the beach intact in the last 30 years.

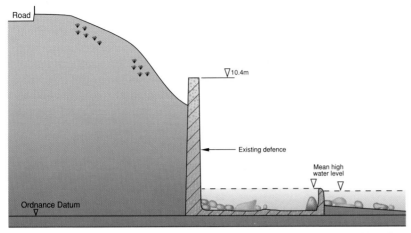

Figure 8.19 Section through the original wall. This wall is shown in Figure 8.20. It is the highest wall in the centre of the bay and is showing signs of ageing.

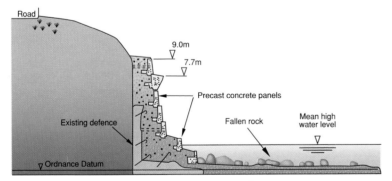

Figure 8.20 Part of the original wall.   **Figure 8.21** Part of the northern wall.

Figure 8.22 Section through part of the northern wall.

Reviewing the defences at Collywell Bay

The old part of the wall at Collywell Bay is deteriorating. As part of a summer vacation job, you are working for a firm of engineers contracted to advise on the best way to protect the cliff at Collywell Bay. Your job is to write a report about the best strategy for renewing the defence, using the data available to you as follows:

a) Figure 8.17 which shows wind speeds and direction during the year along this stretch of coast;

b) Figure 8.18 which shows gale speeds and direction during the year along this stretch of coast

c) Figure 8.19 which shows a section through the original wall

d) Figure 8.20 which shows a photograph of the original wall

e) Figure 8.21 which shows a photograph of newer wall along the northern end of Collywell Bay

f) Figure 8.22 which shows a section through the northern wall.

Write a report of about 1000 words, with annotated diagrams, showing which designs you consider most effective, and why. In your report, state:
- the problems faced at Collywell Bay
- the effects of previous attempts to protect the cliff foot
- your evaluation of the success of these attempts
- possible alternatives to protect the cliff
- reasons why you accept some of these alternatives, but not others
- your preference for cliff management at Collywell Bay.

Sandy Bay

Sandy Bay is located north of Blyth, between the estuary of the River Wansbeck and Newbiggin-on-Sea. This part of the Northumberland coast is shown in Figure 8.23. The caravan park at the southern end of the bay brings a seasonal influx of tourists who stay in the area, and Newcastle-upon-Tyne is only 20 minutes drive away for daily visitors.

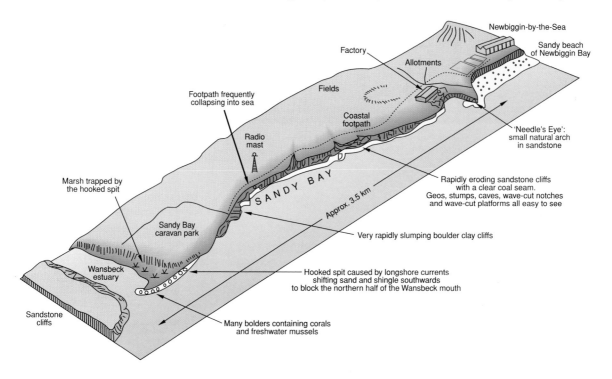

Figure 8.23 The Sandy Bay area.

1 What evidence is there of sediment movement in the Sandy Bay area? In which direction does it seem to be taking place?
2 How far does the direction of movement correspond with what you would expect from the rose diagrams in Figures 8.17 and 8.18? How would you explain this?
3 Where and what is the source of the sediment?
4 Using Figure 8.23, compare the land use in this stretch of the coast to that at Collywell Bay.

Why is erosion so serious at Sandy Bay?

Study Figure 8.24. It shows a succession of shore lines at Sandy Bay since 1853. In the last 150 years the coast has receded rapidly, much more so than at Collywell Bay. The geology is significant. Here, like Collywell Bay, the main rock types are sandstone, shales and coal seams. The sandstone has been undercut and repeated collapse has caused the upper parts of the cliff to retreat. Further south, the cliffs consist of boulder clay, the resistance of which is low. Once removed, the boulder clay particles remain suspended in sea water rather than being deposited on the beach; some are carried south towards the spit. This is the area where the caravan site is situated.

Figure 8.24 Cliffs at Sandy Bay.

1 Study Figure 8.25. Measure the distance eroded on the cliffs between:
   a) 1853 and 1897   b) 1897 and 1922
   c) 1922 and 1959   d) 1959 and 1978.
2 Find the average rate of erosion per year in each of the four periods. Plot this on a graph.
3 Is erosion constant, speeding up, or slowing down? Predict the location of the coastline by the time you are aged 40.
4 Does the photograph in Figure 8.24 give any clues as to why this is happening? Refer to the Theory box *Erosion processes and cliff formation* on page 99–100.

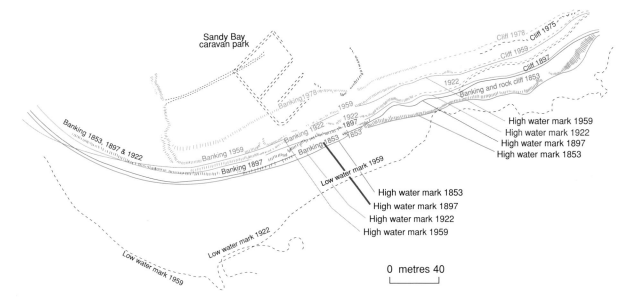

Figure 8.25 Shorelines at Sandy Bay since 1853.

Erosion and human activity

Three factors have increased the rate of erosion significantly along this stretch of coast. Large areas of Northumberland have been subjected to coal mining during the last 200 years; Sandy Bay has been especially affected. There is much surface subsidence here, caused by the abstraction of coal in deep seams which have been left to collapse into each other. In some cases, this has accelerated the rate of erosion. Sea water is thus able to invade a beach which has been subject to subsidence, and reach the cliff foot.

Field drainage on farmland has contributed to the problem. Where drains reach the cliff, the outflow of water from the fields has run down the cliff face and caused the cliff to retreat. This situation has been made worse by saturation of clays which have slipped when lubricated by the run-off.

Further north, at Newbiggin-on-Sea, coastal defences have reduced sediment supply to Sandy Bay, especially sediment of larger grain size. The larger grain size and mass is important as it provides greater beach stability than that offered by fine-grained boulder clay.

What can be done?

Little was done by the borough council to combat erosion until recently. Some boulders have been placed, or have fallen, at the cliff foot and help to protect the cliff. However, longer-term solutions are necessary to protect the cliffs at Sandy Bay. Figure 8.26 shows a proposed scheme which uses a technique known as 'rip-rap'.

Cost-benefit analysis meant that the possibility of protection was shelved for future reassessment. The caravan site owners were not happy with the delay and, after consulting English Nature, erected rock armour revetments on the beach (Figure 8.27). The borough council will soon carry out a complex cost-benefit analysis to see if any further protection work is needed at Sandy Bay. This is consistent with their Shoreline Management Plan (SMP), which intends to hold the line along selective stretches of coast. Figure 8.28 is an economic appraisal of the SMP, and illustrates costs and benefits from the three options.

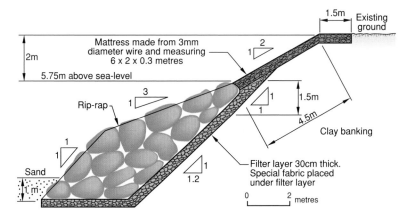

Figure 8.26 Possible protection scheme at Sandy Bay.

1 Study Figure 8.26. Describe the 'rip-rap' structure.
2 Identify its component parts. How does it attempt to solve the problems of:
  a) cliff-foot erosion
  b) storm-wave energy
  c) excessive land drainage and saturation of clay
  d) cliff slumping?
3 The cost of this is scheme is about £700 per metre at 1999 prices. Possible strategies for this 3.5km stretch of coast include:
  • protect the whole 3.5km stretch
  • protect certain parts of it
  • protect none of it.
Form groups of two or three people. Carry out a cost-benefit analysis, described in Chapter 3 on page 40.
  a) Which option seems most feasible?
  b) How do your recommendations compare with those made by other groups?
  c) How might the owners of the caravan-site react to your recommendation?
  d) How should the caravan-site owners defend their case for protection work to be carried out at a time of spending cuts in local government?

Figure 8.27 Rock-armour revetments at Sandy Bay

Case	Damages (£)	Benefits (£)	Costs (£)	B/C Ratio
Do Nothing	(596,190) 604,958	–	–	–
Hold the Line (Throughout Unit)	– –	(596,190) 604,958	982,000	(0.61) 0.62
Hold the Line (Selectively)	(15,009) 5,010	(581,181) 599,948	409,000	(1.42) 1.47

Figure 8.28 Results of Economic Appraisal for Management Unit 37, which includes Sandy Bay

Figure 8.29 and 8.30
Managed Retreat between Collywell Bay and St Mary's Lighthouse.

Managed retreat between Collywell Bay and Sandy Bay

The land between Collywell Bay and St Mary's Lighthouse at the northern end of Whitley Bay is of relatively little economic value. It mostly comprises privately owned rough pasture. A footpath runs along the top of the cliff which is popular with ramblers. Access to the beach is restricted to a set of steps halfway between Collywell Bay and the lighthouse. The cliffs along this coast are subject to erosion (Figures 8.8 and 8.9).

Costs of protecting these cliffs would be considerable using either hard or soft engineering, and cost-benefit analysis shows both options to be impractical. The preferred option on many stretches of coastline now is that of 'managed retreat', where the situation is monitored and safety structures are maintained. The photographs in Figure 8.29 and 8.30 show that, as cliffs erode, the fence and footpath in Figure 8.29 will be moved back towards the land.

Managed Retreat: controversy versus realism

1 Managed retreat policies are controversial. Many people disagree with the principle and see it merely as a government money-saving ploy.

   a) Read the three statements below. Highlight those parts of the statement that are fact and those that are opinion. How easy is it to distinguish between fact and opinion?

   b) Consider the possible backgrounds of the people who might have made these statements.

   STATEMENT 1: 'Managed retreat is a cop-out. Money always seems to be the important factor. People's land, England's land is being destroyed and it always comes down to 'it costs too much'. I think that there are more important things than money and people's livelihood is one of them.'

   STATEMENT 2: 'Coastal erosion is a fact. It happens, and little can or should be done about it. Cost-benefit analysis shows that some land is not worth protecting anyway, and in this case managed retreat is the only viable option.'

   STATEMENT 3: 'Hard defences will continue to be the only option where important natural or human assets are at risk. But "managed retreat" is suitable where sea walls protect land of low value or quality. Hard defences would only prevent valuable sediment from reaching sand dunes, shingle beaches and salt marshes elsewhere.'

2 Form six groups. Each group should formulate a reply to one of these statements.

   • Group A should agree with statement 1; Group B should disagree with statement 1.

   • Group C should agree with statement 2; Group D should disagree with statement 2.

   • Group E should agree with statement 3; Group F should disagree with statement 3.

3 Now write an essay of about 1 000 words, saying how you rate the 'managed retreat' policy for eroding coastlines.

4 Draw a three-circle Venn diagram (see Chapter 3). Label one circle Collywell Bay, one Sandy Bay and the other, St Mary's Lighthouse to Collywell Bay.

   a) Add labels to show the problems experienced in each of these areas.

   b) Using a second colour, add labels that highlight the human development and activities for each area.

   c) Using a third colour, add labels according to what has been done to combat erosion.

Druridge Bay

Druridge Bay is the largest and most northerly stretch of this part of the Northumberland coast. It lies at a point where the relatively undeveloped coastline of the north meets the more developed and industrialised coastline of the south, eventually leading towards the Tyne and Wear conurbation. The Bay consists of a long, mainly sandy stretch of beach and dune coastline between Cresswell to the south and Low Hauxley to the north. In October 1995, Druridge Bay was awarded 'Heritage Coast' status by the Countryside Agency, an acknowledgement of its status as one of the best undeveloped coastlines in England. It includes sand dune reserves owned by the National Trust, Druridge Bay Country Park, six SSSI, and two nature reserves.

Figure 8.31 Aerial photograph of part of Druridge Bay.

Draw a sketch of Druridge Bay from the aerial photograph in Figure 8.31. Annotate it with the features named in this section. You will add more features to this sketch later, so it make it at least A4 size.

Human activities at Druridge Bay

The number of features of ecological interest led to the Bay being awarded its Heritage Coast status. There are nonetheless significant pressures on the area, including visitor pressure, agriculture, coal extraction, sand extraction and nuclear energy.

Figure 8.32 Tourist pressure during the summer at Druridge Bay.

Visitor pressure

Druridge Bay is only 25 minutes' drive from Newcastle-upon-Tyne. Figure 8.32 shows Druridge Bay on a day in mid-June. While the beach does not attract the crowds of people who might visit a major resort, the numbers of visitors are significant. Car parks vary in distance from the main beach, but involve a walk of some 200–300m through the sand dunes. This has resulted in erosion of the dunes by trampling.

Figure 8.33 Dune erosion. Bare exposed patches of sand typify serious erosion at a fairly early stage. What begins as isolated and small patches of bare sand soon expands, as sand is removed by wind, and other species of plant become threatened.

Dune erosion is cumulative; once areas of dune are exposed, the removal of sand by wind threatens the survival of other plants close by as the surface lowers. Thus dunes quickly become scarred by large patches of bare sand (Figure 8.33). Though this can be managed, and paths can be fenced off during re-growth, dunes become highly mobile once vegetation cover is lost, as Chapter 10 explains. Strong on-shore winds accelerate the process, as shown in Figures 8.17 and 8.18.

The Druridge Bay Country Park has been developed partly as a safety valve for the Bay. By concentrating visitor pressure in this one area by providing facilities and activity possibilities, pressure is less in other areas.

Agriculture

Most land adjacent to Druridge Bay has little agricultural value. However, overstocking of cattle and sheep on grazing lands adjacent to the dunes has affected the ecosystem by disturbing the balance of plant communities in the area. The use of hay as a winter feed has introduced grass seeds of varieties not found in Druridge Bay, and these are becoming dominant in some grazing areas.

Coal extraction

The decline of the coal industry in north-east England during the mid-1980s and early 1990s has closed all the deep mines within 25km of this area. Opencast extraction of coal still takes place, however. East Chevington site is the largest adjoining the bay. Opencast mining always leaves an impact on the landscape. Ladyburn Lake is a former site, now a habitat for 38 species of water fowl. On the other hand, site restoration also affects soil characteristics; restored farm land is never the same as it was before mining began. It may also have an impact upon erosion rates of the beach and dunes at Druridge Bay.

Sand extraction

Study Figure 8.34. It describes an issue involving Druridge Bay, which seems to threaten the beach and dunes.

This issue is complex because sediment movement is complex, as Figure 8.35 shows. Sediments have sources – cliffs from which they originate – and sinks, where they are deposited. Beaches, coastal spits, dunes and off-shore bars all act as points where sand is deposited. The depletion of one, such as the beach, has an effect upon the others. Beach erosion, or sand extraction from it, would lower the beach level, bring the sea further inland and threaten the dunes, starve the spit of sand, and thus deplete the off-shore bar. These both develop and affect water movement and sediment transfer around Druridge Bay. Sand extraction at Druridge Bay has now ceased but the effects are still felt – the beach is narrower than it would have been without extraction having taken place.

BEAUTIFUL BEACH CARTED AWAY FOR BUILDING SAND

One of the most beautiful and unspoilt beaches in northern England is disappearing – carted away by the lorry load to be turned into roads, shops and buildings.

The pure white sand of Druridge Bay, an eight-mile stretch of Northumberland coast lined with dunes, lagoons and nature reserves, is the victim of a 30-year-old planning decision described as legal nonsense.

In the early 1960s, the government overruled local planning authorities and gave the go-ahead for sand extraction at Druridge on condition that only a 'small mechanical digger' was used. The permission was granted to a small firm for local building needs. But in the 1970s, the rights were bought by Northern Aggregates, a subsidiary of Ready Mixed Concrete (RMC), and the rate of extraction accelerated.

Planners blame the loose wording of the original permission, typical of the days when environmental awareness was low. Today's 'small mechanical digger' can scoop up two tons in its bucket. And the demand for building materials has grown substantially.

According to planners at Castle Morpeth Borough Council, the condition gives RMC 'carte blanche to take as much sand as they like'. At a rate of 40 000 tons a year, about 1.5 million tons of sand have been extracted from Druridge, causing 'extreme' erosion. Beach levels have dropped, the dunes are narrowing by a metre every year and underlying clay, rock and a fossilised forest are regularly exposed as the sand thins.

The erosion also jeopardises plans by Northumberland Wildlife Trust to create new habitats for otters, marsh harriers and now the rare bittern. One of the threatened dunes, an SSSI, has a colony of scarce marsh helleborine orchids.

Ironically, RMC prides itself on its environmental record and is a corporate member of the Yorkshire, Cleveland and Durham wildlife trusts, though not of the Northumberland trust, which rejected its application. The firm argues that its contribution to erosion is 'minimal'. Local councils say they cannot afford the estimated £500 000 to compensate RMC for revoking the permission.

In 1994, RMC agreed to cut extraction by a quarter and has said it will stop using Druridge if similar sand can be found elsewhere. An alternative site has been earmarked but will not be available for at least a year.

Many protesters claim this will be too late for Druridge as the bay is a 'closed' system and cannot generate its own sand. Ian Douglas, reserves manager for the wildlife trust, said the risks where pointed out in the 1960s. 'Thirty years later, we are still talking'.

Figure 8.34 From *The Independent*, 15 January 1995

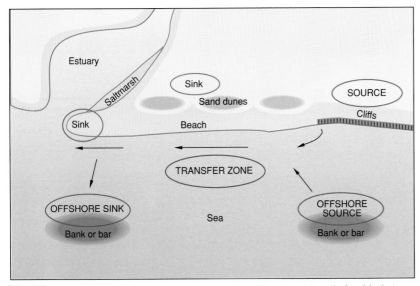

Figure 8.35 Sediment circulation around a coastline. This shows the relationship between sediment sources from cliff erosion, transfer along the coast by longshore drift towards four sinks, or points where sand is deposited. If sand extraction continues, the beach becomes depleted, water depth increases as does wave energy, and the beach suffers accelerated erosion.

Add further labels to your sketch of Druridge Bay to show the chief pressures on the landscape. Use different colours to highlight which you see as major threats to the landscape, medium threats, and minor threats.

Nuclear energy

Since the 1980s, Druridge Bay has been subject to proposals to build a nuclear power station in the southern part of the bay close to Widdrington. This proposal continues to be the subject of debate. National policies con-cerning nuclear energy have become somewhat fluid since the Chernobyl disaster in 1986, and probably depend upon the decision whether or not to privatise the nuclear industry. Following pri-vatisation, it is highly likely that the new company would seek to expand. Druridge Bay has remained a potential site since its character-istics – on a low flat coast with access to a major conurbation – were first identified.

Theory

Sand dune formation

Four requisites for sand dune formation have been identified:
- a large supply of readily available sand, dry enough to be transportable by wind;
- a reasonably flat or low-gradient surface on which the dune can develop;
- a stabilising agent to control the form of the dune once it has begun to develop, usually in the form of vegetation;
- strong onshore winds.

Sand is moved by the process of saltation. The amount of sand moved depends upon variables such as grain size, shape, strength of the wind, and any debris on the beach. Initially, sand ripples develop as a result of friction between air layers close to the beach. The reduced velocity causes deposition, which in turn creates an increase in deposition, thus increasing the ripple effect. Inland from the high water mark, colonisation of sand by salt-tolerant grasses, such as marram, stabilise sand so that it creates an obstacle around which wind must pass, causing air turbulence. The turbulence generated causes further deposition on the side away from wind exposure. Marram is capable of maintaining its growth to suit the increased dune height; its roots may be very long. Dunes rely upon vegetation to form; without such vegetation they would not occur. For further information on dune ecology, read Chapter 10.

Construct a sequence of annotated diagrams to show the development of sand dunes and their vegetation.

Technique

Conflict matrices

Issues about the use of space arise because different people have different perspectives about how the space should be used. This creates conflict. A conflict matrix helps to identify uses of land which conflict with each other and which are not easily resolved. For instance, in the case of Druridge Bay, you may decide that there is a conflict of interest between 'Conservation' and 'Extraction of sand'. In this case, you would shade in the box where these two intersect.

Some conflicts are complex, however. 'Recreational use' may conflict with 'Extraction of sand', but only at certain times of the year. In this case, you might shade in a different colour to indicate that it is a more moderate, or perhaps irregular, conflict.

If you complete the matrix for Druridge Bay, it will help you to analyse the situation. You may identify certain land-use types which are not compatible with others; you may decide that some uses should be stopped or prevented. It will also help you to develop ideas for how to manage Druridge Bay and which features require most support and attention.

	Recreational use	Flora and fauna	Agricultural land	Coal mining	Nature reserves/conservation	Commercial extraction of sand	Nuclear power plant	Structural works elsewhere on coast
Recreational use								
Flora and fauna								
Agricultural land								
Coal mining								
Nature reserves/conservation								
Commercial extraction of sand								
Nuclear power plant								
Structural works elsewhere on coast								

Figure 8.36 A conflict matrix. The three colours are used to show degrees of conflict. Red is used to show that the conflict is marked and consistent, orange is used to show that conflict may be significant but perhaps temporary, while yellow is used to denote light conflict.

Interest groups in Druridge Bay

Can different groups have the same interests at heart? Consider the list of people in Figure 8.37. Are they likely to see a conflict matrix in the same way?

1 Individually, select one group from the list in Figure 8.37. Complete a conflict matrix as you think that group would perceive it.

2 Compare your matrix with another group's matrix and find one that agrees with yours. For example, RSPB might agree with the Northumberland Wildlife Trust.

3 Repeat the exercise until you have established groups of like-minded people. Has a conflict situation emerged in your class? How can the future of Druridge Bay be managed?

4 In your groups, decide upon a plan that answers the question: 'Should Druridge Bay be protected from further development as part of English Heritage Coast, or should it be left for development?'

- **Northumberland County Council**
- **Alnwick District Council**
- **Castle Morpeth Borough Council**
- **National Trust**
- **Northumberland Wildlife Trust**
- **English Nature**
- **Royal Society for the Protection of Birds**
- **Countryside Agency**
- **Druridge Bay Campaign**
- **Friends of the Earth**
- **Local parish councils**
- **Alcan Farms**
- **Environment Agency**

Figure 8.37 Interest groups in Druridge Bay.

A policy for the south Northumberland coast?

This chapter has shown how the south Northumberland coast varies considerably in character, and the issues that each part faces (Figure 8.38). A holistic view of the South Northumberland study area needs to be taken, both of coastal processes and coastal protection measures. It has been shown that processes in one place, both human and natural, have impacts on processes elsewhere. Figure 8.38 shows the areas of sediment transport, erosion and deposition along Collywell Bay, Sandy Bay, and Druridge Bay. Consider whether or not this area should be treated as one, or whether it should be managed in its different component parts.

Look at Figure 8.11 which shows organisations responsible for managing Britain's coast, the concept of Shoreline Management Plans (SMP), and Figure 8.12 showing major sediment cells of England and Wales. Which is the best way of organising coastal management? Should it be done locally, regionally through cells or nationally? Perhaps there is a case for international management, involving EU funding. Chapter 9 explores this further.

Discuss in 1000 words whether it is realistic to manage coastlines at a scale of 'sediment cells' shown in Figure 8.12, or whether Collywell Bay, Sandy Bay, and Druridge Bay should each be managed differently and separately?

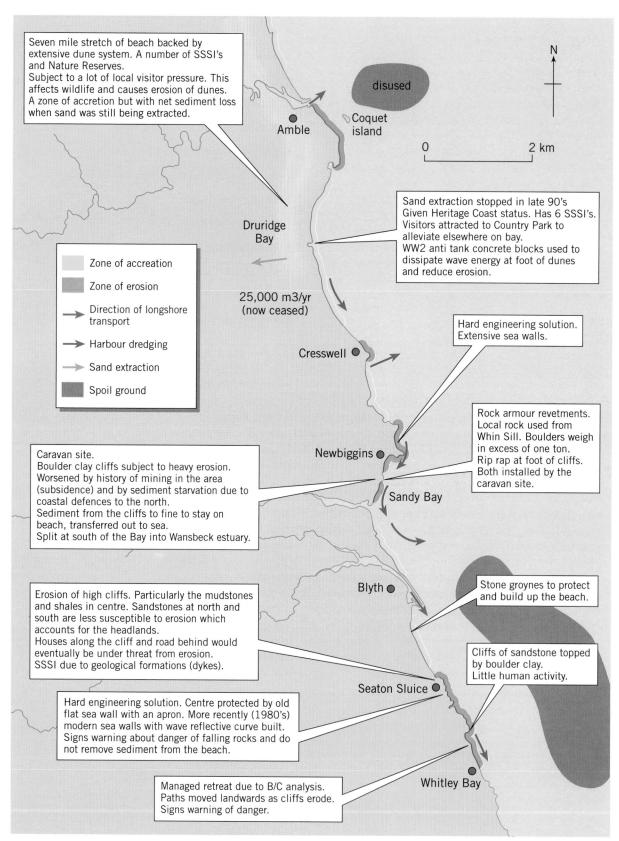

Seven mile stretch of beach backed by extensive dune system. A number of SSSI's and Nature Reserves.
Subject to a lot of local visitor pressure. This affects wildlife and causes erosion of dunes.
A zone of accretion but with net sediment loss when sand was still being extracted.

disused

N

Amble

Coquet island

0   2 km

Sand extraction stopped in late 90's Given Heritage Coast status. Has 6 SSSI's.
Visitors attracted to Country Park to alleviate elsewhere on bay.
WW2 anti tank concrete blocks used to dissipate wave energy at foot of dunes and reduce erosion.

Druridge Bay

Zone of accretion

Zone of erosion

Direction of longshore transport

Harbour dredging

Sand extraction

Spoil ground

25,000 m3/yr (now ceased)

Hard engineering solution. Extensive sea walls.

Cresswell

Rock armour revetments. Local rock used from Whin Sill. Boulders weigh in excess of one ton.
Rip rap at foot of cliffs. Both installed by the caravan site.

Caravan site.
Boulder clay cliffs subject to heavy erosion. Worsened by history of mining in the area (subsidence) and by sediment starvation due to coastal defences to the north.
Sediment from the cliffs to fine to stay on beach, transferred out to sea.
Split at south of the Bay into Wansbeck estuary.

Newbiggins

Sandy Bay

Erosion of high cliffs. Particularly the mudstones and shales in centre. Sandstones at north and south are less susceptible to erosion which accounts for the headlands.
Houses along the cliff and road behind would eventually be under threat from erosion.
SSSI due to geological formations (dykes).

Blyth

Stone groynes to protect and build up the beach.

Cliffs of sandstone topped by boulder clay.
Little human activity.

Hard engineering solution. Centre protected by old flat sea wall with an apron. More recently (1980's) modern sea walls with wave reflective curve built.
Signs warning about danger of falling rocks and do not remove sediment from the beach.

Seaton Sluice

Managed retreat due to B/C analysis. Paths moved landwards as cliffs erode. Signs warning of danger.

Whitley Bay

Figure 8.38 The south Northumberland coastal system.

Summary

You have learned that:

- The coastal zone is a complex scene of many different and often interdependent natural processes. These need to be understood if effective management is to take place.
- The coastal zone is an area where natural and human processes meet. These have an impact on each other and changes can result because of this.
- Many different groups use – and have responsibility for – the coastal areas of England and Wales and this can cause conflicts of interest. Careful management is therefore necessary.
- There are different approaches to managing coastline and many strategies within these approaches. Management varies from place to place. We need to be able to choose the correct management strategy for each individual case.
- Action on one part of the coastline can have major impacts, often adverse, on another. Coasts in England and Wales are managed in major cells with this in mind.
- Coastal erosion and deposition are natural processes and although we can influence them, we cannot hope to completely control them. We must bear this in mind when planning management schemes.

Ideas for further study

Work in groups of four. Each group should choose one of the following briefs and research a case study to match it. Use information from geographical journals such as Geographical Review, from the Internet, or from other text books.

Produce a case study card (one side of A4 only) to circulate amongst your group. You could use a computer package to produce your case study card, such as Microsoft Word. Be careful to include the important and relative details only; the space is limited.

- A stretch of coastline where 'soft' engineering techniques have been used
- A small area where coastal erosion is a major problem and a lot of money has been spent on coastal defence
- A study of ineffective coastal protection
- Management of a depositional feature such as a spit.

References and further reading

Bishop, V and Prosser, R, *Landform Systems*, Collins Educational 1997

Bray, M and Hooke, J, 'Managing the Wessex Coast: towards an integrated approach', in *Geography Review*, September 1996

Hill, M, *Advanced geography Case Studies*, Hodder and Stoughton, 1999

Holiday, A, 'Managing the Wessex Coast: Coastal defences at Preston Beach Road', in *Geography Review*, September 1997

Manuel, M, McElroy, B and Smith, R, *Coastal Conflicts*, Cambridge, 1995

Prosser, R, *Natural systems and human responses*, Nelson, 1992

Younger, M, 'Will the sea always win?' *Geography Review*, May 1990

International coastlines: the care of West Africa

Chapter 8 has shown how different processes occur that affect people along stretches of coastline, and that management strategies are needed to deal with each process. This chapter looks at West Africa, in order to show how international issues can arise when coastal processes affect more than one country. It also explores how coastal management issues may be approached differently in LEDCs.

The West African coast

Study the two news stories in Figures 9.1 and 9.2. One is from 1987 and describes erosion along the coast between Ivory Coast and Nigeria (Figure 9.3). The second shows how the issue was still alive in 1998. Find the places mentioned in Figure 9.3, which shows the coastline concerned. Erosion has been occurring at a rapid rate here, often up to 8m per year, along a 2 000km stretch of coast from Ivory Coast, through Ghana, Togo and Benin to Nigeria.

Why is there an erosion problem in West Africa?

Most of the West African coast is made up of a shallow, sandy coast over which breaking waves crash daily. Long beaches separate the shoreline from the sea, as shown in Figure 9.4. The coast consists of large stretches of sand embankments, created by sand blown on-shore, which have closed off and separated from the shore a chain of lakes, lagoons and mangrove swamps. These lakes and lagoons are areas of water enclosed by sand bars

> Ocean currents sweeping along the coast of West Africa are washing huge chunks of coastland into the sea, threatening settlements, tourism and industry in the region. A new dam is being blamed. The Guinea current, one of the strongest in the world, is nibbling away at 2000km of coastline between the Ivory Coast and Nigeria. It carries off some one and a half million cubic metres of sand each year. The consequences for Ghana, Togo and Benin are potentially catastrophic.

Figure 9.1 From the *New Scientist*, 15 January 1987.

> ## EROSION THREATENS AFRICA'S COAST
>
> The assistant executive secretary of the Scientific, Technical and Research Commission of the Organisation for African Unity, Mbaye Ndoye, has said Africa is experiencing 'very widespread' coastal erosion on its western ... shores. These are caused by both natural and human factors, Ndoye told an environment forum held in Dakar, Senegal. 'The intensity varies from one region to another,' he said in a paper entitled, Problems of Coastal Erosion In West Africa. 'Among natural causes of erosion are very low coastal topography, intense waves and high winds and weak soils.'

Figure 9.2 From Pan African News Agency, 2 November 1998 by Aly Coulibaly.

deposited by longshore drift. The area is prone to storms, but the fetch – some 4 000km to South America – means that high waves easily erode the fine sands of the coastal bed almost permanently.

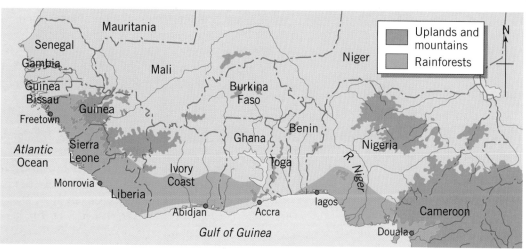

Figure 9.3 The West African coastline.

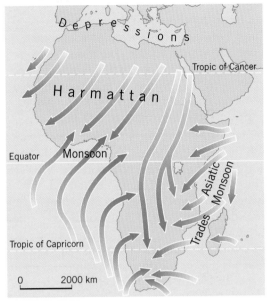

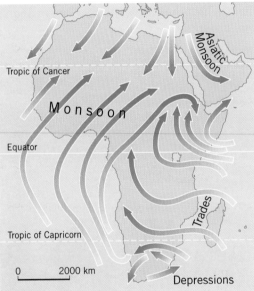

Figure 9.5 Wind patterns over Africa. Notice how winds blow from the same direction along the coast of West Africa throughout the year. This establishes a permanent east-west current, which is one of the natural causes of the erosion problem.

Draw a large sketch map of the coastline stretching between Ghana and Benin as shown in Figure 9.3. Annotate your sketch map with causes of the erosion problem in West Africa.

Figure 9.4 A stretch of the West African coastline. The West African coast is generally characterised by low-level sandy shores. Cliff features are generally uncommon.

These are driven by almost permanent on-shore winds that blow across the Gulf of Guinea. Figure 9.5 shows how in both January and July, monsoon winds blow on-shore from the south-west, creating a series of waves that establish a year-round current from west to east.

Few natural harbours or bays exist in the African continent, and along the coast of West Africa, harbours built to encourage and support economic expansion have been artificial. This is part of the coastal erosion problem. In early days of trading between West African countries and Europe, ships had to dock off-shore, and 'surf boats' – shallow-bottomed wooden craft designed to ride the surf – would carry unloaded import goods into the shore, or take export produce beyond the shore to ships for loading. The process was hazardous and slow, and limited the volume of goods that could be transported between countries. Economic development in the 1960s and since has been dependent on the creation of artificial ports, such as Tema and Takoradi in Ghana (Figure 9.3), designed to accommodate ships that can reach in-shore by means of dredged channels.

The problem is therefore one of rapid erosion of sands. Three West African countries – Ghana, Togo and Benin are experiencing particular problems.

Issues arising from coastal erosion in Ghana

Study Figure 9.3 as you read this section. One of the places most affected by erosion along the Ghanaian coast is Keta, on the eastern edge of the country, close to the border with Togo. Ghana's Minister of Works and Housing referred to Keta in 1998 as '*hitherto a busy coastal town, now a shadow of its former self as a result of years of sea erosion. A visit to Keta will shock you with pain at the sight of buildings in the sea. It has been going on all these years and it has not stopped*'.

Keta is located on the eastern edge of the Volta River Delta, to the east of a major shoreline sediment delta called Cape St Paul. The Cape has developed as the result of a major realignment of the delta of the Volta River, following the construction of the Volta Dam at Akosombo in 1961. Erosion is occurring at Keta because sand moving from west to east by longshore drift – which shifts up to 1 million cubic metres each year – is deposited first at the Cape. This creates a sediment deficit in Keta, where the shoreline is retreating at rates of up to 8m per year. In fact, a drift of sediment is slowly advancing along the shoreline toward Keta, but its rate of movement is too slow to save it and adjoining towns from erosion, hence the necessity for the sea defence project.

The effect of sustained erosion shown in Figure 9.6 has meant loss of property and the major road between Keta and Togo. Formerly a town of 70 000 people with a thriving fishing industry, Keta is now decimated in size, and the fishing industry has been destroyed.

Draw a large sketch map of the coastline stretching between Ghana and Benin as shown in Figure 9.3. Annotate your sketch map with details of the erosion problem in Ghana. Use one colour to identify social issues that have resulted from erosion, a second to show economic issues, and a third to shown environmental issues.

long into the sea. Like Ghana, the coast is low-lying and sandy, and the creation of a port has involved dredging a channel. Unless protected by a breakwater, continual drifts of sediment by longshore drift from west to east would have silted up the channel.

Meanwhile, at Kpeme, 30km east of Lome, a jetty which is part of the port development is about to collapse. In 1987, the *New Scientist* reported that this would lead to the collapse of exports of phosphate, which then accounted for half of Togo's overseas earnings.

Figure 9.6 Coastal erosion near Keta, Ghana.

Figure 9.7 Breakwater in Togo. A similar barrier has been constructed at Lome

Coastal erosion in Togo

Study Figure 9.3 as you read this section. In Togo, hundreds of metres of the main coastal road between Ghana and Togo's capital, Lome, have disappeared. At Tropicana, a tourist village 9km east of Lome, the rate of erosion in the late 1980s was 100m in five years. The development of the port of Lome in the 1980s led to the construction of a breakwater 1 300m

Coastal erosion in Benin

A pilot project to control the erosion problem along the Benin coast was initiated between April and June 1998 on a beach close to Cotonou, the second city. It costs US$80 000. The project aims to control the rate of erosion along a 125km stretch from the Togolese border to Nigeria.

Country	Population (million)	Infant mortality	Life expectation	GNP per capita in US$ (1995)	Aid per capita	Debt owed overseas as % of GDP	People per doctor 1995
Ghana	18.1	66	56	390	-1.2%	53.7%	22 970
Togo	4.7	91	57	310	-1.1%	66.7%	11 385
Benin	5.9	98	54	370	+0.1%	41.7%	14 216
UK	59	6.2	77	18 700	+1.8%	0%	611 (in 1980)

Explanation of terms
- Infant mortality – the number of infant deaths per 1000 births
- GNP in US$ (1995) – Gross National Product, an indicator of average earnings per person across the whole country
- Aid per capita – the total amount of aid received from overseas donors, divided by the number of people
- Debt owed overseas as % of GDP – the percentage of a country's total earnings that are taken up by debt repayments

Figure 9.8 Datafile showing economic status of Ghana, Togo and Benin.

The Keta Sea Defence Project – a solution to coastal erosion?

Chapter 8 showed how processes of erosion and deposition occur along a small stretch of coastline, and how local councils manage this using grants from national government. How can this work:
- where more than one country is involved?
- where the countries concerned are much poorer and lack resources by which to solve the issues through engineering solutions?

The data in Figure 9.8 show that there is little money for coastal protection or expensive engineering work. It was estimated in 1987 that to save Togo's coastline would cost £1.2 million per km. Some aid has been targeted at coastal protection, including donations from the

1 Continue to add annotations to your map of the coast between Ghana and Benin, labelling erosion issues from **a)** Togo **b)** Benin.
2 How has your understanding of coastal processes from Chapter 8 enabled you to explain:
  a) the threat of silting of the channel into the port of Lome?
  b) the likely cause of the collapse of the jetty at Kpeme?
3 Does your map show that the main issues arising from coastal erosion in West Africa are social, economic or environmental?
4 To what extent would you say that the causes are **a)** natural or **b)** the result of human activity?

CONCERNING COASTAL PROTECTION WORK FOR KETA IN GHANA

Construction works on the Keta Sea Defence project will commence after an agreement with the contractors, Messrs Great Lakes Dredge and Dock Company of United States, is signed on 12 March, 1998. The Minister of Works and Housing said in Accra that government has completed negotiations on all major issues for construction to start in a few weeks. 'But the commencement of construction is (dependent) on the availability of an Eximbank Loan Facility of about $70.3 million and the Government's contribution of $12.4 million,' the Minister said at a press briefing on the project.

4 March 1998

$94 MILLION LOAN FOR GHANA'S KETA SEA DEFENCE PROJECT

The United States Export-Import Bank (Exim Bank) has approved a $94 million long-term loan to enable Great Lakes Dredge and Dock Company, located in Illinois, USA, to sell equipment and services in Ghana to build a sea wall to protect the Keta Beach and Lagoon from erosion. The project will prevent catastrophic impacts on agriculture, fisheries, commerce and living conditions of the inhabitants of Keta and its surrounding areas.

The Chairman of Exim Bank, James A. Harmon, declared: 'This project will support vital infrastructure development in the important Ghana market which will lead to economic growth and at the same time create many US jobs.'

Mr Valley, Project Manager, said that his company would hire 15 to 30 supervisors and about 20 US engineers, mechanics and other personnel to supervise, operate and maintain the equipment during the building of the Keta Sea Wall.

It is also to reclaim 300 hectares of land from the lagoon for inhabitants to build on in the areas of Keta, Vodza and Kedze.

11 March 1999

Figure 9.9 From the Pan African News Agency.

1 Consider Figure 9.9. What was the time lag between first approval of the project and money becoming available? Why should it take so long?
2 Who financed the deal and why did it cost so much? Who gained from the deal?
3 What would have happened to costs had Ghanaian workers been employed? Why do you think this option was not used?
4 Make a large copy of the table below; complete it for the Keta Sea Defence project, using all details from this study.

	Benefits	Costs
Social Economic Environmental		

EU. However, the Ghanaian government has earmarked expenditure for construction of the Keta sea defence wall in its budget for 1999. Government revenue is limited, however. Consider the two news extracts in Figure 9.9, concerning the finance for a sea defence project to protect Keta, in Ghana.

The project has four components:
- sea-defence works to limit further erosion.
- land reclamation from the lagoon adjacent to the town of Keta, providing an area for local inhab-itants to rebuild homes that were lost to erosion
- construction of a 8.3km road or causeway at Keta Lagoon between Keta and Havedzi, re-establishing a road link between these townships lost to erosion
- flood control for Keta Lagoon, providing relief from extreme flooding conditions for people living around the lagoon.

It is usual to complete an Environmental Impact Assessment when considering major work of this kind. It is this kind of work which is sometimes by-passed when money is short in LEDCs, as Figure 9.10 shows. However, the coastal defence work will have impacts elsewhere.

Follow the progress of this project using the web page reference at the end of this chapter.

An Environmental Impact Statement (EIS) has been submitted in order to obtain the necessary approval to begin construction of the Keta Sea Defence Works (KSDW) project. The EIS presents an evaluation, analysis and prediction of the impacts of the KSDW project, and recommendations are made where appropriate.

The project is scheduled to begin in early 1998. There is very limited environmental information on the Keta Lagoon, the ocean shoreline, or the environment near the project site. Due to the urgency of completing the work to prevent further unnecessary loss of property and help protect the livelihood of the local inhabitants, there is not sufficient time to undertake detailed, long-term environmental studies.

Figure 9.10 Environmental Impact Statement for the Keta Sea Defence works project.

1 Why should an Environmental Impact Assessment (EIA) be difficult to obtain for this project?
2 What information ought to have been available if a proper Environmental Impact Assessment were to be carried out? Why was it not available?
3 How might the governments of Togo and Benin react to the work being carried out at Keta? Why?
4 Why was the government of Ghana not bound to consult them? Should there be a process by which governments have to consult? If so, who should monitor it?
5 You are employed during a gap year to work with Great Lakes Dredge and Dock Company, located in Illinois USA, in compiling a proper EIA. Your job would be to monitor and evaluate the success of the Keta project. Write a proposal of about 750 words to show:
   a) what information is needed to assess whether or not the Keta project is a success
   b) how this information should be obtained, and by whom
   c) what you would regard as measures that would prove that the project had been a success in three years time, ten years time, and 30 years time.
Consider Togo and Benin in your proposal, as well as Keta.

Summary

You have learned that:
- Erosion along coastlines can occur at international or even continental scales.
- The causes of erosion may be both natural and human.
- The effects of erosion are especially marked in LEDCs.
- Economic development may have an impact on natural systems operating along coastlines.
- Proposals to solve erosion problems may involve large-scale engineering work, which is often outside the ability of LEDCs to pay.
- Costs of remedial work along coastlines are often high, reflecting costs in employing MEDC companies to do the work.
- Impact studies are needed to assess the possible outcomes of environmental projects.

Ideas for further study

1 Monitor progress on the Keta project, using the Panafrican News Agency web site, whose address is given below. Include a progress watch on proposals by Togo and Benin to address their erosion problems.
2 Contrast this with major coastal defence works occurring elsewhere in the world, e.g. Louisiana, where there is also an erosion problem. You can locate materials about this on the internet. Use a search engine to find references for key phrases such as 'Louisiana coastal erosion'.

References and further reading

'Keta Sea Defence Project Begins Soon', Accra, Ghana, 4 March 1998
http://www.ghana.com/headlines/h52.htm
'Keta environmental impact statement', Research Planning Inc., South Carolina, USA, Home page address http://www.researchplanning.com/envir/keta.htm
'Ghana awards US company sea defence contract', 13 March 1998, Panafrican News Agency, on web page http://www.africanews.org/PANA/news/19980313/feat10.html
'$US94 million loan for Ghana's Keta Sea Defence Project', 11 March 1999, Panafrican News Agency, on web page http://www.africanews.org/environ/stories/19990311_feat7.html

This chapter looks at coastal ecosystems in two parts of Britain. The first part of the chapter looks at sand dune ecosystems at Studland Bay in Dorset. Here, the natural ecosystem is being disturbed by human activity, though in this case, largely from tourism.

The second part, a study of the salt marsh ecosystem at Alnmouth Bay in Northumberland, shows how plant and animal communities develop over time along and around estuaries. These environments are under threat. Many salt marshes have been lost to industrial and port developments, such as those at Teesside and along the Thames estuary. Now, large areas of salt marsh are still being lost to major building projects such as the tidal barrage in Cardiff Bay and the construction of the second Severn Bridge.

Studland Heath, Dorset

Studland Beach stretches for about 6 km south-east from the entrance to Poole Harbour in Dorset, shown in Figure 10.1. It is owned and managed by the National Trust. Extensive sand dunes have formed at the back of the beach, shown in Figure 10.2, and these form part of the Studland Heath

Figure 10.2 Studland sand dunes.

National Nature Reserve. Although owned by the National Trust, it is managed by English Nature. In summer it is popular with visitors. Many come to enjoy the long sandy beach but some explore the nature reserve by following nature trails. Others pass through walking the South Western Coastal Footpath.

Plant succession on Studland Bay dunes

Sand dunes provide an excellent place for studying plant succession. The process is shown in Figure 10.3. It is surprising that plants ever grow on sand; it is often dry and any moisture is salty. Dry sand is mobile, and begins to form dunes along the strand line. This process is described fully in Chapter 8. Small **embryo dunes** form and are easily destroyed, unless the sand is sufficiently moist and stable enough to be colonised by plants. At Studland, embryo dunes are colonised by sea couch grass and sea lyme grass which are salt resistant. Their spreading roots help to bind the sand, allowing the embryo dune to grow. Marram

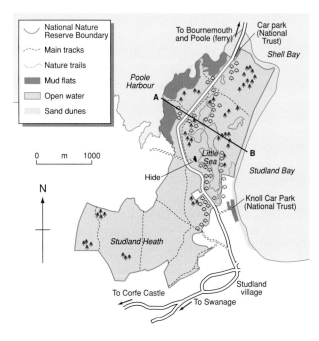

Figure 10.1 Studland Heath Nature Reserve.

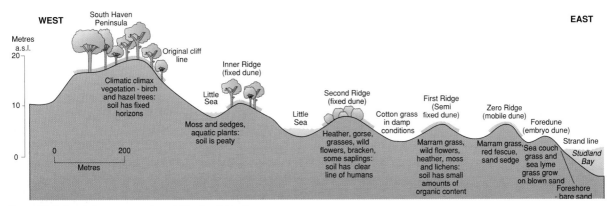

Figure 10.3 Sand dune succession at Studland Heath. See transect line A-B on Figure 10.1.

grass becomes established once the dune is about a metre high. Marram has long roots, enabling it to obtain water, and it grows with the dune. It can withstand dry weather but not salt water. Once established, sand is trapped and dunes can grow by up to a metre a year.

Embryo dunes grow until they are large enough to be called **mobile** or **yellow dunes**. At Studland, Figure 10.3 shows that the mobile dune is named Zero Ridge. It has been established for about 50 years and is colonised by red fescue and sand sedge

as well as marram grass. Few animals are found on the embryo dunes, but Zero Ridge supports snails and insects as well as rabbits, lizards and sea birds. The air behind sand dunes is calmer allowing hollows, known as **dune slacks**, to develop.

When marram dies and decays, nutrients are added to sand enabling the dunes to support other plants that can survive in dry sandy conditions. As more organic matter is added, the soil increases in depth and moisture content, which encourages different species to invade and colonise. In this way,

Theory

Plant succession

Bare ground such as exposed rock or wasteland in a city does not remain bare for long. Similarly, sand dunes, mud flats or shallow water at the edge of a pond are gradually colonised by plants able to survive in these conditions. These are **pioneer species** and mark the beginning of the process of plant succession, whereby other plants invade and take over until a balance is reached. If bare ground has never previously been colonised, the process is known as **primary succession** and may take centuries to complete. If the land has been exposed by human activities then the process is known as **secondary succession** and may be more rapid.

Pioneer species survive on exposed sites without competition. They modify the environment by binding soil and adding nutrients when they die and decay. Creeping plants or those with leaf cover help the soil to retain moisture. These changes allow other species to colonise, at the expense of pioneers which can no longer compete. The invaders, in turn, modify the environment by providing shade as well as improving soil. Birds and insects also start to find food and shelter.

Further modifications attract different species until stability is reached. If there are no limiting factors, the final community will be adjusted to the climatic conditions of the area and is known as the **climatic climax community**. Each stage in the process of succession is known as a **seral stage**. The whole process is shown in Figure 10.4.

Often plant succession does not reach the climatic climax, because something prevents full succession taking place. This may be soil conditions, relief, drainage characteristics, or the human management of the land for agriculture. Such factors are known as **arresting factors**, and result in the development of a **sub-climax community**.

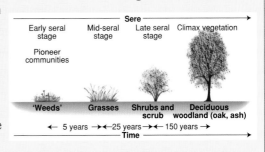

Figure 10.4 Plant succession in an abandoned field.

the mobile dunes become more stable and are known as **semi-fixed** and, later, **fixed dunes**. Studland First Ridge is colonised by dandelions, sea bindweed and other wild flowers, together with heather, moss, and lichen. Marram grass is still common and there is little bare sand which has been trampled. Animals too are more varied. Butterflies, meadow pipits, grass snakes and lizards are found here. Second Ridge is covered with heather and gorse, and there is a little marram grass. A thin black line of humus is found here, showing the first stage of soil formation. This allows more varied plant and animal species to inhabit the dunes.

Inland from Second Ridge, some parts of the dune slacks contain stretches of brackish water, the largest of which is Little Sea. Small trees, such as birch, are found as well as water-loving plants such as bog myrtle, and aquatic animals, ducks and other water birds. Inland, gorse and heather give way to birch and hazel trees. This is the climax community of Studland Heath, and it would take over the area if trees were not cut to maintain the diverse habitats of the heath.

Pressures on Studland Heath and Studland Bay

On a busy day in summer up to 25 000 people visit Studland. The vast majority come to the beach but some visit the sand dunes and nature reserve. The number of visitors is increasing each year, and problems for both beach and dunes are also increasing, as Figure 10.5 shows.

The main problems for the area are:
- dune erosion caused by people walking through the dunes to the beach or for shelter, threatening plant and animal species
- traffic congestion in car parks and roads leading to the area
- over *12 tonnes* of litter per week left by visitors. Outside the bins provided, it is dangerous to small animals and birds
- at least once a year, heath fires destroy plants and animals. The most common cause is discarded cigarette ends. Lizards and snakes can escape by burrowing, but may not escape predators once vegetation cover has gone.

Management strategies at Studland

Both the National Trust and English Nature have introduced schemes to protect Studland Heath and the dunes, while allowing the public to visit the area. Since 1982 the National Trust has:
- enlarged the four main car parks and increased their capacity by 800
- built a visitor centre with shop, cafe and information point
- increased the number of toilets, adding facilities for the disabled
- closed some paths and fenced parts of the sand dunes
- planted marram grass
- placed litter bins on paths and at the back of the beach
- placed fire beaters on the heather and gorse heath as well as fire breaks and water hydrants
- erected information boards to educate people about the area.

Figure 10.5 The use of car parks in Studland.

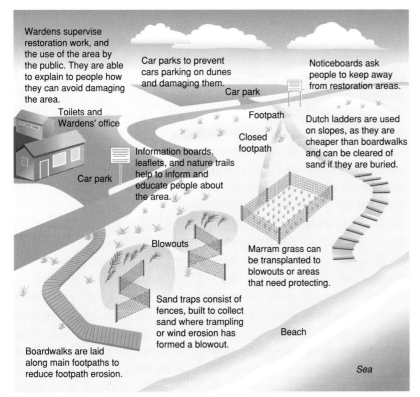

Wardens supervise restoration work, and the use of the area by the public. They are able to explain to people how they can avoid damaging the area.

Car parks to prevent cars parking on dunes and damaging them.

Car park

Noticeboards ask people to keep away from restoration areas.

Toilets and Wardens' office

Footpath

Closed footpath

Dutch ladders are used on slopes, as they are cheaper than boardwalks and can be cleared of sand if they are buried.

Information boards, leaflets, and nature trails help to inform and educate people about the area.

Car park

Blowouts

Marram grass can be transplanted to blowouts or areas that need protecting.

Sand traps consist of fences, built to collect sand where trampling or wind erosion has formed a blowout.

Beach

Boardwalks are laid along main footpaths to reduce footpath erosion.

Sea

Figure 10.6 Some ways of managing sand dunes.

The National Trust and English Nature differ in their approach to management. The beach is managed by the National Trust and the Heath and Nature Reserve by English Nature. The National Trust is a charity funded by subscription. The Trust is keen for people to visit the beach and use the facilities they have provided. These facilities bring about £0.5 million a year to the Trust, which is used for improvements to the beach and its facilities, and for conservation projects in this part of Dorset. However, English Nature, a government agency, prefers to restrict access to the Nature Reserve to protect and conserve this environment. The use of the beach by the National Trust encourages more people to visit the Nature Reserve which can cause problems. A policy is needed that combines conservation with public access.

Options for Studland Heath and Bay

1 What are the conflicts between the ways that the National Trust manage the area, and how English Nature would see it managed? How can the differences between the approaches be addressed?
2 Are the motives of the National Trust the same as those of English Nature?
3 In groups, consider the following options for Studland Heath and Bay. Decide on points in favour of and points against each option.
   a) Restrict access to a maximum of 15 000 people on any day.
   b) Charge admission to the Heath and Bay.
   c) Remove any management strategies at all, and leave the Heath and Bay to the public.
   d) Close all areas of heath and dunes except by means of board-walks and laid out paths.
   e) Establish Studland Heath and Bay as a Site of Special Scientific Interest, and leave it to be preserved for study only.
4 Which option is preferred by **(a)** you, and **(b)** the rest of the group?

Techniques

Transect studies through sand dunes

Taking a transect through the sand dunes from the strand line to the dunes is a useful way of collecting data. An example of how you might do this is shown in Figure 10.7. A transect of 300-500 metres will take you about 3-4 hours fieldwork.

Once you have chosen a transect, decide what data to collect and how to do it. Some ideas for this are found below and in the Techniques box on page 129. Before choosing methods of data collection, think about what data you need for your study and how you are going to record the information you collect. You cannot record information along the whole transect, so choose a method of sampling. Three methods are shown: systematic, random, and stratified.

Systematic sampling

The easiest method is to sample systematically, which means sampling along the transect at regular intervals. Fifty-metre intervals are usual, so for a transect of 500 metres, choose eleven sites at which to collect data.

Note that this method might result in some features being omitted.

Random sampling

Random sampling means choosing points at random along the transect using random number tables, shown in Figure 10.7. Before selecting sample points, decide how many sites you need in the time available and which will give you the information you want. Like systematic sampling, this method could result in some features being omitted.

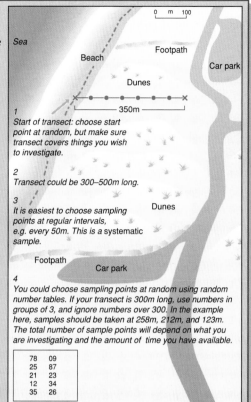

1
Start of transect: choose start point at random, but make sure transect covers things you wish to investigate.

2
Transect could be 300–500m long.

3
It is easiest to choose sampling points at regular intervals, e.g. every 50m. This is a systematic sample.

4
You could choose sampling points at random using random number tables. If your transect is 300m long, use numbers in groups of 3, and ignore numbers over 300. In the example here, samples should be taken at 258m, 212m, and 123m. The total number of sample points will depend on what you are investigating and the amount of time you have available.

78	09
25	87
21	23
12	34
35	26

Figure 10.7 Selecting sample sites for a transect through a sand dune.

Stratified sampling

Stratified sampling is more complex, but, some would say, gives more accurate results. An example might be in a retail survey in which you are sampling people who are shopping. If you know that 46% of the population is male in your study area, you might want to make sure that 46% of your sample are male. If one-quarter of all men are over the age of 65, you should ensure that, of your 46%, one-quarter are men over the age of 65.

Therefore, on a sand dune, if you know that your sample area includes a mobile dune, semi-fixed and fixed dunes, then you should ensure that you include these. Your survey area might therefore include a mobile dune, the slope and crest of each of the fixed dunes, and each of the slacks behind the dunes. This is called stratified sampling.

Techniques

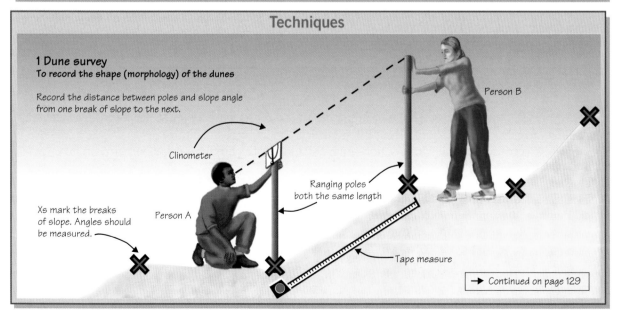

1 Dune survey
To record the shape (morphology) of the dunes

Record the distance between poles and slope angle from one break of slope to the next.

Clinometer

Person B

Ranging poles both the same length

Xs mark the breaks of slope. Angles should be measured.

Person A

Tape measure

→ Continued on page 129

Techniques

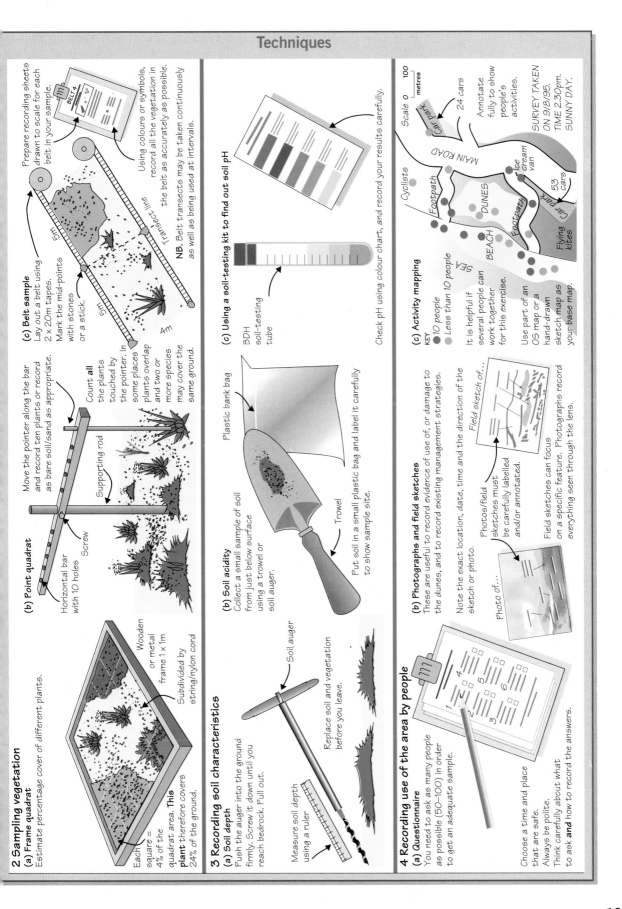

2 Sampling vegetation

(a) Frame quadrat
Estimate percentage cover of different plants.

Wooden or metal frame 1 x 1m

Subdivided by string/nylon cord

Each square = 4% of the quadrat area. **This plant** therefore covers 24% of the ground.

(b) Point quadrat

Move the pointer along the bar and record ten plants or record as bare soil/sand as appropriate.

Count **all** the plants touched by the pointer. In some places plants overlap and two or more species may cover the same ground.

Supporting rod

Horizontal bar with 10 holes

Screw

(c) Belt sample
Lay out a belt using 2 x 20m tapes. Mark the mid-points with stones or a stick.

5m 5m 4m

Transect line

Prepare recording sheets drawn to scale for each belt in your sample.

BELT 4

Using colours or symbols, record all the vegetation in the belt as accurately as possible.

NB. Belt transects may be taken continuously as well as being used at intervals.

3 Recording soil characteristics

(a) Soil depth
Push the auger into the ground firmly. Screw it down until you reach bedrock. Pull out.

Soil auger

Measure soil depth using a ruler

Replace soil and vegetation before you leave.

(b) Soil acidity
Collect a small sample of soil from just below surface using a trowel or soil auger.

Plastic bank bag

Trowel

Put soil in a small plastic bag and label it carefully to show sample site.

(c) Using a soil-testing kit to find out soil pH

BDH soil-testing tube

Check pH using colour chart, and record your results carefully.

4 Recording use of the area by people

(a) Questionnaire
You need to ask as many people as possible (50–100) in order to get an adequate sample.

Choose a time and place that are safe.
Always be polite.
Think carefully about what to ask **and** how to record the answers.

(b) Photographs and field sketches
These are useful to record evidence of use of, or damage to the dunes, and to record existing management strategies.

Note the exact location, date, time and the direction of the sketch or photo.

Photo of...

Photos/field sketches must be carefully labelled and/or annotated.

Field sketch of...

Field sketches can focus on a specific feature. Photographs record everything seen through the lens.

(c) Activity mapping

KEY
● 10 people
○ Less than 10 people

It is helpful if several people can work together for this exercise.

Use part of an OS map or a hand-drawn sketch map as your base map.

Scale 0 — 100 metres

Car park
24 cars

Annotate fully to show people's activities.

SURVEY TAKEN ON 9/8/95.
TIME 2.30pm.
SUNNY DAY.

Cyclists
MAIN ROAD
Footpath
Ice cream van
DUNES
53 cars
Car park
Footpath
BEACH
Flying kites
SEA

129

Figure 10.8 OS map of Alnmouth and the salt marsh around the estuary.

© Crown copyright

Alnmouth salt marsh, Northumberland

Figure 10.8 shows an Ordnance Survey map of a small salt marsh at the mouth of the River Aln in Northumberland, where it enters the North Sea at the village of Alnmouth. This is one of Britain's most important salt marshes because of the diversity of plant and animal life found there. Some plant species found here are at, or close to, the northern limit of their distribution in Britain. The site was declared a Site of Special Scientific Interest (SSSI) in 1988. It is the largest salt marsh on the north-east coast of England between Lindisfarne in North Northumberland and the Tees Estuary, and is managed by the National Trust.

Figure 10.9 Alnmouth salt marsh.

The salt marsh lies to the south of the river. It is surrounded by mud flats to the north and east but lies next to sand dunes to the south. Much of the area consists of mud flats which are intertidal – covered by the sea for several hours a day. The mud flats are dominated by cordgrass, which feeds birds such as red shank, curlew, snipe, dunlin and other waders. The transition from water to land is a gradual one and occurs in three stages:

- the lower parts of the marsh, which are most frequently flooded, are dominated by sea purslane and some rare free-living forms of seaweed
- on the higher marsh, covered by the sea for only a few hours a day, vegetation is more open with thrift, salt marsh grass, rushes and sea aster
- eventually, salt marsh gives way to dry sand dune, producing a distinct open habitat which supports grasses and wild flowers.

1 a) Using Figure 10.8, draw a sketch map that shows the location and extent of Alnmouth salt marsh. Include the river, the village of Alnmouth, the mud flats, the salt marsh and the sand dunes to the south. Use colour to identify mud flats, the lower salt marsh (covered at high water) and the upper salt marsh (covered only by the highest tides). Include a key.

   b) Label your sketch map to show the different plant and animal communities found in the area.

2 What evidence is there that this area is popular with visitors? Why should this be?

3 Suggest reasons why the village of Alnmouth was built on the north and not the south side of the river.

Theory

Understanding salt marshes

Salt marshes are areas of flat, silty sediments that accumulate around estuaries or lagoons. They develop in sheltered areas where deposition is possible, where salt water and fresh water meet, and where there are no racing tides or strong currents that might prevent sediment accumulation. They tend to be covered by water at high tide, and exposed at low tide. Marsh plants, capable of tolerating both fresh and salt water, develop as plant communities on the surface. This may be in a lagoon, along the edge of a beach or in an estuary. Salt marshes are common features around the coast of Britain and important breeding grounds for birds, fish and crustaceans.

Estuaries are constantly changing on a daily and seasonal basis as a result of tides. Tidal processes are the main cause of transport and deposition of sediment in estuarine areas. Twice each day the tide flows in to produce a high tide, and then ebbs to produce a low tide. The difference in water level between high and low tide is called the tidal range. The two high tides are usually slightly more than 12 hours apart and produce changes in water level in areas of shallow coastline, such as estuaries. Constant fluctuation of tides results in large amounts of silt being deposited both by the ebbing tide and by the river as it meets the sea.

The development of a halosere

In time, tidal sediments fill the estuary to produce mud flats, barren expanses of silt and clay exposed at low tide but covered at high tide. Salt-tolerant plants may then colonise the mud flat. The development of plant communities on salt mud flats is an example of a **halosere**. It is another example of plant succession (see page 52).

The first plants to colonise are green algae and eel grass, an important winter food for migrating wild fowl. These pioneer plants can tolerate being submerged by the tide for almost twelve hours and they stabilise the mud flats so that other plants can grow there. Two other plants, cordgrass (spartina) and glasswort (salicornia), may soon become established. These are **halophytes**, plants that can tolerate saline conditions. They grow on mud flats with up to four hours exposure to the air in every twelve. These plants help to trap further sediment which is deposited daily.

As sediment accumulates, the surface becomes drier and different plants can colonise to form the **sward zone**, shown in Figure 10.10. The survival of this zone depends upon less submersion beneath water. Sea asters and sea meadow grass may form a thick turf in the middle zones, which are only submerged for a few hours each day. However, vegetation is not continuous and hollows remain where sea water is trapped. Creeks divide up the marsh, created by stream water flowing across the estuary at low tide. Evaporation leaves behind salt pans which are too salty for plants to survive.

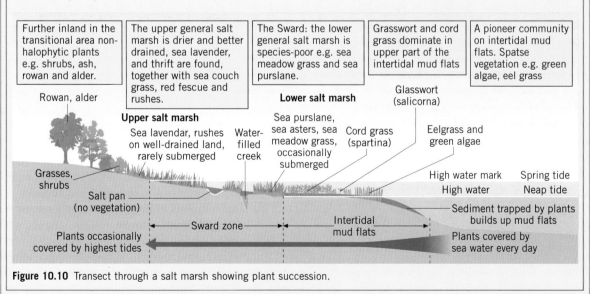

Figure 10.10 Transect through a salt marsh showing plant succession.

As the surface is raised further by deposition, sea lavender, rushes and thrift may be established higher on the marsh (the upper sward) or on well-drained places. This part is only submerged by the highest spring tides. Further inland still, non-halophytic grasses, shrubs and eventually trees will take over. These form high-water roosts for birds described below, and nesting areas during the breeding season.

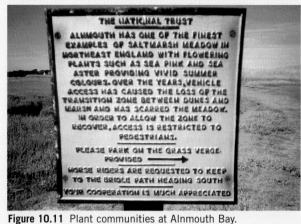

Figure 10.11 Plant communities at Alnmouth Bay.

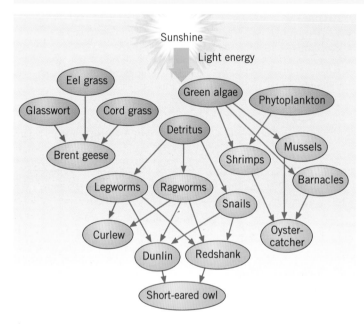

Figure 10.12 Generalised food web for a salt marsh.

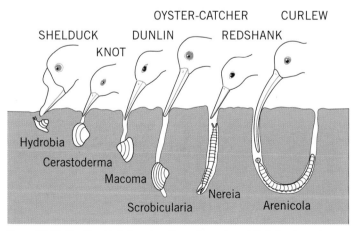

Figure 10.13 Feeding birds found on the salt marsh and their food supply.

The structure and functioning of salt marsh ecosystems

A complex food web develops that supports both salt- and fresh-water animals. A simplified version of a salt marsh food web is shown in Figure 10.12. Areas closest to the sea are almost constantly submerged and contain nutrient-rich sediments from which plants, such as eel grass or glasswort, feed directly. There is little oxygen, and the silty water limits the activity by phytoplankton, which depend on sunlight.

The grasses provide both food and habitat for birds and insects. Insect life is varied, including butterflies, moths, grasshoppers and crickets. There are a dozen kinds of spiders which live only in Britain. The wolf spider, for instance, clings to the submerged stems of spartina at high spring tides and waits for the ebb. While these rely upon living plants, other insect life such as the woodlouse live beneath driftwood or flotsam.

Many birds feed on salt marsh grasses, including grazers such as Brent geese, red shank and shelduck, which feed on snails, shrimps, rag worms and lug worms. All are adapted to live on different parts of the salt marsh. Figure 10.13 shows the relationship between feeding birds on the marsh and their food supply; beak length and shape helps to determine their food supply. On the sediment surface, shell fish such as mussels and oysters – which feed by filtering estuary water – attract sea birds such as oyster catchers.

Theory

Human impact on salt marshes

Salt marshes are valuable to people as well as to wildlife and are threatened in many ways. Farmers may use the land for grazing livestock including cattle, sheep or horses. This leads to selective removal of certain species such as red fescue. If chemical fertilisers are added this alters the natural composition of the marsh. In some areas salt marshes have been reclaimed. The soil in established marshes is rich in nutrients and farmers may drain and plough them so that crops can be grown. Large areas of salt marsh have also been lost to major building projects, such as the tidal barrage in Cardiff Bay and the construction of the Severn Bridge.

Recreation is another way in which salt marshes can be threatened, especially where they are found along attractive coastlines or adjacent to sand dunes and popular beaches. Both trampling and vehicle tracks can cause damage.

Management of salt marshes can be difficult because several different interest groups may be involved. If the land owner is unsympathetic to the needs of the wildlife, the marsh may be drained for agriculture or sold for development. In some cases the declaration of salt marshes as SSSIs has enabled protection; elsewhere the National Trust or National Park Planning Boards have been able to prevent damaging changes from taking place.

Management of grazing on the marshes is often an important step; if grazing cannot be stopped completely it is often possible to reduce stocking densities. Discussion with local farmers about the damage excessive chemical sprays inflict can lead to a reduction in the amount of chemicals reaching the salt marsh. Salt marshes located in estuaries used for industry need to be protected from pollutants in industrial effluent. This is the responsibility of the Environment Agency which has the power to take polluters to court if excessive pollution occurs.

The human impact on salt marshes at Alnmouth Bay

Salt marshes are sensitive environments and are easily scarred by human activity. They are valuable to people as well as to wildlife, and are threatened in many ways along the Bay. These threats are mainly from farming and tourism.

Farming

The main threats to this salt marsh are grazing, horse riding and vehicles. No livestock grazing is allowed in the area because this leads to the selective removal of certain grass species, such as red fescue. However, rabbits graze the sand dunes where they are an important and beneficial part of the ecosystem. Where rabbits spill over on to the marsh, they damage the vegetation and have a dwarfing effect on plants.

The addition of chemical fertilisers to surrounding farmland also alters the natural composition of the marsh. In some areas of Alnmouth Bay salt marsh has been reclaimed. The soil in established marshes is rich in nutrients and they form a tempting resource for farmers to drain and plough them for crops. Estuaries are sensitive to change in agricultural practice in surrounding areas, particularly an increase in chemical fertilisers or pesticides which might find their way on to the salt marsh. Similarly, any plans to drain parts of the area or to alter the course of the river by re-aligning its banks or dredging it could have a significant impact.

Tourism

Recreational pressure is also affecting this area. The growth in tourism – both day trips and longer stays – is affecting the Bay. The attractiveness of this part of the Northumberland coast (discussed in Chapter 8) is leading to increased trampling through to sand dunes and local beaches. Both trampling and vehicle tracks are causing damage. Vehicles are a threat because cars gain access to the beach by using the track that runs along the southern edge of the salt marsh. Horse riders and walkers also cross over the marsh at times, trampling the vegetation underfoot.

---

1 Draw a simple diagram to show nutrient flow in a saltmarsh ecosystem. Annotate your diagram to show how use of nitrogen fertilisers in the surrounding area would affect nutrient flow.
2 Explain how bait digging on the salt marsh could disturb energy flows in the ecosystem.
3 If you were responsible for caring for this part of the National Trust's property,
   a) how would you know whether or not the salt marsh was being damaged over time? Suggest how damage to the ecosystem over time could be measured.
   b) how would you try to dissuade visitors from taking their cars, riding horses or walking across the salt marsh?

Managing Alnmouth Bay

Management of the salt marsh is difficult because several different interest groups are involved. Alnmouth Bay is owned by the Duke of Northumberland, but it is leased to the National Trust under a protective covenant. The whole area is part of the Northumberland Coast Area of Outstanding Natural Beauty and it was declared a Site of Special Scientific Interest (SSSI) in 1988, because of its size, location and the diversity of plant and animal life found there.

At present the National Trust management is non-interventionist. In other words, the Trust has decided not to interfere with the natural ecosystem as long as no serious damage has been done. No grazing by domestic animals is allowed and signs indicate to visitors the delicate nature of the ecosystem. Regular surveys are carried out by both the National Trust and English Nature, who are responsible for the conservation of SSSIs, and if any deterioration in quality of the ecosystem is found then more active steps will be taken in order to protect the marsh. Possible steps could include:

- encouraging seed germination by disturbing the soil to allow a more varied range of flora
- tackling problems of pollution in the River Aln, such as excess nitrates from fertilisers, and pollution on the sea shore
- preventing vehicle access to the beach close to the salt marsh.

In Chapter 8, different interest groups were identified along the Northumberland coast including:
- National Trust
- Northumberland Wildlife Trust
- English Nature
- Royal Society for the Protection of Birds
- Countryside Agency
- Friends of the Earth
- Parish, District and County Councils
- Environment Agency.

a) Why would these groups also have interests in how Alnmouth is managed?
b) How and where might they conflict?
c) What suggestions would you make to resolve conflicts?

Summary

You have learned that:
- Ecosystems develop and change over time through the process of plant succession, until they reach equilibrium. This can be upset by natural or human factors.
- Sand dunes, such as Studland Heath, and salt marshes, such as Alnmouth, are ideal places to study plant succession.
- Ecosystem management may include repairing damage and strategies to prevent further damage, or it may be non-interventionist.
- Decision-makers must consider specific local factors before drawing up appropriate management strategies.

Ideas for further study

1 Compare the structure and functioning of a wetland ecosystem, such as Sweat Mere, with the saltmarsh ecosystem such as Alnmouth salt marsh.
2 Make a study of the north Northumberland coast from Lindisfarne to Druridge Bay. Study:
- the variety of landscape, geology, flora and fauna found along the coast
- the pressures on the coast caused by recreation and tourism
- the ways in which the area is being or could be managed.

References for further study

Bellamy, David *Coastal Environments*
R.S.P.B. (1994) *Ecosystems and human activity*, Collins Educational
Stoff, T, Hindon, J and Crump, R (1994) *Sand dunes: a practical conservation guide*, FSC publication 25

11 Managing coral reefs

This chapter looks at coral reefs, one of the most biologically productive and diverse of all natural ecosystems. It shows what coral reefs are, why they are important, and how they are under threat. It focuses upon Ban Don Bay in Thailand, where reefs are under threat as a result of economic development and the lack of effective controls on its impacts. It examines how coral reefs might be managed more effectively in the future.

Where are coral reefs located?

Coral reefs are found within the tropics, and between latitudes 30° north and 30° south. However, they are not found throughout the tropics, but at specific locations that fulfil their requirements. To survive, coral polyps need water with a high temperature year-round; they cannot survive in water cooler than 18°C. They also require clear water, usually less than 30m deep, for sunlight to filter through; any sediment in the water can block sunlight which polyps need for photosynthesis. Hence, most reefs are located along the narrow edges of continental shelves, before water depth increases. The largest by far is the Great Barrier Reef which runs in a 2000km stretch, parallel to the eastern coast of Australia.

Within this distribution, there are different types of reef:

- **fringing reefs** are close to the coast and are usually only separated from it by a narrow stretch of water
- **barrier reefs** are large and generally continuous reefs, parallel to the shore and separated from land by shallower water. They are occasionally discontinuous, emerging as small islands, and sometimes with much deeper water between. The Great Barrier Reef in Australia is 2000km in length but is not always continuous in its form
- **atolls** are found on or near the surface of the sea when islands that are surrounded by the reefs subside. Atolls can be found either in deep sea or on the continental shelf.

Figure 11.1 shows these kinds of reef. Figure 11.1a shows a coral island emerging above the sea surface. Such islands consist of small coral fragments that have been broken during storms and which gradually erode into small sand-like fragments. Where fragments accumulate, small islands form, known as coral cays. Most of the country of Kiribati in the Pacific Ocean consists of such islands, less than 5m above sea level.

Figure 11.1 Different types of coral reef. Photograph A shows a part of the Great Barrier Reef in Australia, and Photograph B shows an atoll complete with an enclosed lagoon. Note how the Great Barrier Reef is not continuous but varies in width and thickness, sometimes emerging above the sea as coral cay islands.

Why coral reefs are important

The World Conservation Strategy of 1980 identifies coral reefs as one of the 'essential life-support systems' necessary for food production, health and other aspects of human survival and sustainable development. Although coral reefs cover only 0.2% of the ocean floor, they are particularly important because:

- they support 25% of marine species
- they protect coastlines from erosion that would otherwise suffer as a result of storm surges, and act as natural recycling agents of carbon dioxide from sea water and the atmosphere
- they contribute to the formation of sandy beaches – through disintegration of coral shells – and sheltered harbours
- they are a source of raw material, such as corals and coral sands, of building material, of black coral for jewellery, and of stony coral and shells for ornaments
- increasing numbers of reef species are being found to contain compounds with medical properties.

The diversity of coral ecosystems has led to their potential as a major source of food, supporting both subsistence and commercial fisheries. Reefs contribute vast amounts of money to local and national economies. In some parts of south-east Asia and the south Pacific, local and island economies would collapse without coral reefs.

Coral ecosystems

1 a) Construct a food web for a coral reef, using information in Chapter 5 to help you.
   b) Identify different trophic levels, using the food web to help you.
   c) Construct annotated energy cycles, using information in Chapter 5 to help you.
2 How might the following affect this food web:
   a) global warming, where sea level might rise by up to 5m?
   b) clearance of rain forests on land, leading to changes in river discharge and sedimentation?
   c) increased local economic activity in tourism?

Theory

What are coral reefs?

Coral reefs consist of billions of coral polyps, which live together in reefs, or colonies. Each coral polyp is an animal and belongs to the same group as jelly fish and sea anemones. Each is about 2 to 3cm long and feeds on minute organisms in the sea. Colonies of polyps establish themselves upon suitable surfaces, which fulfil their requirements for survival. Obtaining minerals from their food supply (minute organisms such as larvae and algae, or plankton) enables each polyp to secrete calcium carbonate continuously around the lower half of its body, forming a protective chamber, like a shell. This occurs at a slow rate – about 1cm per year.

What makes corals unusual is their relationship with the plant community.

Figure 11.2 A typical coral reef scene. This photo was taken on the Great Barrier Reef in Australia and shows a variety of coral. The branching corals are 'antler' corals, helping to provide a network of protection for small fish. The varied colours are due to the presence of different algae in the corals.

Three-quarters of all living tissue within a coral reef is vegetable. Within each polyp are several yellow-brown granules, consisting of algae similar to those that swarm in plankton. The vast array of colours in coral reefs is due to the presence of different algae, reflecting different colours in the sunlight. They absorb waste from the polyp. They convert its phosphates and nitrates into proteins, and, with sunlight, use carbon dioxide to produce carbohydrates through photosynthesis. In this process, they release oxygen, which is what the polyp needs to survive. The relationship is therefore a symbiotic one, where each is suited by the arrangement with the other, and each needs the other for survival.

Polyps reproduce sexually (by means of fertilised eggs) and asexually (by budding). When eggs hatch, new polyps swim independently and settle elsewhere, thus continuing the process of coral formation. Where budding occurs, filaments grow from the coral polyp to form buds, which begin to secrete their own calcium carbonate on top of the parent. Being gradually buried, the parent polyp dies. The colony therefore consists of layer upon layer of vacant limestone chambers which accumulate as reefs or atolls. The process of building coral reefs is thus a slow process, and if a reef ecosystem is disturbed either naturally or through human interference, it may never recover.

In fact, reefs become self-sustaining communities. The constant breaking waves over the edge of the reef on the ocean side creates a vast reserve of oxygen-rich sea water, itself supporting a range of sea life. Corals are not only important for the mass of polyps that they support, but for the diverse marine life which is dependent upon the reef for food or security. Vast ranges of species of fish inhabit the coral reefs. Some, like parrot fish, feed directly upon polyps in the coral, tearing off coral to extract them. Starfish produce digestive juices which they use to squirt into the polyp casings, then suck out the polyp as a soup.

Other species use the protective network of coral on which to base or defend themselves. Clams settle on the coral bed and filter plankton from the water. Eels live within the network of coral, pouncing on unsuspecting small fish that pass. Damsel fish live within reach of branching 'antler' corals, filtering organic material from the sea while sheltering within the forest produced by the corals. Small hermit crabs, carnivorous fish, sponges and sea horses all contribute to a hugely varied community.

What are the threats to coral reefs?

A number of impacts can adversely affect coral reef systems. Natural disturbances, such as hurricanes, storms and sea level fluctuations have always placed temporary pressures on coral reef systems, but there has been a recent increase in the effect that humans have on reefs. This impact is more significant because much of the damage is lasting, rather than temporary.

Tourism – the world's biggest growth industry

Long-haul travel has become increasingly accessible to more people. It costs 40% less in real terms to travel from the UK to Australia's Great Barrier Reef in 2000 than it did in 1990. Coral reefs offer many recreational opportunities and are a magnet for tourists. This adds significantly to local and national revenues but it also creates additional problems.

The coral reef is a system of interdependent components. The complex nature of the relationships with light, temperature, water clarity, salinity and oxygen means that damage to one component can trigger a negative chain reaction that affects many different organisms in the system. There are thus rules and guidelines for tourists in coral reef areas (Figure 11.3).

Fishing

Increasing populations in coastal areas result in greater subsistence fishing, adding to the effect that commercial fisheries have on the reefs. Removal of fish leads to imbalance in the ecosystem. With fewer fish to control their population, algae become more dominant.

1 Make sure when you enter the water that you are wearing sneakers or some other form of footwear, in order to protect your feet if you step on sharp coral or sea urchins.

2 In the water make sure you swim, not walk, in order to avoid damaging the coral. When swimming, make sure you do not damage the coral in any way, by touching it or knocking it with a fin. Remember that such damage remains visible for years.

3 When heading for the open sea, cross the reef only by the route marked by floats and ropes. Do not cross the reef anywhere else.

4 It is strictly forbidden to stand or walk on the reef, which causes major damage.

5 All rocks and animals in the reserve – both in the water and on the shore – are protected natural assets. Do not touch them and on no account remove them from the water.

By following these rules and keeping the reserve clean, you will have an enjoyable visit, as well as preserving its unique beauty for other visitors to enjoy too.

Your co-operation is appreciated.

Figure 11.3 Safety measures designed to protect coral reefs in Israel on the Red Sea.

A number of damaging fishing practices are taking place in coral reef areas. Although illegal, dynamite and cyanide fishing still take place. Muro-ami and kayakas fishing require swimmers to drive fish into nets by striking the coral. Trawl fishing vessels drag nets along the sea bed on metal wheels, which break coral and drive fish into the nets. These practices are increasingly used to produce greater yields.

Economic activity

The growth of coastal populations and economic development of ports and other coastal urban areas has had an enormous impact on coral reefs. Pollution through sewage and industrial emissions, oil spillage, and transfer of pesticides into coastal waters from rivers are all common. Chemical emissions from industry have adverse effects upon the coral ecosystem, together with outflow pipes from power stations that discharge hot water into the sea. Pollution of reefs is not always terminal; once pollutants are removed, the reef recovers relatively rapidly. Other results of coastal development are putting increasing pressure on coral reefs. As construction continues, the amount of freshwater run-off increases, transporting sediment into the sea, and decreasing the amount of sunlight that reaches the coral. Deforestation has similar results.

Global warming and ozone depletion

Global warming is a relatively recent phenomenon that may affect coral reef ecosystems increasing sea temperatures and raising sea level, thus affecting the amount of sunlight reaching the coral. This could threaten the world's coral reefs. Depletion of the ozone layer means that damaging UV-B rays will reach the coral ecosystem. The newspaper article in Figure 11.5 highlights how in 1999, two islands in the Seychelles have disappeared.

Corals hit by fossil fuels

The world's coral reefs, already damaged by record sea temperatures, are threatened by rising carbon dioxide levels, an international team of scientists has determined.

Their study has shown that the burning of fossil fuels is raising levels of carbon dioxide, causing a reaction that erodes the reefs.

Calcium carbonate is the foundation of coral reefs, producing the tiny reef-dwelling creatures called polyps. However, rising levels of carbon dioxide in the atmosphere are causing the sea to become more acidic as the gas is absorbed by the water. This tends to dissolve calcium carbonate, making reef formation more difficult.

In the next century, levels of carbon dioxide in the atmosphere are expected to reach double the level that they were before the Industrial Revolution, when human beings first began to burn fossil fuels on a big scale. The result may be that live reefs become increasingly fragile and may stop growing.

Charles Arthur, The Independent

Figure 11.4 Threats to coral reefs.

Classifying pressures on coral reefs

1 Form groups of two or three people. Classify the pressures on the world's coral reefs (Figure 11.4) into between three and eight categories. Share your findings with other groups. Explain how you decided to classify these pressures.

2 How would this classification help to plan management of coral reefs?

Managing coral reefs in Thailand – Ban Don Bay

Many LEDCs find that coral reefs are suited to their tourist industry and are a means of economic development. The tourist industry has therefore expanded rapidly in these areas, especially with the rise of long-haul tourism. Thailand is now a popular destination from Europe, in spite of the 13-hour flight. While many travel to cities such as Bangkok, many now travel to islands close to Phuket, on the west coast of Thailand, and Ban Don Bay on the east.

The Ban Don Bay area of Thailand comprises three groups of islands, known as Koh's: Koh Samui, Koh Phangan, and Mu Koh Ang Thong. Figure 11.6 shows these and the relative quality of their coral reefs. All of them possess coral resources of value both to the Thai fishing industry and its rapidly growing tourist industry. Koh Samui was the first island to be developed for tourism, followed by Koh Phangan and later, Koh Tao. In Mu Koh Ang Tong National Park there are few residents and visitors are not allowed to stay overnight.

Koh Nang Yuan (Figure 11.7), is an island off north-west Koh Toa and is described in Trailfinders Autumn 1999 magazine.

'We chartered a boat to go snorkelling and sailing and ended up at Koh Nang Yuan (which changes from one island to three and back again with the tide), where the underwater world of pinnacles, caves and coral gardens teems with life.'

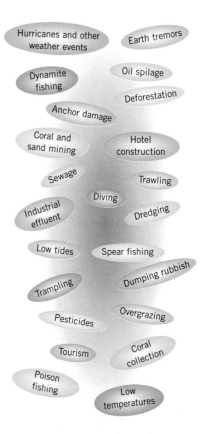

Figure 11.5 Pressures on coral reefs.

What is the spatial distribution of reef quality?

Use Figure 11.6 to describe and offer explanation for the spatial distribution of coral reef quality in Ban Don Bay.

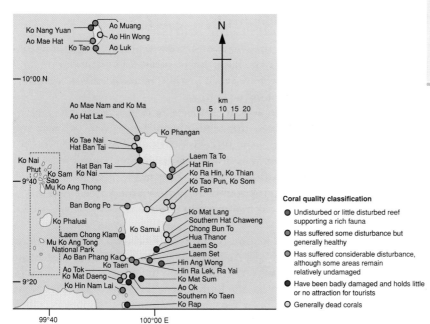

Coral quality classification

- Undisturbed or little disturbed reef supporting a rich fauna
- Has suffered some disturbance but generally healthy
- Has suffered considerable disturbance, although some areas remain relatively undamaged
- Have been badly damaged and holds little or no attraction for tourists
- Generally dead corals

Figure 11.6 Locations and relative qualities of coral reefs in Ban Don Bay.

The impact of tourism

The growth of tourism in Thailand has had a major impact on its coral reefs. The sheer number of visitors is having an effect, despite careful management and restrictions in some areas, as is the collection of coral memorabilia. Visitor pressure results in increased numbers walking on the reef where waters are shallow. It is a paradox that the more beautiful the place, the more people visit, and the more rapid is the deterioration of the environment. In searching for new locations, the process affects an increasing number of places. Koh Nang Yuan (Figures 11.6 and 11.7) off Koh Tao is suffering from this.

Tourist development continues on Koh Samui. It is now one of the most popular resorts in Thailand, with daily flights from Bankok, and is on the itinerary of many UK tour agents, whose brochures

Figure 11.7 Koh Nang Yuan off north-west Koh Tao.

Figure 11.8a The first stages of another tourist development on Koh Samui.

Figure 11.8b Next stage – the developers move in!

actively promote this once isolated and quiet island. The photographs in Figure 11.8 show a now more common scene in Koh Samui. Forest and vegetation clearance is resulting in increased run-off and therefore sedimentation in Koh Samui's seas, a contributing factor in the deterioration of its coral reefs. The local population has also declined as more and more land is taken by tourism.

1 Study the photographs in Figures 11.7 and 11.8. On an A3 page, write the following headings: Tourism; Development; Environment: Economy; Sustainability. Use the five Ws (what, where, when, who, why) to develop questions using the photographs as stimulus. Once the questions have been formulated try to answer them.

2 How are the headings linked? Do they go hand-in-hand or do they reveal conflict?

Managing the tourist threat

The coral reefs of Thailand and of much of south-east Asia have been subject to exploitation and damage in recent years. The main causes include:
- intensive fishing
- the use of destructive fishing methods such as dynamite fishing, where areas of reef are subjected to explosions of dynamite
- sedimentation due to deforestation
- pollution from urban and industrial areas
- tourism in heavily frequented areas
- boat moorings in reef areas.

In response to this degradation of coral reefs in south-east Asia, the Association of South East Asian Nations (ASEAN) which includes Thailand, Brunei, Malaysia, the Philippines, Singapore, and Viet Nam, have been working on strategies to preserve their coral reef resources. Some experimentation is taking place with marine parks and reserves, such as Muh Koh Ang Tong National Park in Ban Dong Bay. The islands of Ban Don Bay in Thailand have been a pilot for coral reef management and the strategy for this area will provide the focus of this study.

Despite blast or dynamite fishing methods being banned, the practice continues, mainly by people from outside the area. The situation is worsened by an influential few who receive a share of the catch from such practice, for ignoring it.

The tourist boom in the area has meant increased human contact with the coral reefs and has also resulted in increased silting and pollution, especially waste water from the coastal zone development. By 1999 Thailand earned US$ 6 billion in foreign exchange from tourism. Seven million people now holiday in Thailand, a 64-fold increase since 1969.

Managing the Ban Don Bay coral reefs

Coral reefs require careful management if sustainable usage is to occur. An increasingly popular approach is to protect certain areas and to develop marine parks, where fees are charged for entrance, souvenirs, and boat permits. The income derived from this allows both development of reefs as resources and allows a sustainable form of management. Lack of funding has proved a problem and as a result, effective management is difficult.

At present there is little positive management in Ban Don Bay, and certainly no co-ordinated management plan for the reefs. There are regulations governing destructive fishing techniques and coral collection. Apart from this, management has been left to governors at provincial level, the quality of which varies enormously.

Degradation of the coral reefs of Ban Don Bay continues (see Figure 11.6). There is:
- a low public appreciation of the value of the coral reefs as a resource
- little knowledge of workable management options
- a lack of both funds and people to police coral reefs effectively
- a lack of co-ordination at different levels.

When developing a management plan for Ban Don Bay, two approaches were adopted by local government. Firstly, it was thought that a bottom-up approach was needed; management could only be successful with the support and involvement of the local people. Secondly, a management plan was necessary that would allow sustainable use of coral reefs. Rather than prohibit their use of the reef, local people could continue to reap the benefits of the reefs in a sustainable way.

There are four types of management programmes:
- recreation and tourism
- fisheries management
- water quality maintenance
- public awareness and education.

Recreation and tourism

In order to control the damaging effect that tourism is having upon the reefs, without damaging the potential income from tourism, a system of zoning has been proposed. This creates:
- sanctuary zones where any destruction of coral is prohibited
- conservation zones where tourist activities are allowed within guidelines and restrictions, e.g. where plastic bottles or cans are not permitted on to islands
- a general use zone where there are only guidelines for habitat protection.

Fisheries management

Particular attention is drawn to the role of reefs in fisheries and in the tourist industry. The reefs provide fish and crustaceans on which many coastal communities depend for their livelihood. Fishery management seeks both to limit and prevent damage to the reef through the prevention of destructive and illegal fishing practice, and to avoid damage caused by coral and shell collection. It also seeks to preserve traditional culture.

It has been recommended that:
- coral reef fisheries are maintained at sustainable levels. The World Conservation Strategy considers this to be a global priority. Most countries have legislation for fishing
- fisheries are maintained in a sustainable way through stock enhancement and development of under-used varieties of coral life such as sea cucumbers and molluscs
- the over-exploitation of shells for ornamental purposes be stopped by developing breeding grounds specifically for this purpose
- further research into coral reefs takes place; a vital step if effective management is to take place.

Water quality maintenance

Reef management needs to be considered in terms of the whole coastal zone and must include river catchment areas. Water quality is to be monitored and legislation drawn up to prohibit damaging construction, especially in areas adjacent to sanctuary zones. Coastal zone management involves putting sewage outlets down-flow from coral reefs, discharging thermal effluent into deep water, and generally taking the reefs into account when planning land use. Environmental Impact Assessments, described in Chapter 3, should be undertaken whenever development may affect the reefs.

Public awareness and education

Raising public awareness and education of both local people and visitors is considered essential to the success of any management plan. Figure 11.9 shows a means of educating tourists on how to conduct themselves in the area. For example, tourists are no longer allowed to bring plastic bottles or cans to the island.

Information centres are to be set up for visitors and a trained team will visit villages on an educational campaign. In time, the importance of coral reefs would also be incorporated into the schools' curriculae. Schools have a major role in educating children about the importance of reefs as essential resources and as areas in need of protection and careful management.

International management

ASEAN (the Association of South East Asian Nations, which includes Thailand, Brunei, Malaysia, the Philippines, Singapore and Viet Nam) is seeking to establish a network of Heritage Parks and Reserves. International collaboration is important in developing man-agement strategies for coral reefs; reefs do not follow political boundaries and are hence a collective responsibility.

Figure 11.9 Tourist education on Koh Nang Yuan.

Establishing priorities for managing Ban Don Bay

1 Prepare a conflict matrix for Ban Don Bay, referring to Chapter 8. Identify different interest groups such as local government, tourists, fishing families, etc.
2 What difficulties does the conflict matrix suggest exist?
3 Suggest ways of managing these difficulties.
4 Figure 11.10 shows management programmes for Ban Don Bay. The final column in the table is empty. Working in pairs, copy and complete the table, giving each programme a priority on a scale of 1 to 12 (1 = maximum priority). Give reasons for your choices.
5 Compare your rankings and reasons with another pair. How and why are they similar or different?

Program	Recommended projects
Recreation and tourism	• Demonstration of mooring buoys at Ko Taen • Park zoning and boundary extension to Ko Tao • Sanctuaries at Ko Taen and Ko Phangan • Conservation zones at Ko Samui and Ko Phangan • Multiple-use zoning of Mu Ko Ang Thong National Park
Fisheries management	• Marine sanctuary at Ko Tuen • Management scheme for reef fish • Upgrading of law enforcement on illegal fishing
Water quality Maintenance	• Mandatory EIAs for beach resort construction • Guidelines for mitigating the situation of reefs • Water quality monitoring
Public awareness and education	• Information and visitor service centre • Exhibition and media campaign • Training tour guides and operators • School curriculum on marine ecology and conservation

Figure 11.10 Summary of priority management programmes.

1 Read through this study about the attempts made to manage coral reefs in Thailand. In your opinion, which aspects are well-managed? Which are not?

2 Decide upon actions that are necessary to improve management of the reefs. Draw up a plan of 750-1000 words and justify your ideas.

3 Imagine that all your proposals are adopted. How would you know whether or not your ideas have been successful in five years time? How would you evaluate the success of your plan?

Summary

You have learned that:

- Coral reefs are a type of ecosystem. It is a very diverse ecosystem and is important in many respects, economic, social and environmental.
- Coral reefs form only under specific conditions.
- In Ban Don Bay there are many and varied pressures on the coral ecosystems.
- They are very fragile and slow-growing ecosystems and need careful management.
- The coral reefs of the world are under threat.
- Economic development and environmental protection do not always go hand-in-hand. Sustainable development is not always the primary target, even if it should be.

Ideas for further study

Ban Don Bay is an example of management of coral reefs systems in an LEDC. Research a case study of management in an MEDC. The Great Barrier Reef off the coast of Australia or the coral reefs of Florida are good examples.

Compare them to each other. Use a large, two-circle Venn diagram to compare the characteristics of the reefs and the ways in which they have been managed.

References for further study

Garland, Alex (1999) The Beach, Penguin Books

Manuel, M, McElroy, B and Smith, R (1995) Coastal Conflicts, Cambridge

Middleton, N (1995) Coral reefs: Ecosystems under pressure, Geography Review, January 1995

White, A T, Coral reef management in the ASEAN, US Coastal Resource Management Project

Woodfield, J (1994) Ecosystems and human activity, Collins Educational

Any novels by Carl Hiassen

Coastal flooding in Bangladesh

So far, rivers and coastlines have been presented as separate entities, which work mainly independently. This final chapter of coastlines explores coastal flooding, where the interaction between rivers and coasts presents people living in low-lying coastal areas with continual problems. It focuses upon Bangladesh; annual news reports remind us that flooding is a feature of life every year.

Water management in Bangladesh

Water is central to the lives of the people of Bangladesh. Approximately 80 per cent of the country is made up of the floodplains and delta of the Ganges, Brahmaputra (Jamuna), Meghna and a number of smaller rivers which form one of the largest delta systems in the world. The complexity of the river system is shown in Figure 12.1. Each river is huge. The average peak flow of the combined rivers in the lower Meghna alone is about 2.5 times that of the Mississippi! The three main rivers and their tributaries drain an area of more than 1.5 million square kilometres.

Water plays a dominant role throughout the year – either too much, with flooding common in the monsoon season, or a lack of it, with drought often occurring in the spring. The climate is predominately tropical monsoon and the three main rivers have a marked seasonal flow, reflecting monsoon rainfall distribution, shown in Figure 12.2. Not only is the flow of water affected by rainfall, but other factors also contribute to sudden surges of water at different times of the year.

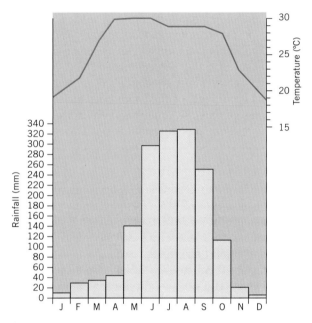

Figure 12.2 Annual rainfall graph for Dhaka. Note the surge in rainfall during the summer months, May-September.

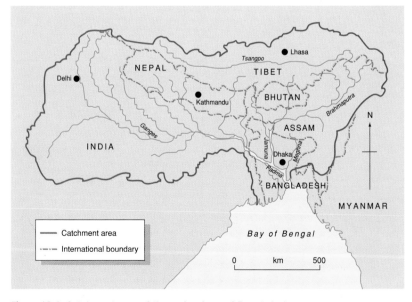

Figure 12.1 Catchment area of the major rivers of Bangladesh.

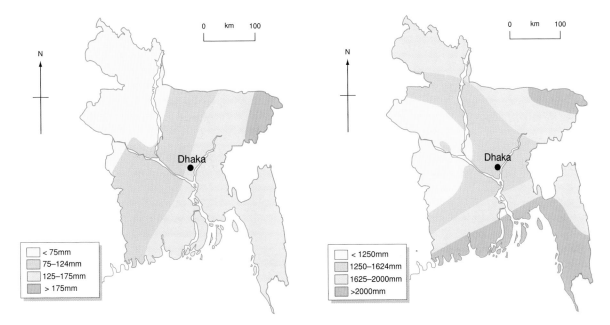

Figure 12.3 Seasonal rainfall distribution in Bangladesh **a)** dry season, November-March, **b)** rainy season, June-September.

Hydrological background

1 Use Figure 12.3 to describe the annual pattern of precipitation in Bangladesh.

2 **a)** Make a large copy of the catchment area from Figure 12.1.

  b) Shade in the country of Bangladesh and estimate the percentage of the catchment area that Bangladesh makes up.

  c) How many other countries are affected by river flow in these rivers? How might this present problems for Bangladesh?

3 Use an atlas to identify features of the catchment area such as relief, land use and vegetation. How might any of these features affect the flow of the rivers during the year?

Theory

What is a delta?

A delta is a low, almost flat area of land at the mouth of a river where it flows into a quiet body of water, such as an ocean, sea or lake. At the point of entry, the velocity of the river is reduced and its load deposited. Estuaries and deltas differ; deltas carry greater supplies of sediment and therefore deposit far greater volumes than estuarine rivers, where sediment volumes are less and tidal currents and river discharges keep channels clear. Figure 12.4 shows the delta area of the Ganges, Brahmaputra (Jamuna), and Meghna rivers. Notice its shape; on the satellite image, dark pink tongues of sediment drift out into the sea, creating a broadening area of sediment which gradually builds up to form new land.

The formation of deltas

At the point of entry to a sea or lake, river velocity slows and halts, and as a result deposits much of its load, called **alluvium**. This partially blocks the river channel, and forces the river to seek another route, which itself in turn becomes clogged. The process is repeated as new channels form, creating a maze of semi-parallel channels, known as **distributaries**, through which the river flows towards the sea. As a result of distributaries, a delta grows laterally, or sideways, 'fanning out' as it reaches further into the sea or lake.

Figure 12.6 shows a cross-section through a delta, and shows how a delta grows further out to sea. It grows sea-wards as further layers of sediment are added. At the point where sediment is deposited on the edge of a delta, rapid sub-surface currents form on the steep drop towards the ocean or lake floor. These are known as **turbidity currents**. The surface of the delta is built up by continued deposition so that it is partially exposed above sea level, especially at low tide. As the alluvium is rich in nutrients, vegetation growth is prolific. Roots bind the alluvium together and this provides a more secure base for further expansion of the delta. In tropical areas of the world, particular plant communities form mangroves along river deltas and estuaries.

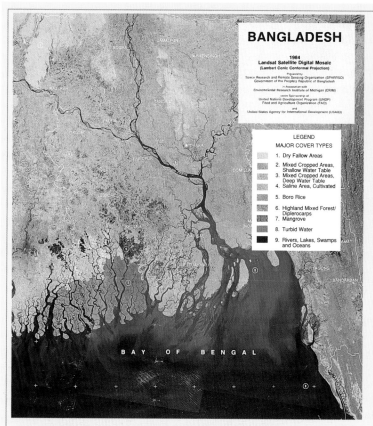

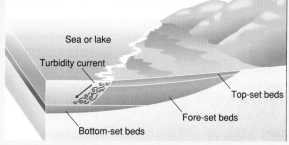

Figure 12.4 A satellite photograph of Bangladesh and of the delta formed by the confluence of the Ganges, Brahmaputra (Jamuna), and Meghna rivers. Note that these colours are filtered and enhanced to show contrasts in land, sea, and land cover (forest, crops, etc.) The key is shown on the photograph.

The size and shape of a delta depends on the load of sediment, the rate of flow of the river and the wave power and tidal range of the ocean. In some deltas, the shape is modified as a result of a balance between river deposition and the removal of sediments by ocean waves and currents. The Mississippi delta, for example, exhibits a 'bird's foot' shape, which results from large quantities of sediment carried into quiet water and the volume of discharge from the Mississippi River creating permanent distributaries. Tides and currents cannot remove all the sediment carried, and so the delta grows out to sea.

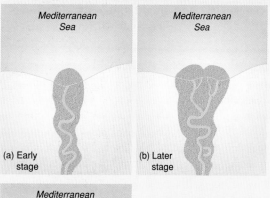

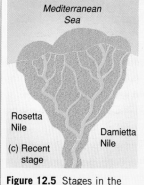

Figure 12.5 Stages in the formation of the Nile delta.

Figure 12.6 The formation of a delta in cross-section.

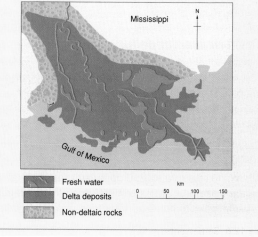

Figure 12.7 The Mississippi Delta

Life on the floodplains

Although catastrophic floods tend to dominate the way we see Bangladesh, floods are essential to the livelihoods of most of the 123 million people and to the economy of the country. Bangladeshi way of life is closely adapted to the floodplain environment: settlements are on floodplain ridges, with individual houses on raised earth mounds; roads and railways are on embankments; and fish production and fishing are important activities, particularly in the monsoon season.

Agriculture is the main economic activity, with rice occupying 80 per cent of the cropped area. Rice is grown in three seasons, two (*aus, aman*) mainly under rain-fed and flooded conditions in the *kharif* (wet) season; the other (*boro*), mainly with irrigation in the *rabi* (dry) season. Over the centuries, farmers have selected thousands of rice varieties to suit local micro-environments, including deep-water varieties for areas flooded up to 5m deep. Individual farmers generally grow several varieties adapted to different depths and duration of flooding, which helps to spread risks, labour use and market opportunities.

Floodwaters cover between a third and half of the country during the monsoon period, and annual floods are generally beneficial. They are an integral part of the floodplain system and play a key role in supporting agriculture; groundwater aquifers are recharged and floodplain soils are enriched, enabling agricultural production without the extensive use of fertilisers. Floodwaters also play an important role in the migration and breeding of fish which are caught by a large proportion of rural families: 75 per cent of dietary animal protein in Bangladesh is supplied by fish, and the fisheries sector of the economy contributes about 12 per cent of annual export earnings.

Continual changes in channel routeways mean that original river channels may dry up. The Brahmaputra (Jamuna) river, shown in Figure 12.9,

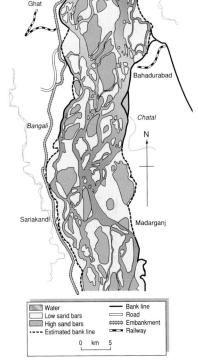

Figure 12.9 Braided channel of the Jamuna River. Notice how the sediment has choked the channel so that it has split into sections, a process known as 'braiding'.

Figure 12.8 Settlements on the floodplain ridges.

is braided during the dry season and its channels change position as water levels rise during the following monsoon. This affects the formation of sand bars which may become high enough to form islands, known as **chars**. In a poor country where there is a shortage of land, chars are often settled by the poorest, sometimes with disastrous results when the next changes take place. They may become eroded and flood, displacing people and their livelihoods.

Theory

Characteristics of flooding

It helps to distinguish between 'normal' floods in Bangladesh, to which farmers' cropping practices are well adapted, and 'damaging' floods which occur when water rises earlier, higher, more rapidly or later than farmers expect. High or sudden floods may also breach road, rail and flood embankments, and submerge settlements and industrial sites. Floods – both 'normal' and 'damaging' – can be classified into four main types: flash, river, rainwater and storm-surge.

Flash floods result from exceptionally heavy rainfall over neighbouring hills and mountains. They occur mainly in the north and east, where rivers from India enter the country. Damage to crops may occur and large amounts of sediment is deposited, which leads to silting up of river channels.

River floods result from snow melt in the high Himalayas during summer, to which are added heavy monsoon rains in the catchments of the three major river systems. In general, this type of flooding is beneficial, because it brings alluvium, but it may cause damage to crop

growth if floods arrive early (June, mainly along the Brahmaputra and Meghna) or late (after mid-August, along all rivers). Flooding is particularly serious when peak flows in the Ganges and the Brahmaputra coincide, as happened in 1988 and 1998.

Rainwater floods are caused by heavy and prolonged rainfall within Bangladesh. Heavy pre-monsoon rainfall in April-May may cause local run-off to collect in depressions, and this is later trapped on the land by rising river levels. In effect, the water table rises above the ground surface. As with river floods, the extent of crop damage depends on the time when the flood occurs. Unlike flash and river flooding, this type does not deposit new alluvium but may wash topsoil from ridges into adjoining depressions.

Storm surges are associated with cyclones that move north from the Bay of Bengal. Incoming surges last only for a few hours at a time with high tides, but the return outflow may get trapped behind roads and embankments. The area affected is usually limited to within 4–8km of the coast, but the impact can be devastating, with great loss of life and huge areas of cropland inundated with damaging saline water.

The 1988 and 1998 floods

Severe and damaging floods in 1988 and 1998 made headline news in the international media. Each was reported at the time as being the worst on record. By studying the hydrology of the rivers concerned as shown in Figure 12.10, we can compare them in order to understand how and why the two floods were each severe, but different.

Study Figure 12.10.
1 Copy the tables below. On each one shade in the months when flooding occurred for each of the three rivers, Ganges, Brahmaputra (Jamuna), and Meghna in:
   a) 1988
   b) 1998.

1988

	June	July	August	September
Ganges				
Brahmaputra				
Meghna				

1998

	June	July	August	September
Ganges				
Brahmaputra				
Meghna				

2 a) Which of the three rivers had the worst flood in 1988?
   b) Why did flooding occur in September 1988, but not in July?
   c) In what ways was 1998 different from 1988?
   d) Which of the three rivers had the worst flood in 1998?
   e) Why did Bangladesh suffer flooding in September 1998 as well as in July-August?
   f) On this basis, which was the worse flood – 1988 or 1998?
3 Does Figure 12.11 confirm your judgement about which was the worse flood? Explain your answer.

a The Ganges

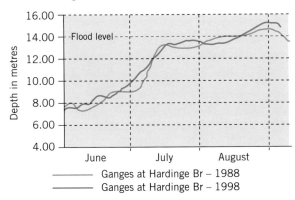

Ganges at Hardinge Br – 1988
Ganges at Hardinge Br – 1998

b The Brahmaputra (Jamuna)

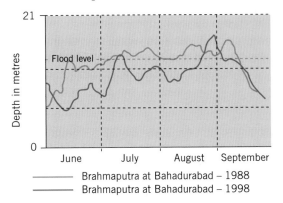

Brahmaputra at Bahadurabad – 1988
Brahmaputra at Bahadurabad – 1998

c The Meghna

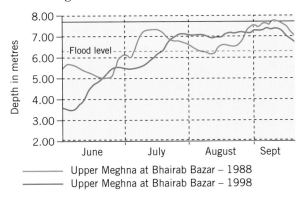

Upper Meghna at Bhairab Bazar – 1988
Upper Meghna at Bhairab Bazar – 1998

Figure 12.10 Hydrology of the rivers flowing through Bangladesh in the period June – mid-September, 1988 and 1998.

The 1988 floods

These were predominately river floods. Heavy early monsoon rainfall in the north-east of the country brought the Meghna and Brahmaputra rivers above danger level temporarily in July. Heavy rainfall at the end of August brought the Brahmaputra to flood peak on 30 August, and the Ganges and Meghna to record high levels on 2 and 7 September, respectively. The sudden, simultaneous rise of these rivers caused them to spill over onto adjoining floodplains. Official sources estimated that about 60 per cent of the country was submerged, including two thirds of the capital, Dhaka. An estimated 45 million people were affected and over 2000 deaths reported. Roads and railways were severely disrupted by broken bridges and breached embankments for several weeks. Crop losses were at their worst on floodplains adjoining major rivers.

The 1998 floods

A combination of timing, duration, and size of flood in Bangladesh made 1998 unlike any other year. Flood waters receded at a much slower pace than usual. Water levels did not begin to recede until late September.

- More than 2 months of flooding in July and August were triggered by heavy monsoon rains, and caused 75 per cent of Bangladesh and 50 per cent of the capital city, Dhaka, to become submerged.
- In September, continuous rainfall in Bangladesh and in the river catchment areas outside Bangladesh led to increasing water levels along the Brahmaputra, Meghna and Ganges.
- To make matters worse, flash flooding hit the Chittagong and Cox's Bazar region three times.

Flood	Station and river name	Danger level of river (m)	Peak level of river (m)	No. of days above danger level
1988	Bahadurabad, *Brahmaputra*	19.50	20.62	15
	Hardinge Bridge, *Ganges-Jamuna*	14.25	14.87	22
	Bhairab Bazar, *Meghna*	6.25	7.66	76
	Dhaka (Capital city), *Buriganga*	6.00	7.58	23
1998	Bahadurabad, *Brahmaputra*	19.50	20.37	57
	Hardinge Bridge, *Ganges-Jamuna*	14.25	15.19	23
	Bhairab Bazar, *Meghna*	6.25	7.33	54
	Dhaka (Capital city), *Buriganga*	6.00	7.24	51

Figure 12.11 Danger levels along the three major rivers and Dhaka, the capital, in 1988 and 1998.

- Further monsoon rains and water from melting snow in the Himalayas made the situation worse still.
- Normal flow of floodwater through rivers was hindered due to existence of high tide in the sea. The high tide was measured at a record 5.52m above the sea level on 10 September 1998.

Item	1988	1998
Duration of floods	21 days	65 days
% of country affected	60%	75%
% of capital city covered by flooding	67%	50%
Area flooded in km	2 282 000 km²	Over 1 million km²
No. of people affected	45 million	31 million
Houses totally or partially damaged	7.2 million	980 000
Human lives lost	2379	1050
Livestock lost – cattle and goats	172 000	26 500
Rice production lost	2 million tonnes	2.2 million tonnes
Km of trunk roads damaged	3000	15 900
Flood embankments damaged	1990 km	4528 km
Industrial units flooded	>1000	>5000
Schools flooded	19 000	14 000
Rural hand tubewells (for irrigation) flooded	240 000	300 000

Figure 12.12 Comparing the 1988 and 1998 floods

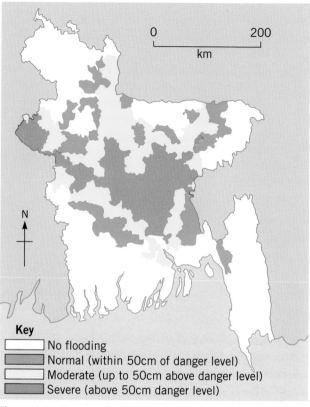

Key
- No flooding
- Normal (within 50cm of danger level)
- Moderate (up to 50cm above danger level)
- Severe (above 50cm danger level)

Figure 12.14 The extent of flooding in Bangladesh in September 1998.

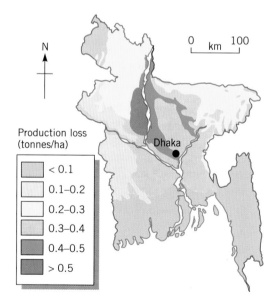

Production loss (tonnes/ha)
- < 0.1
- 0.1–0.2
- 0.2–0.3
- 0.3–0.4
- 0.4–0.5
- > 0.5

Figure 12.13 Loss of monsoon season rice in 1988.

Figure 12.15 Flooded Dhaka in 1988. There were similar scenes in 1998.

1 a) In pairs, make two large copies of Figure 12.1. On one, annotate the causes of the 1988 flood, and on the second, the causes of the 1998 flood.

   b) In another colour, annotate each map with the impacts that occurred in different parts of the country. Use text information and Figures 12.11 – 12.15.

2 Using the information provided, contrast the problems presented by the floods that a) a rural farmer and b) an urban dweller in Dhaka would had to cope with during these two severe floods.

3 a) In pairs, read all the information in this section, and list all the impacts of flooding in Bangladesh, for both 1988 and 1998.

   b) Draw a large copy of the Venn diagram (Figure 12.16) on to a sheet of A3 paper. Take each of your impacts and decide whether it is social, economic or environmental, or a combination of these. Classify your impacts by placing them on the diagram.

   c) Are the impacts mainly social, economic or environmental?

4 Form groups of three or four people. Using all the information, decide which was the worse flood – 1988 or 1998? Decide your criteria, and argue your case with other groups.

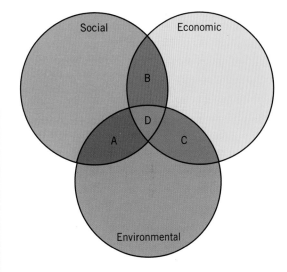

A = Social and Environmental
B = Social and Economic
C = Economic and Environmental
D = all three

Figure 12.16 Venn diagram.

Theory

Flood control in Bangladesh

Different schemes are used to manage flooding in Bangladesh.

- **Controlled flooding** occurs where the spread of water over the floodplain is regulated with mechanisms, such as sluice gates, which are built into embankments. This controls depth of water and helps to ensure that damage is limited.
- **Compartments**, shown in Figure 12.17, are created when embankments are built to enclose depressions into which flood water may be directed. These enable flood water to be contained inside compartments.
- **River training** attempts to direct river flow in such a way as to minimise the bank erosion and scouring of river beds.

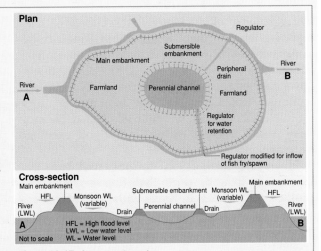

Figure 12.17 A compartment.

- In **flood forecasting**, data are gathered about rainfall, run-off and river levels upstream. They are used to assess the impact on the river downstream, which helps to predict the volume of floods on their way.
- **Flood proofing** limits the potential damage of floods. The development of irrigated agriculture in the dry season reduced dependency upon the flood season for crops. In this way, the Bangladeshi people are less dependent upon the success of the monsoon floods.

Controlling flooding

The Flood Action Plan

Major flood control, drainage and irrigation projects have been undertaken in Bangladesh since the 1960s, with varying degrees of success. In 1989, following the 1988 floods, the World Bank held an international conference to co-ordinate all flood control studies and efforts. The Flood Action Plan (FAP – Figure 12.18) was drafted. It consisted of 26 projects, with the aim to identify, plan and implement technically, economically and environmentally sound flood prevention measures. The first phase was completed between 1990 and 1995, costing approximately $150 million. By 2005 a range of projects will be undertaken, costing up to $500 million. The FAP is funded by donor countries, and work will be carried out by agencies of the Bangladeshi government, donor country governments, by international and local construction firms, by university academics, and national and international NGOs (Non-Governmental Organisations).

The FAP's overall aim is to provide flood protection to the following flood frequency levels:

- Dhaka and other major cities, 500-1000 years
- district towns, trunk roads and main railways, 100-500 years
- main river embankments, 100 years
- agricultural land, 10-100 years.

The main proposal is to construct new embankments along upstream sections of the Brahmaputra and the Ganges within Bangladesh. Construction will proceed downstream in stages, allowing river channels to adjust to increased flood flows and sediment load from the newly confined sections upstream. As embankments are completed, compartments will be created behind them, using internal embankments and existing embankments where possible.

However, total flood control will not be attempted. Instead, controlled flooding will be allowed up to 'normal' levels with which farmers are familiar and can cope with. The FAP will also investigate flood forecasting and disaster preparation including provision of boats for escape and shelters on raised ground.

The Jamalpur Priority Project Study

The possible effects of embankment construction were investigated in one community, (Action Plan 3.1, Figure 12.18). Project staff collected detailed information from 500 households, and met representatives of local councils and members of

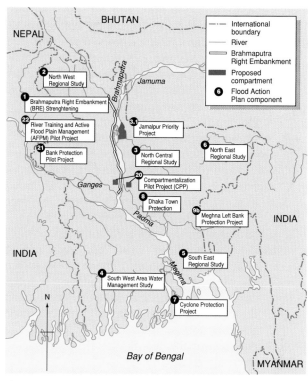

Figure 12.18 Components of the Flood Action Plan.

the public. They discussed four options with the local people:

Option A floodproofing and drainage improvements

Option B controlled flooding for the entire area, with some compartmentalization

Option C controlled flooding of only half of the project area

Option D full-scale compartmentalization of the whole area to exclude all river flooding.

Options C and D were fairly quickly dropped on economic and social grounds. Option A would benefit the landless, non-farming and fishing households. Option B was expected to increase economic growth but would improve only the living standards of farming and land-owning households. Some of those who would lose threatened to breach the embankments, an action known locally as 'public cuts'. These have been carried out elsewhere when embankments have prevented the flow of floodwater back into rivers.

There was also conflict between those who lived inside and outside the embankments. Predictions made by project staff indicated that Option B was likely to increase flood depths outside the project area; over half a million people in the chars of the Brahmaputra would be threatened. The population within the project area is the same size. A decision has still to be made.

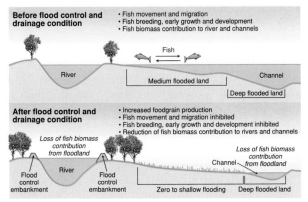

Figure 12.19 Impact of embankment construction on fish.

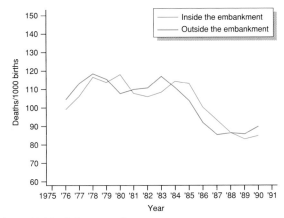

Figure 12.20a Infant mortality.

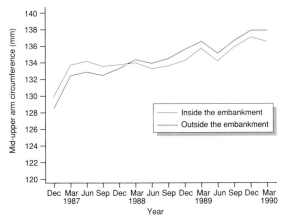

Figure 12.20b Mid-upper arm circumference.

---

1 Is there any evidence in Figure 12.20 that living inside embankments creates either health problems or advantages?

2 Summarise the advantages and disadvantages of **a)** building embankments close to the river, **b)** building them further away.

The Meghna-Dhonagoda Irrigation Project

A study was undertaken to examine the impacts of one of the existing embankments, the Meghna-Dhonagoda embankment, on the health and well-being of the population living inside it. The embankment is located in Matlabthana, 50km south-east of Dhaka. It was constructed between 1982 and 1987 and provides flood control and drainage to 17 000 hectares and irrigation to 13 800 hectares.

There is widespread belief in Bangladesh that projects like this cause health problems associated with water quality inside the embankment. Traditional practices of human waste disposal and the selection of drinking water have developed so that the annual flood would flush the area of waste. It is not clear how well these practices have adapted to the new situation. If water quality suffers, there will be increased health problems amongst children. Figure 12.20 shows two indicators of the well-being of those children.

The debate about embankments
Positioning of embankments

There is strong pressure within Bangladesh to build embankments close to rivers, in order to protect as many people and agricultural land as possible. However, proximity to the river increases the likelihood of erosion, particularly because the rivers have unstable, braided and meandering channels. The channels of both the Ganges and Brahmaputra have shifted significantly in recent centuries. Some existing embankments have had to be rebuilt further from the river a number of times, and at some distance.

By confining flood flow between embankments, river levels and flow speeds are increased, which increases channel instability. A previous study showed that building embankments close to rivers and defending them with groynes would cost twice as much (US$10 billion) as building them 5km away, and their operation and maintenance costs would be $180 million versus $165 million each year. However, embankments set further back would leave 6.3 million people exposed to flood risk, compared with 1.3 million if close to the river.

Time-scale

Even a comprehensive plan to build embankments would take 20-30 years before all major floodplains were protected. In the meantime, many people will still be exposed to unregulated floods.

Impact of embankments on river processes

In places where embankments have been constructed on major rivers (such as the Mississippi and along three rivers in China), the level of the river bed had been raised above the adjoining countryside. The consequences of breaching embankments in a flood would be catastrophic. Another effect of speeding up the river flow between embankments upstream may be increased discharge and sediment downstream, creating further difficulties.

Drainage

Some existing embankments have trapped flood water and prevented it from returning to the river once its level has dropped. This leads to health problems, and affects crop production and fish movements. People affected in this way have been known to breach the embankments deliberately, allowing flood water to drain from their land. If this happens within a compartment it can lead to difficulties between neighbouring communities, as a 'public cut' to drain one area may flood the land next to it.

Breaching of embankments

Damage inflicted when embank-ments are overtopped or breached by flood waters may be more severe than in unprotected areas, because flooding is more likely to happen suddenly.

Criticism of the Flood Action Plan

The Flood Action Plan has also been criticised by some for not including enough 'people's participation' in the whole process. Criticisms suggest that the poorer people of Bangladesh – the majority of the population – are not being heard and are having to live with the effects of projects that they do not want.

Human lives shrugged off in flood plan

Protection of human lives against floods caused by cyclones in the Bay of Bengal, such as the one which killed more than 100 000 people last week, has been a low priority for successive Bangladesh governments and aid agencies. A 26-project Flood Action Plan unveiled by the Bangladeshis and the World Bank in London 18 months ago included just one project aimed at protecting people from tidal waves whipped up by cyclones. The other projects were aimed at reducing the effects of river floods, which damage property rather than kill people.

Of the estimated US $500 million to be spent on the plan, coastal embankments are expected to cost around US $70 million. Most of the remainder will go on improving inland river embankments to protect cities and farms from floods on the rivers Ganges and Brahmaputra. These two great rivers inundated up to half the country in 1987 and 1988.

In an interview with New Scientist in Bangladesh last November, Wybrand van Ellen, a Dutch hydrologist, conceded that coastal flooding had a low priority. 'If one took human life as a yardstick for expenditure, this would be the problem to solve,' he said. 'Cyclones are regular and can kill hundreds of thousands of people, whereas the 1988 river flood was a once in 150 years event and still only killed a few hundred. The difference is that cyclones don't hit the capital, and don't flood the lawns of foreign ambassadors.'

Ainum Nishat, professor of water resources at the Bangladesh University of Engineering and Technology in Dhaka and a member of a local overseeing panel, said: 'Yes, there is a priority for inland protection, where the risk to human life is much less. The reason is that the national priority is to increase agricultural production.' The government wants to protect its cities and its 'green revolution' farms which grow high-yield varieties of rice that are more vulnerable to flooding than traditional varieties.

Coastal embankments 'are not designed primarily to protect against cyclones, but to prevent salt water entering fields during normal high tides', Nishat said. Most embankments are 5 metres high, but last weeks tidal wave was at least 7 metres high. Several heavily populated island at the mouth of the wide coastal delta, such as Urichar Island, which bore the brunt of the death toll last week, have no banks.

A similar flood in 1970 killed between 150 000 and 500 000 people. No one knows the true figures because the inhabitants of these islands are landless migrants, the poorest of Bangladesh's 100 million poor. Cyclones also ravaged the coast four times in the 1960s and three times in the 1980s.

Evaluating the construction of embankments

1 Read Figure 12.21. List the different criticisms made of the FAP.
2 What measures are needed to protect the people of Bangladesh more effectively?
3 Should the priority of FAP be to protect agricultural land or lives?
4 Form groups of three or four. You have been asked to produce a report for flood management workers in other parts of Bangladesh, to let them know the effects so far of the FAP. Your report should include written and spoken presentations.
  a) Explain in two minutes the causes of flooding in Bangladesh.
  b) Summarise the arguments for and against embankments.
  c) Evaluate the effectiveness of embankments on the evidence so far.
  d) Recommend the role that embankments should play (if any) in the FAP.
  e) Suggest how future developments might be monitored.

Figure 12.21 From *New Scientist* 11 May 1991.

1 VILLAGERS

'Give us the power and the resources that the Water and Power Development Authority has, and we would do things better than them economically and technically.'

'If I were to be consulted, what would I say? You see, I'm just an ordinary man. I don't know anything. All I know is that one has to have meals every day.'

'No, no. The Water and Power Development Authority-wallahs have never bothered about these things. They aquire the alnd first, start construction work for the embankments and then notify us that this land has been aquired. People give up the land whether they have food to eat or not . . . If anyone had siad that I won't give my land, then the WAPDA officials would have brought in the police and roped him in . . . '

'If they had contacted the public before doing things then we would have stopped the Water and Power Development Authority from taking the trouble of building these embankments. We have said foremost that rather than building embankments you should try to do something about the river. Make it deeper. But they go on building embankments. Each one goes into the river, then they build another. They are making money out of this, while its the public of the area who are being killed off.'

'Oh yes, the foreigners were here one day last month. But they only went to the school and spoke English. We are not educated. We could not understand.'

2 MEMBERS OF NON-GOVERNMENT ORGANIZATIONS

'When I spoke to the FAP consultants in Dhaka I felt there was some scope for participation. But here in the field I see there is no scope.'

'What is significant by its omission in any debate is the fact that while the Food Plan Co-ordination Organization recieves its mandate from the donar consortium meeting held in London 1989, the government of Bangladesh at that time had no such mandate to enter into such a monumental agreement from the people of Bangladesh.'

3 BANGLADESH WATER DEVELOPMENT BOARD

'With a low literacy rate and limited exposure to the outside world, rural people are not adequately equipped to find/suggest solutions to all of their problems. On the other hand, they may be suspicious of solutions given by experts.'

4 FAP CONSULTANTS

'If you want to consult everybody and wait for a solution agreed by all interested parties, we'll never finish.'

'Without people's participation you can't sell the project!'

'Another new idea from the social scientists. Only slogans! First "poverty alleviation". Then "women" and "environment". Now "people's participation"! It's just a new fad!'

'Oh yes, but you have to consult my socio-economist, not me. I have not time for my people's participation. I'm working 12 hours every day on the project.'

Figure 12.21 Comments on people's participation in the FAP.

1 Which indicators in Figure 12.12 show that, although flooding was more extensive in 1998,
   a) damage or loss was less?
   b) damage or loss was more?
2 a) How do you explain the differences?
   b) Do these indicators point to the success or failure of the FAP (together with other measures) so far?
3 In pairs, carry out the following:
   a) Read through the quotes in Figure 12.21. Identify the different points being made about 'people's participation' in the FAP. Classify them into those in favour and those against.
   b) Within each list, what do the individuals who have made the comments have in common with each other? What might this tell us about the process of development?
   Compare your findings and views with another pair.
4 Summarise, in 500 words: 'FAP – success or failure?'

Exploring the causes of flooding outside Bangladesh

Deforestation in the Himalayas

It is widely thought that flooding in Bangladesh has become worse recently, as more forests have been cleared in the Himalayas of India and Nepal. Increased run-off would increase the amount of top-soil eroded from the land, which in turn would carry it downstream and deposit it on the river beds, thus reducing the carrying capacity of the river. However, some studies challenge this view. Ives and Messerli have shown that:

- clearing forests for agriculture in the Himalayas has occurred for many centuries and could not account for any recent change
- more soil may be lost from a forested slope than from a slope terraced for agriculture
- high erosion and run-off rates occur naturally as a result of climate and landscape
- there is no published evidence that flood magnitude and sediment load have increased recently.

Global warming

There is concern that global warming may create further problems for the people of Bangladesh. As yet there is no evidence that rainfall is increasing; a review of the rainfall records of Calcutta, of the past 150 years, show no upward trend. (Bangladeshi records were only started in the 1960s.) Also, there is no existing evidence that sea levels are rising, which could aggravate inland flooding by raising river base levels. However, should sea levels begin to rise, even by a small amount, they would have considerable impact on Bangladesh.

Looking to the future – a Ganges barrage scheme?

Of all areas, the worst-affected regions in Bangladesh have been close to the coast. Floods can be controlled by both structural and non-structural measures. The Bangladeshi government has so far constructed 8613km of embankments, of which 4041km are located in the coastal region.

While these have largely proved beneficial, their effects on silting of rivers and on riverbank erosion are enormous. The government is now considering dredging river beds and raising the levels of buildings and roads above the highest flood level.

As well as the massive river dredging programme, and the proposal to reclaim marshy lands beside the rivers, a 'Ganga Barrage' is being considered by the present government. It would need overseas aid. The barrage – on the Ganges river – is estimated to cost US$1.7 billion and a feasibility study would be completed by 2001. This project is only possible because a 30-year Ganges water-sharing treaty has been signed with India. The proposed barrage would irrigate 1.35 million hectares of farm land and protect another 1.4 million hectares from floods. Bangladesh plans to start dredging the Gorai river in the eastern region, and the government is taking measures to protect major towns and cities. Dhaka already has a network of embankments which surround it. However, the breaching of these in 1998 caused major flooding.

Whatever the outcome, flooding in Bangladesh is likely to occupy global attention for years to come. You should follow some of the *Ideas for further study* to update your knowledge of these projects.

Figure 12.22 Potential impact of sea level rises.

1 In groups, evaluate the arguments for and against a 'Ganga barrage'.
2 How far do you believe that Bangladesh should **a)** receive money as part of an international aid package to finance flood control, or **b)** be granted loans to pay for such work?

Summary

You have learned that:

- Flooding affects coastal regions, particularly deltas where several rivers may converge and meet the sea.
- Flooding in coastal environments usually has several causes; severity of flooding usually depends upon a balance of these. These causes include rainfall patterns and intensity, river regimes, and storm surges.
- The impact of floods may be social, environmental or economic. They affect both urban and rural dwellers in different ways.
- Poorest countries are least able to manage flooding, especially where 'hard' solutions may be required as sea defences.
- Aid is likely to be required where countries are unable to provide necessary assistance to manage problems such as flooding.
- Actions taken in managing floods may not meet with approval from all people.

Ideas for further study

1 Investigate one project that attempts to manage coastal flooding in another part of the world. Possibilities include major projects where embankments have been constructed along the course of major rivers such as the Mississippi, or, on a smaller scale, localities in the UK such as Towyn.

Follow an enquiry, such as:

a) What are the causes of flooding there?

b) What threats do such floods pose? For whom?

c) What attempts are being made to manage flooding? By whom? With what effects?

2 What are the possible flood threats posed by global warming to different areas of the world?

References for further study

You can obtain up-to-date data about flooding in Bangladesh in 1998 and subsequently through the following web pages:

http://bangladeshflood98.org/research/

http://www.drik.org/flood98

Changing coastal environments: summary

Enquiry questions	Key ideas and concepts	Chapters in this book
Introducing coastal systems 1.6a What are the key features of coastal environments? 1.6b What systems operate within coastal environments?	• Coastal environments contain a wide variety of landscapes and ecosystems. • Coastal environments are an interface between land and sea. Their holistic study includes the wider coastal zone. • Coastal environments contain dynamic systems operating at a variety of scales. Single storm events cause dramatic changes.	• *Chapter 8 (The Northumberland coast)*
Process and change in coastal environments, landforms and ecosystems 1.7a What physical factors and processes influence coastal landforms?	• The nature of the coastal environment is the result of a number of interacting factors. • Coastal landforms result from the interplay of marine and sub-aerial processes. • Short-term and long-term changes. Contrast dynamic nature of coastal accretion/erosion with longer-term changes of sea level (eustatic/isostatic)	• *Chapter 8 (The Northumberland coast)* • *Chapter 9 (West African coastline)*
1.7b What processes lead to change in coastal ecosystems?	• Changes in coastal ecosystems result from natural change (concept of succession) and the direct and indirect impact of human activities.	• *Chapter 10 (Studland Bay and Alnmouth Salt Marsh)* • *Chapter 11 (Coral reef coastline in Thailand).*
Environment – people interactions 1.8 How can change in coastal landforms have an impact on people's daily lives?	• Short-term impacts of rapid erosion, changing deposition, coastal flooding. • Long-term impact of global climate change on rising sea levels.	• *Chapter 8 (Northumberland coast)* • *Chapter 9 (West Africa)* • *Chapter 11 (Coral Reefs)* • *Chapter 12 (Bangladesh and coastal flooding)*
1.9 How have human activities, some of which may be conflicting, influenced coastal environments? What are some of the consequences?	• **An overview** of coastal land uses on an extended coast. Potential conflicts between activities. • ***Choose one*** of the following to illustrate key questions in 1.9: – urban and industrial development and issues of water quality – recreational and tourism pressures – land reclamation schemes.	• *Chapter 8 (Northumberland coast) – recreational and tourism pressures* • *Chapter 10 (Alnmouth salt marsh) –- recreational and tourism pressures* • *Chapter 11 (Coral Reefs) – recreational and tourism pressures*
Management and the future 1.10 In what ways does the management of coastal environments pose a continuing challenge for people?	• Human pressures on coastal environments create the need for a variety of coastal management strategies. • Coastal management strategies may be short-term or long-term; sustainable or non-sustainable. Successful management requires an understanding of coastal processes and systems. • Management involves policy, planning and practice. Countries develop contrasting policies for coastal management. • The demands for coastal management are likely to increase, with rising sea levels, more frequent storm activity and continuing coastal development.	• *Chapter 8 (Northumberland coast)* • *Chapter 9 (West African coastline)* • *Chapter 10 (Alnmouth salt marsh) – recreational and tourism pressures* • *Chapter 11 (Coral Reefs) – recreational and tourism pressures* • *Chapter 12 (Bangladesh and Flood Action Plans) – Contrasting policies and strategies in the UK and West Africa / Bangladesh*

Introducing rural environments

'The land of England is overwhelmingly countryside, while the great majority of people live in urban areas. Over half the population would like to live in the countryside, and about one third are concerned about the quality of the countryside.'

Thus begins The State of the Countryside 1999, a report published by the UK Countryside Agency, which is responsible for development in rural areas. This section will explore how most land within a country may be rural, while the pressures that it faces may come from urban areas, such as demand for food, for leisure space and in many LEDCs, for new supplies of labour to sustain industry.

What are rural areas?

Definitions of 'rural areas' vary. Look at the photographs in Figure 1, taken in different parts of the world. Consider whether or not each represents a rural scene. Different groups use different criteria in defining 'rural'. This introduction looks at four ways of defining rural, by:

- settlement size
- an index of rurality
- regional context
- land use.

Figure 1 Rural areas in different parts of the world.

Photograph C Dispersed farms in Grisedale, Yorkshire Dales National Park.

Photograph D Kenyan cattle herders.

Photograph A Aerial view of a settlement about 6 miles east of the edge of Perth, Western Australia.

Photograph E Wadebridge, Cornwall – town or village? Rural or urban?

Photograph B Outback Australia.

Photograph F A village in rural India.

Defining by settlement size

In this case, settlement size means 'population'. The following will show how different governments do not agree on what we mean by 'rural' populations. Even the Countryside Agency and the UK government use confusing definitions.

- They define 'rural areas' as any area or postcode which is outside settlements with 10 000 or more people. On this basis, 9.3 million people – 20 per cent of the UK population – live in rural areas.
- However, of this group, 6 million live in settlements of between 1000 and 10 000 people. To many, these may be large settlements that are urban in nature, or that local people describe as 'towns'. If 'rural' is defined as settlements of fewer than 1000 people, just over 3 million – or 7 per cent of people in the UK – live in rural areas.

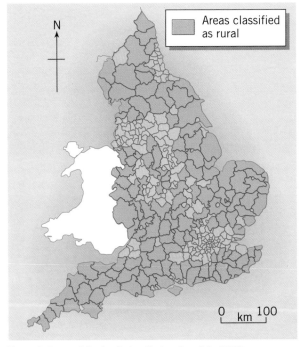

Figure 2 Areas of England classified as 'rural' in 1999.

- Rural parishes are defined by the Countryside Agency as 'parishes that have a resident population of 10 000 or under'. How, therefore, is a settlement of 9500 people within a parish of 10 500 people to be defined?

Definitions vary from one government to another.
- In Italy and Spain, settlements with a population greater than 10 000 are urban and every other settlement is rural by default.
- In France the population threshold of an urban settlement is 2000.
- In Denmark and Sweden the threshold is only 200 inhabitants.
- The United Nations has its own definition – an urban settlement is one having a population greater than 20 000. In this and many other situations, rural settlements are described by default; they are 'not urban'.

Defining by regional context

Defining 'rural' on the basis of population may be problematic. It can also be confused with definitions of 'town' (which is an urban term) and 'village' (which is a rural term). It can include settlements that some people might consider quite large. For instance, Cornwall – considered by most people to be a largely rural county – contains several 'towns' which do not have as many as 10 000 people (such as Wadebridge, with about 5800 people – Figure 1, photo E). Even better-known ones such as the resort of Newquay have only about 18 000 people.

Defining whether a place is a 'town' or a 'village' may therefore depend on the area in which the settlement is located. Consider the following cases.
- Hawes in North Yorkshire, within the Yorkshire Dales National Park, has a number of banks, a large general store, a cattle and retail market once a week, clothes and gift shops, including an outdoor wear shop, as well as several small hotels and pubs. Yet its population is 900.
- Wilcannia in New South Wales, Australia, has a population of about 200. Its location is shown in Figure 3. It lies in the 'outback' of New South Wales, in semi-desert country along the main highway that links Adelaide in South Australia with Broken Hill and Sydney in New South Wales. As isolated as any settlement in New South Wales, it has a range of functions, such as a bank, post office, petrol station, motel, bakery and a range of shops (Figure 4).

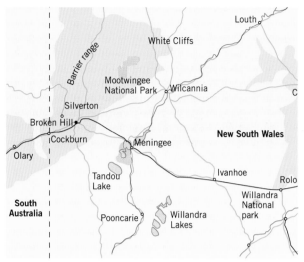

Figure 3 Location of Wilcannia, New South Wales, Australia.

Figure 4 The main street in Wilcannia.

In each case, the settlement is defined in terms of its location as a 'town'. Local people refer to each one as a 'town'. Each place is located in an area of low population density, where the nearest other settlements are farms or isolated houses. By contrast, more recent suburban villages close to larger towns or cities have fewer functions than might be expected, because they are dominated by a local town.

It helps to think of the settlement within the area in which it is located. As well as providing for the few people who live there, the functions in both Hawes and Wilcannia cater for two other groups of people: those who live within reach of it, and those who are passing through. In Hawes, tourists are responsible for a large proportion of purchases made within the town, whereas in Wilcannia, passing lorries and cars form a bulk of the trade. Places that might under any other circumstances be known as 'villages' become important as towns where they are regional centres.

Defining by an index of rurality

Previous definitions have shown that it is not possible to think of rural and urban as absolutes. It may help, therefore, to think of them as a graduation between one and another. Figure 5 shows England and Wales according to five graduations of rural and urban produced in 1977 by P Cloke, a geographer, who devised a scale known as the **Cloke index**. He tried to measure the extent to which an area could be said to be urban or rural. He took a number of characteristics of rural areas and used census data of rural indicators. These are shown in Figure 6.

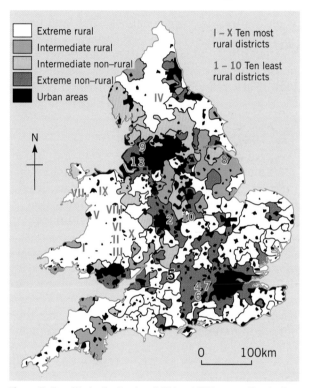

Figure 5 Rurality in England and Wales 1971, according to the Cloke scale of extreme rural to extreme urban.

Female population aged 15 – 45
Occupational structure
Population change 1961 – 71
Commuting-out pattern, i.e. number of people
   working outside the settlement
Distance from an urban centre of 50 000 people
   or more
Population over 65
Household amenities
Population density
Inward-migration over the previous 5 years

Figure 6 Indicators of rural life used by Cloke to calculate the index for the map in Figure 5.

161

Defining by land use

The Countryside Agency defines rural land use as 'agriculture, forestry, open land and water, minerals and landfill, outdoor recreation and defence. All housing development is classified as 'urban'. Refer back to the photographs in Figure 1 and consider the extent to which this definition may be helpful.

The rural – urban continuum

It is perhaps more helpful to see 'rural' and 'urban' as different ends of a continuum. It is sometimes difficult to define where urban areas begin or end. Rural lifestyles are changing. Mechanisation of farming has meant that large numbers of agricultural jobs have been lost in the UK since 1945. At the same time, urban dwellers have moved into villages within easy commuting distance of the city. Some rural places are dominated by tourists and second-home owners from cities. Many villages have therefore lost their agricultural function and have become housing estates – or dormitory suburbs – for cities, giving them an urban function. There is therefore a **rural-urban continuum**.

Figure 7 Two views showing the process of urbanisation on the urban 'fringe' of Perth, Western Australia.

Photograph A shows a sign marking out new development.

Photograph B shows new building on the 'rururban fringe'.

On the continuum, cities differ between the core and the outer-edge where they are sometimes known as the **'rururban fringe'** – the vague boundary between countryside and city. Figure 7 shows how urban areas expand and absorb what was once open countryside by a process of urban spread. Towns and villages surrounding it become suburbanised and adopt urban functions, such as homes for those who commute into the city each day, or use it for shopping, schools, health and leisure facilities. Beyond the fringe lies a more sparsely populated rural area. People here may still use and depend upon the city, but on a more infrequent and irregular basis. Now, many countries are concerned to protect the rural edges of cities. The transition between rural and urban always makes rural areas more urban, never the other way around.

Urbanisation and counter-urbanisation

In the UK, many villages have reversed a long-term population decline since about 1970. Until then, loss of opportunities and employment in rural areas led to declining populations. Grisedale, shown in Figure 1 (Photograph B), had a population of sixteen families and working farms in 1881. In 1981, one last farmer was left. Now, a reversal of fortunes has taken place. Purchase of barns and former farmhouses has led to re-population of the dale, this time by urban dwellers looking for holiday or second homes, and, in a few cases, by retired former urban dwellers. A process of **counter-urbanisation** has occurred, reversing the drift to the cities. It results in an increase in the number of people who live in the rural area, due largely to urban to rural migration. This is discussed more fully in Chapter 20.

On the other hand, the movement of people from rural to urban areas is commonly found across the world.

- It occurs in the UK, so that a teenager in Cornwall is likely to move away at the age of 18 to go to University or to train as a nurse, for instance; for the foreseeable future, Cornwall has no university nor teaching hospital. Younger people often try to move away, especially for further education. Figure 8 shows the views of one student who grew up near Skegness in Lincolnshire and is now studying at Brunel University in west London.

- It occurs in LEDCs, so that the drift of people into cities such as Bangkok or Sao Paulo is constant and adds several tens of thousands to city populations each year. This movement, which results in increasing proportions of people living in cities, is known as urbanisation. Urbanisation is discussed in the final section of this book.

Living in rural communities

Many aspects of rural life are perceived as very positive. Communities in rural areas support a number of organisations that provide both leisure opportunity and chances for people who live together to mix and know each other. Figure 9 shows the range of clubs and societies available. In addition to these, 45 per cent of rural communities in 1997 had a village newsletter, 74 per cent a church newsletter, and 93 per cent had a public notice board. There are nearly 9000 village halls in rural England, many of which have been enhanced by lottery money since 1994.

The place was like a ghost town. It is a nice place, with low crime rates and a good atmosphere in winter, and it is a nice place when you are young, but when you become a teenager there is very little to do. In the summer there are many tourists who come to the town, causing problems such as an increase in crime rate, traffic jams and a lack of food in the supermarkets nearly all the time. The only advantage of the tourists to me was the fact that I had a seasonal job where I earned quite a bit of money, but Skegness has few advantages because it is very isolated. The nearest decent shopping centre is one hour away in Lincoln, and the roads to Lincoln are poor.

I came to London basically because I wanted to live in London for part of my life. There is so much to do and so many people to meet in London. I thought that now, as a student, would be the best time to move into London, because it is a very difficult area to move into when you are older due to high house prices and jobs, but I knew that if I didn't like the lifestyle then I could move out of the area easily. I just had to try it and I'm glad I did. I love it – the diversity of lives and cultures all mixed together is excellent.

Phil Ringsell, student at Brunel University

Figure 8 Images of growing up in Lincolnshire, close to Skegness.

The image of rural areas among young people, especially those in urban areas, is that they offer very little. In fact, the reverse is often the case. Young people in rural areas are often allowed more freedom, as parents perceive greater safety in rural areas compared to urban. Figure 10 shows the lifestyle of a family in Gnosall, Staffordshire, in 1999. Far from being isolated, the range of activities in rural areas is often great, and most families interact with their local towns.

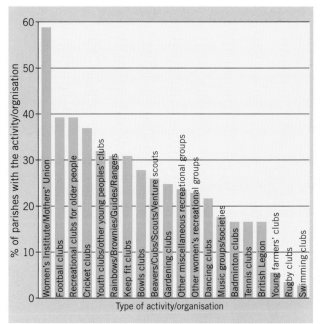

Figure 9 Community activities in rural localities in the UK, 1997.

Mike's cub camp went off successfully and he now has his camping warrant. Rhian got me to dye her hair this morning, prior to going shopping for school disco gear at Telford.

Sian has been going shopping with mates & to McDonald's & skating! Tonight is a swimming gala, then off walking tomorrow with Scouts. She went to see Les Miserables on Wednesday at Birmingham, played netball for the school on Tuesday & cross country for them on Wednesday. I can't keep up with this pace!

Figure 10 Life in rural areas – the Loveless family in Gnosall, Staffordshire.

References for further study

Countryside Agency, *The State of the countryside Report 2000*, available on the Countryside Agency website http://www.countryside.gov.uk

13 Rural change in Ashwell

This chapter is about rural change and the issues that arise out of change. It focuses upon Ashwell, a village in northern Hertfordshire, about 40 miles north of central London. During the twentieth century, many villages such as Ashwell, both in the UK and in other MEDCs, have seen considerable changes. Most of these are part of broad changes in society or in the economy, but are felt locally. This chapter shows how changes in the economy and employment, lifestyle and national politics have all made their mark on people's lives.

Ashwell

Figure 13.1 shows north Hertfordshire, near the county boundary with Cambridgeshire. It shows Ashwell in relation to other towns such as Baldock and Royston, and transport routes such as the A1(M) motorway and the A10 Old North Road. Most of the roads leading to the village are narrow. The surrounding rural area is largely farmland. According to the 1991 census, 1629 people lived in Ashwell and the population had risen 23 per cent since 1971. It continued to rise during the 1990s. Many people in the village are concerned about its changing character as demand for housing continues. What kind of village might it be in future?

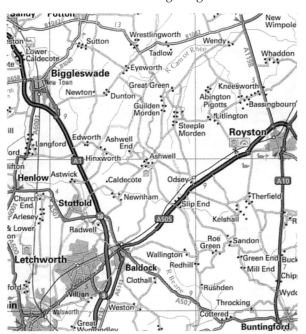

Figure 13.1 The location of Ashwell, Hertfordshire.

1901	1911	1921	1931	1951	1961	1971	1981	1991
1281	1284	1163	1132	1258	1336	1360	1612	1629

Figure 13.2 Population of Ashwell during the twentieth century. No census was taken in 1941.

The changing village

Read Figure 13.3, extracts from the Ashwell Year Book, an annual publication about the village, including Parish and County Council reports. Concerns focus upon the rapid growth of housing in Ashwell and the effects on the village community.

One viewpoint, Extract 2, concerns the local aquifer, a water-bearing rock. The local rock type is chalk, which holds reserves of water underground that are pumped to supply most of the village. The reserves of water are shown in Figure 13.4. The impact of growing demand of housing and water supply is one of many concerns expressed by people living in the village. Development in Ashwell may not be 'sustainable' for much longer; village life may not be able to continue without threatening future water supplies altogether. Read more about sustainability in Chapter 6, page 82.

The issues that Ashwell faces are not just about resources such as water. Much attention has been focused upon the character of Ashwell and ways in which this is changing. What is the 'character' of a settlement? How can it be assessed?

The changing character of Ashwell

While a stereotyped view of an English village might be of thatched cottages (of which there are relatively few) and a close community, most places depend upon the balance of features. Together, these help add to the character of a place. 'Character' includes environmental, economic and social characteristics, rather than just appearances, for example:
- ages and styles of the buildings
- the form or shape of the settlement
- people and housing in the village
- local economy and employment
- transport and traffic
- landscape, wildlife and open spaces.

Extract 1
Many are basically happy with the village as it stands, but a lot of attention is being given to its future, its character and size, given the pressure being exerted to build many more houses in North Herts.

Extract 2
Low level of water in the springs: Mrs Reddaway said that the level was now completely dependent upon being constantly recharged from the … pumping station, but was still very low. When the … pump fails, the level drops even further. The aquifer was still in a poor state.

Extract 3 *From Minutes of the Chairman's report to the Ashwell Annual Parish Meeting, 23 March 1999.*
He … explained that the Village Character Workshop was to be held this Saturday … when parishioners will have their chance to say what they like about Ashwell. This can then be put forward as a … guide to builders and the Planning Authorities, rather than having to react, sometimes angrily, to plans submitted, which has been the position in the past.

Figure 13.3 Extracts from the Ashwell Year Book 1999, showing some of the concerns of people in the village.

Extract 4 *From Minutes of the Ashwell Annual Parish Meeting, 23 March 1999.*
Mrs Cooke, parishioner, asked where the figure of an increase in dwellings of 15% had come from. Mr Short explained that this took the number of dwellings up to 950. This had seemed a reasonable increase, but allowed extra population into the village. … Over the last few years, the rate of development in the village had increased considerably. … (but) a 15% increase in dwellings would not necessarily mean a 15% increase in the number of people living in the village. …the increase was required to support the services such as the shops and the school.

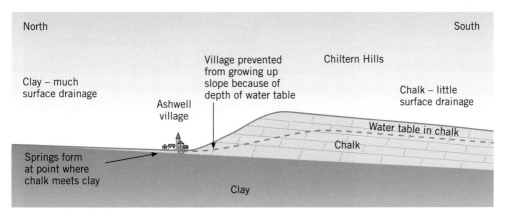

Figure 13.4 Cross-section through the local geology to show the water-bearing rock strata beneath Ashwell. Rocks that hold water in this way are referred to as 'aquifers'.

Issues in Ashwell

1 Draw a sketch map to show the location of Ashwell in relation to other settlements. Use Figure 13.1 and an atlas of south-east England to help you. Show the road and rail links with other places, including London.
2 Classify each of the concerns expressed in Figure 13.3 according to whether they are social, economic or environmental.
3 Read Figure 13.3. Annotate your map to show the social, economic and environmental concerns facing Ashwell during 1999.
4 Why should there be, in Figure 13.3:
   • pressure to build many more houses in North Herts (Extract 1)?
   • concern about water supply from local springs and aquifers (Extract 2)?
   • situations where local people 'react, sometimes angrily, to plans submitted' for new development (Extract 3)?
5 Why is it that 'a 15% increase in dwellings would not necessarily mean a 15% increase in the number of people living in the village' (Extract 4)?
6 Why should there be concerns about the survival of the shops and the school in Ashwell (Extract 4)?

The growth of Ashwell

Figure 13.6 shows a map of Ashwell in 1903. Although long settled by then, clues about the importance of water can be seen. The surrounding area, shown in Figure 13.1, is part of the Chiltern Hills, a range of chalk uplands that run north-east to south-west through this part of eastern England. Chalk is porous and allows the passage of water through it, reducing surface water. Points where water can be extracted are few and are often restricted to springs rising wherever the water table reaches the surface. For many years Ashwell's main water was from the spring shown in Figure 13.5. This is the source of the River Rhee, which later becomes the Cam. It was not the only source of water; below ground surface, water could be obtained by wells at shallow depth (shown as W in Figure 13.6) or later, by means of mechanical pumps (shown as P in Figure 13.6). Note the distribution of wells and pumps in different lines or axes through the village. The maximum depth to which a pump could draw water was 32 feet, so until mains water came to the village in 1910, the growth of the village was restricted.

How and why did Ashwell grow?

Ashwell developed into a market town during Anglo-Saxon times. Domesday Book in 1086 showed that it was one of five boroughs in Hertfordshire, along with St Albans, Stansted Abbots, Hertford and Berkhamsted. Notice in Figure 13.1 how many local roads lead to Ashwell. These roads are very old, dating back to the Middle Ages and before. Ashwell was a market town, attracting farmers and traders from surrounding villages, who used these routes.

Now, Ashwell is dwarfed by neighbouring towns such as Baldock, Letchworth and Royston. By the late thirteenth century these other market towns had developed, taking trade from Ashwell. Although its population continued to grow until the late nineteenth century, its importance as a market declined, and many people in Ashwell relied on agriculture for an income.

The railway came to Ashwell in 1850 and brought economic growth with it, but its importance relative to other settlements gradually declined.

Figure 13.5 The springs in Ashwell. Here the water table is close to the surface. This is the source of the River Rhee, a tributary of the River Cam, which later flows through Cambridge.

Building ages and styles

Ashwell is probably very old. Most of what can be seen now dates from the fourteenth century or later. Few buildings have survived from before then, other than the church, because building materials and styles were strong enough to withstand the time. It is difficult to imagine many twentieth century houses still remaining in the year 2600!

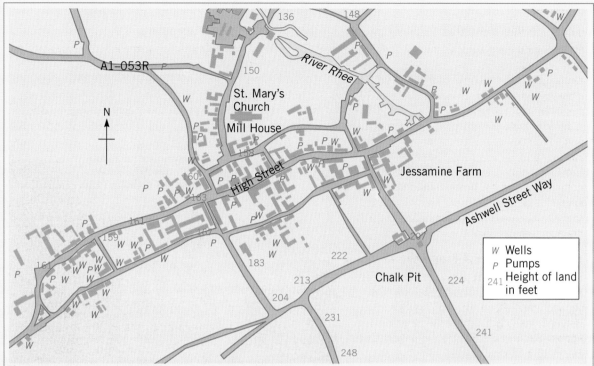

Figure 13.6 Ashwell in 1903. Note the distribution of wells (W on the map) and pumps (P on the map).

However, its decline in population has slowed and reversed during the late twentieth century. Electrification of the railway – which links London's Kings Cross station with Cambridge and East Anglia – now means that the village is 40 minutes away from central London. For commuters, this puts Ashwell within easy reach. At a time of economic growth and when houses in rural areas such as Ashwell are cheaper than those in London, it is tempting for many people to move from the city into villages and commute to work.

Building styles contribute a great deal to the character of a place, since they are all about appearance. Older settlements differ from more recent ones. As well as appearance, building age and architectural style also offer clues about when a place grew and why. The mixture of styles in Ashwell shows how it has developed over a long period. Recent houses can be dated easily because they were carefully planned and permission to build was given by local councils. Ornance Survey maps were first published in 1846 and successive maps may show when housing and other developments have taken place. Older buildings are more difficult to date, because there are rarely written records. The Techniques box below shows how different architectural styles can be used to date buildings and show how a settlement has grown.

Techniques

Using architecture to trace the growth of a settlement

Villages and towns derive their character from their variety of ages and styles of buildings. Most evidence of Ashwell's early growth has been lost but some buildings have survived from the fourteenth century, though often in altered form. It is possible to date when each building was first constructed, using clues from the architecture of the past. Everything about a house, from its roof tiles and chimney to the door, is a clue about its origin. Figure 13.8 shows the clues that can be used to trace the growth of a settlement, using building characteristics. They are indicators for south-east England, which may be helpful for some other parts of Britain. Other areas, such as the central Pennines or Cornwall, have their own building characteristics. It is possible to find local books which will enable you to identify characteristics for your own area.

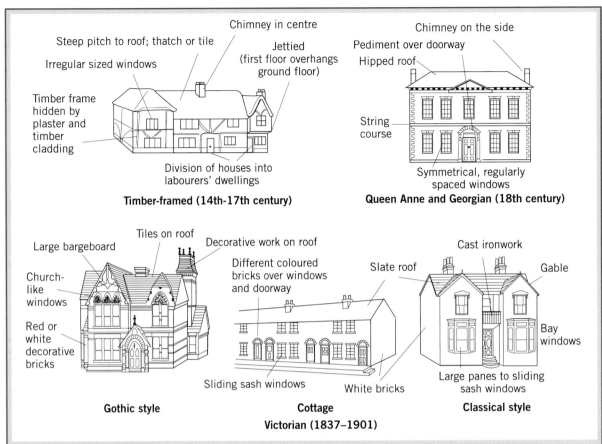

Figure 13.7 Using different 'clues' in Ashwell to determine the age of a building. The clues include chimneys, windows, doorways, shape and overall design.

Figure 13.8 Housing styles in Ashwell from different historical periods.

A timber-frame house, which is a clue about its origin, probably fourteenth-fifteenth century.

An eighteenth century Georgian house, typically symmetrical in design.

In photograph A, the early fourteenth century origins are illustrated by the timber frame of the house. In an area where stone is not especially good building quality, timber was a rich resource. Most early houses open directly on to the street, a feature that has changed in the twentieth century as motor traffic made streets noisier and less safe. As timber supplies dwindled, other materials took over.

Photograph B shows a Georgian house from the late eighteenth or early nineteenth century, typified by the door, positioned centrally to the whole building, and windows symmetrically placed either side. Usually, individual window panes are small, reflecting the cost of glass at the time.

A Victorian house, with gables and large sash windows.

Victorian houses, built from 1837-1901, shown in Photograph C, have much larger panes of glass, reflecting the new industrial methods of producing cheaper glassware. Houses have chimneys and are gabled, usually with slate roofs. The railways in the nineteenth century brought cheap slate to most parts of Britain. Back yards or very small front gardens were most common.

A modern twentieth century house.

The rising cost of land and fuel for heating, and smaller family sizes, meant that twentieth century houses, like those in Photograph D, are usually smaller, set back from roads and often built away from main commercial areas and main roads. Some are built on land that used to form gardens of older properties, so that the amount of open space decreases and the density of housing increases.

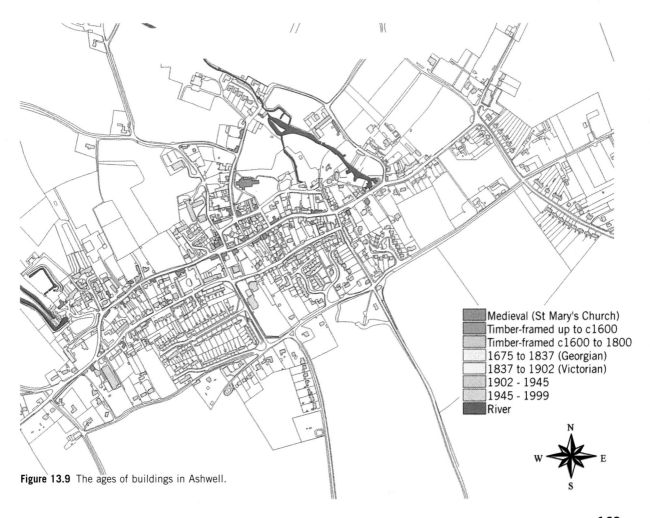

Legend:
- Medieval (St Mary's Church)
- Timber-framed up to c1600
- Timber-framed c1600 to 1800
- 1675 to 1837 (Georgian)
- 1837 to 1902 (Victorian)
- 1902 - 1945
- 1945 - 1999
- River

Figure 13.9 The ages of buildings in Ashwell.

Settlement pattern and form

Figure 13.9 shows a map of the age of properties in Ashwell. Each has been dated according to style and map records, to show the pattern of growth in the village. Each stage of growth takes place in certain ways.

Patterns of settlement growth

Geographers have developed ways of showing how settlements grow and change over time. The diagrams in Figure 13.10 show how settlements may originate and then grow in different patterns.

Diagram A shows a nuclear pattern of growth, which usually occurs around a church or central point, such as a spring. The church was an essential part of this process. Early land holdings rented from the church required farmers to pay a 'tithe' or a tenth of all their produce. The vicar or bishop was rarely able to consume all of this, so a market would develop for the sale of surplus produce. Diagram B shows a linear pattern of growth along a line such as a road. It is possible to see this in Ashwell in Figure 13.10, where houses in the east of the village grew out along the road towards the station in the early twentieth century. Diagram C shows dispersed growth, spread over a wide area. Isolated farm houses sometimes grew away from the village, because to travel out to their landholding from the village would take up too much time.

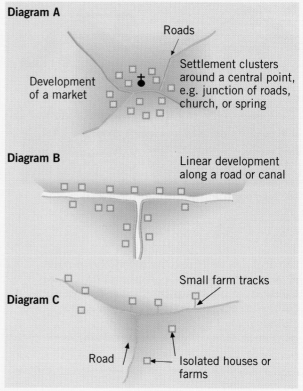

Figure 13.10 Different ways in which villages may originate and grow.

---

1 Suggest as many reasons as you can why, in Figure 13.10, settlements may grow:
- around a central point
- along a line
- in a dispersed form.

2 Which pattern appears to apply to Ashwell in:
  a) the fourteenth to sixteenth centuries
  b) eighteenth and nineteenth centuries
  c) early twentieth century,
  d) growth since 1945?

3 Based on your answers to question 2, is it possible to devise a model of growth for Ashwell? Refer to the box on Urban models in Chapter 17, page 291–292.

4 Use Figure 13.9 to compare Ashwell with theoretical models of settlement growth shown in the Theory box.

Recent housing development in Ashwell

The variety of buildings helps to contribute to Ashwell's character. If demand for housing increases, pressure is put upon space in villages. New buildings may add to or take away some of the character of a place. Since 1971, demand for housing in Ashwell has gradually increased for a number of reasons.

- There is a demand for rural housing from people who wish to move out of cities such as London and Cambridge. Salaries in London are higher than the national average, as shown in Figure 13.12, and many people choose to commute, preferring a rural lifestyle.
- Ashwell has retained most of its village services and functions, so that many people are able to enjoy a high level of amenities, even though they are several miles from a large town.
- Changing family size and an increase in the number of divorces and break-ups has led to a reduction in the average number of people per household. Where 100 people might have been housed in 25 houses in 1970, those same people would now need about 35 houses in which to live.

As a result, there has been substantial change in buildings in Ashwell since 1980. Two kinds of development have taken place:

- Several conversions of older traditional agri-cultural buildings have taken place, such as barns into housing. These are in demand and are expensive.
- New housing has been built, as shown in the key to Figure 13.9. Modern buildings are often cheaper to build than barn conversions, are less individual, less sought after, and therefore cost less than traditional buildings.

Some people in Ashwell argue that the character of the village has been affected negatively by additions of recent housing. Opinions depend upon the kind of development that has taken place. Two developments are shown in Figure 13.11, one, a conversion of a former barn in Ashwell and the second, a 1980s housing development. Recent building has occurred partly by the addition of new buildings on the perimeter of the village, so that it expands in size, and partly by 'in-fill', whereby gaps between older, lower-density houses are used for building. This increases the density of the village core.

People and housing in Ashwell

Ashwell changed considerably in the late twentieth century. Three key issues have affected Ashwell's housing and population in recent years:

- the ageing structure of its population
- the socio-economic character of people in the village
- the increase of commuting, mainly to London.

Figure 13.11 Two developments in Ashwell.
Photograph A shows a conversion of a former mill in Ashwell, creating an individual and expensive house. Photograph B shows a housing development from the 1980s

An ageing population

Ashwell reflected many changes which occurred in the UK as a whole. During the twentieth century, life expectation in the UK has increased from an average of approximately 45 for men and 49 for women who were born in 1900, to 77 for men and 82 for women who were born in the late 1990s. There were over 2 million people over the age of 80 in the UK in 1999. By 2041, this number will have increased to over 5 million.

As a result, Ashwell's residents have changed in age and socio-economic class. One fifth of the village residents are now retired. Statistically, they are better off than most of the rest of the UK. Many of them own a house; they have seen its value grow hugely during the last 30 years. Houses that were worth £10 000 in 1970 may now be worth as much as £250 000. There is a shortage of supply both in south-east England and in Ashwell of older, period-style properties that are unique in style and that have been well maintained. The house in Photograph A in Figure 13.12 is typical. Many retired people are able to sell a large

family house and move to somewhere smaller, thus releasing much of the value in their house when they move.

Socio-economic class

While one section of the community in Ashwell is ageing, another part is also changing. Instead of a rural economy in which most people work in agriculture, a large proportion of Ashwell's working population are now employed outside the village. In addition, there has been a substantial influx of people from London who prefer commuting to living in the city.

Ashwell is also part of the economically prosperous south-east of England which has enjoyed sustained growth. It now has first place on almost all economic indicators in the UK. Figure 13.13 shows an extract from a report by the Halifax Bank in October 1999, which describes economic conditions in the region. A combination of low interest rates and rising incomes now means that people in south-east England spend 23 per cent of income on housing compared to 41 per cent in 1990. Housing has become affordable and a sustained boom in demand for property in villages such as Ashwell has been the result. People in Ashwell, as well as elsewhere in the south-east, have higher incomes and higher measurable standards of living than in other regions of the UK, as shown by the indicators in Figure 13.15.

Figure 13.12 House prices in Ashwell October 1999. These two houses were on sale at £385 000 (Photograph A, a 3-bedroomed thatched house) and £89 950 (Photograph B, a 2-bedroomed flat in a converted block). The flat in Photograph B was the lowest-priced property on sale in Ashwell in October 1999.

The South East has been one of the top performing regions throughout the 1990s. Between 1995 and 1997 real GDP growth averaged 3.8% per annum, compared to 2.8% nationally, with only the South West producing a higher rate. Over the same period, the South East's share of UK GDP rose from 15.5% to 15.9%, leaving GDP per capita 18% above the UK level. The region has seen a further relatively strong performance in 1998 and 1999, with GDP estimated to have risen by more than the national average.

Employment in the South East has risen by 8% over the past three years, although the rate of increase has slowed sharply with a gain of only 0.4% in the past 12 months. The construction sector has seen the fastest increase in employment over the last year with a gain of 2%. Manufacturing employment has continued its downward trend with a fall of 1.6%, while service sector employment has risen by a similar percentage over the period.

The seasonally adjusted unemployment rate in the South East fell to 2.4% in August 1999. The South East has the lowest rate in the country, with a 2% lower rate than the rest of the UK.

Figure 13.13 From the Halifax PLC quarterly housing survey, October 1999.

Part of the reason for increasing prosperity is Ashwell's proximity to London. Figure 13.14 shows average earnings in south-east England compared to the UK. Notice how London's average is higher still; people who have higher salaries are able to afford high housing and travel costs and may choose to commute to work. This is leading to a process known as counter-urbanisation, explained in the theory box on page 178.

Counter-urbanisation

While job opportunities in agriculture have declined as a result of mechanisation, resulting in many people leaving the land, a different process has been going on whereby people from cities have left to live in rural areas. During the late twentieth century, the process of urbanisation has reversed. Cities, such as London and Manchester in the UK, New York City and Chicago in the USA, and Paris, Milan, and Vienna in western Europe, have all shown declining populations. London lost nearly one in five of its population between 1971 and 1991.

During the mid-to-late twentieth century, people moved to live in villages such as Ashwell which are within easy reach of large cities. The rapid increase of settlements on the edge of cities – such as Garforth near Leeds, Ponteland near Newcastle-upon-Tyne – has created major changes within the settlements themselves and the population living there. This process reverses the one in which people move to cities, and is known as counter-urbanisation.

The results of counter-urbanisation are considerable. Some are beneficial, while others cause alarm.
* Villages alter in character as they expand in size. The addition of newer housing estates changes the appearance of the village.
* Communities alter. In-migration has been common in many rural areas as a result of government development grants to mid-Wales, for example. Many former villages now have their own small rural industrial bases, though with small firms dominating.
* The number of shops and services ceases to decline, at least, and may actually increase, as in the case of Ashwell. Many village schools have been saved by the migration into the village of young families. Village services alter; families may shop in supermarkets, which results in closure of the village store, while other services such as restaurants and antique shops may open instead.
* The social and occupational structure is likely to alter. The two most significant groups likely to migrate into villages are retired people and professional families. Both represent better-off socio-economic groups, replacing those from labouring families who are most likely to leave. Instead of a settlement dependent upon farming, a mix of social groups develops.

Figure 13.14 Indicators of prosperity in south-east England in 1999. The south-east was the most prosperous of all 11 UK economic regions in 1999. The national average is shown for each set of data.

Region	Professional / Managerial	Skilled	Semi-skilled	Unskilled
South-East	39.1	40.7	12.3	4.8
UK average	33.6	42.6	14.8	5.7
				Source: HMSO, *Regional Trends* 1999

Figure 13.14a Socio-economic class in south-east England, 1997-98. This table shows the classification of people into different socio-economic groups by occupation. Because it is by occupation, it does not include unemployed people or the armed forces.

Region	% of households who earn		Gross weekly income	
	Less than £150 per week	£600 per week or more	Per household	Per person
South-East	19%	30%	£474	£207
UK average	25%	20%	£408	£171
				Source: HMSO, *Regional Trends* 1999

Figure 13.14b Household income in the UK in 1993.

Region	Dish-washer	Deep freezer	Telephone	CD player	Video	Home computer	Stocks and shares
South-East	31	92	97	69	85	37	37
UK average	22	91	95	62	82	28	29
							Source: HMSO, *Regional Trends* 1999

Figure 13.14c The percentages of households owning certain goods.

Commuting to London

Commuting to London has increased since the late 1980s, when the rail link between London and Cambridge was electrified and journey times improved. Now Ashwell is about 40 minutes from Kings Cross in central London. Based on a 10% sample, Hertfordshire County Council found that 37 per cent of Ashwell's working population commute to London or Stevenage each day.

The impact on local transport is considerable, as

this extract from the Ashwell Parish Council Think Tank showed in 1998.

'There are 2-4 bikes (6-10 in summer) in the station bike park every weekday compared with around 50 cars (about 20 in the station car park and 30 on the surrounding road verges). In addition, about 20-30 people are taken to and from the station at peak hours in the Ashwell taxibus. Another group of about 10-20 are taken or collected by private car to meet each peak hour train.'

For many people, it is a balance of deciding whether to pay extra housing costs in London and be closer to work, or to have a rural environment in which to live and travel from each day. Figure 13.15 shows the cost of getting to London and Cambridge from Ashwell, using an annual season ticket. Remember that all costs normally have to be paid for by the employee, so that an annual Standard class season ticket of £2744 is money earned after tax and National Insurance. Someone paying this amount would therefore have to earn about £4500 just to be able to pay this! On the other hand, housing in London is more expensive than in Ashwell. There are costs and benefits on each side.

Rail Fare from Ashwell to	First Class, per year	Second class, per year
London	£4116	£2744
Cambridge	£1400	£932

[Source: Railtrack, October 1999]

Figure 13.15 The cost of annual rail season tickets in October 1999.

---

1 Draw up a table showing advantages and disadvantages of living in **a)** Ashwell and **b)** London.
2 Which has the greater advantage for you?
3 To what extent might the balance sheet vary according to:
  • your age?
  • your marital or family status (single or married, with or without children, etc.)?
4 Would you prefer to live in London or Ashwell? Justify your answer.

Does everyone benefit?

While prosperity in the village has been increasing for some of the elderly who own their own property and for families who have moved out of London, for others the story is different. As in all situations in which prosperity occurs for some, other people are excluded. There is an increasing gap in Ashwell between those who can afford the rising cost of housing and commuting, and those on low wages, such as farmworkers, or on part-time wages or youth and training schemes who find themselves unable to afford housing. Increasingly, the young and low income earners are excluded.

Even those in professional occupations such as teaching are unlikely to be able to afford housing if they are first-time buyers. Two teachers early in their careers may earn £38 000 – £40 000 between them, which might allow them to borrow about £80 000. Unless they have some savings, Ashwell's cheapest property of almost £90 000 in 1999 will not be affordable.

As a result, there is now pressure in the village for low-cost accommodation. One complex for the

Figure 13.16 Sheltered housing complex for the elderly in Ashwell, 1999.

elderly already exists, shown in Figure 13.16. However, there is an increasing need for smaller properties for those whose families have moved home, and for warden-managed accommodation. There is also a need for low-cost housing for young families and low wage-earners, such as through a Housing Association.

The local economy and employment

Figure 13.18 shows a map of Ashwell in 1924; all the shops and services are marked on the map. Most people worked in agriculture, or in a service which was closely related to it. Until the late 1960s, the majority of workers in Ashwell were either agricultural or worked within the local brewery, which has since closed. Mechanisation in farming has also reduced work opportunities.

People in Ashwell now work in a variety of employment types. Many still work and live in the village. It means that they at least – unlike some commuters – have short journeys to work. Employment is now much more flexible than in 1980 and more people work part-time – both men and women – and the number of people working at home has increased. At a time when people in the UK work longer hours than any other nationality in the EU, and spend longer periods travelling to work, this may mean that some people in Ashwell have a high quality of life, assuming that they earn enough.

Retail shops and suppliers	Crafts	Public houses and restaurants
Baker (1) Butcher (1) Cattery (1) Chemist (1) Craft Shop (1) Dairy (1) Delicatessan (1) Estate Agents (1) General Stores (1) Hair Stylists (4) Kennels (1) Calor Gas supplier, Caravan site and Storage (1) Newspapers (1) Stationery, Confectionery and Garden Requisites (1) Take Away (1) Bank (1)	Basketry/Cane seating (1) Florist (1) Furniture (1) Greetings cards (by hand) and animal photography (1) Pottery (1) Textiles (1) Quiltmaker/ teacher (1) Woodwork (1)	The Bushel and Strike (1) The Jester Hotel (1) The Rose and Crown (1) The Three Tuns, Hotel (1)
		Services
		Dentist (1) Doctors (1) Library (1) Post Office (1) Primary School (1) Taxi and Taxibus (1)

Figure 13.17 Shops and services in Ashwell in 1999. Compare this with the list shown in Figure 13.20 from 1924.

Businesses in the village

Figure 13.19 shows the small businesses in Ashwell, published in the Ashwell Year Book for 1999. Compare it with Figure 13.20, which shows a similar list from Kelly's Directory in 1933.

A range of employment exists in Ashwell. A new complex has been developed from a former range of farm buildings in the village at Dixie's Farm, shown in Figure 13.21. Now, 63 per cent of Ashwell's working population work within the village. Out of a population of about 1700 in approximately 800 households, 30 people work in agriculture, 100 are self-employed, and some come in from outside the village to work in Ashwell. The largest employers in the village include J Cooke Engineering, which employs 35 people, manufacturing metal nuts in different metals for customers in the UK and Europe. The bakery employs about 30 (though several of these are part-time), the surgery about 25 (similarly), and the Field Centre and Community Centre about 6. The pubs in Ashwell employ several people between them, both full- and part-time. The remainder work outside the village in London, Stevenage, Cambridge and local towns of Baldock, Royston, or Letchworth. One person works as far away as Leicestershire.

---

1 Design a classification for the shops and services shown in Ashwell in 1924 (Figure 13.18). Classify each of the shops and services using this system.

2 Using the same classification:
   a) classify the shops and services in 1999, using Figure 13.17 and 13.18
   b) contrast the shops and services in the village in 1999 with those in 1924.

3 Which shops seem absent now compared to 1924? Why should this be?

4 Explain any other differences you have noticed between 1924 and 1999.

5 Do any of the shops or services in Ashwell in 1999 surprise you? Why should this be?

6 How far were the shops and services in 1924 based on local need? To what extent does this seem to be true now?

The range of services in Ashwell

By contrast to Figure 13.18, Figure 13.17 shows a full list of shops and services in 1999. There were actually more services in 1999 than there were in 1924, but they have changed significantly. People rely on services elsewhere, for example, supermarkets in Baldock and Royston (Tesco), Letchworth and Biggleswade (Sainsbury's) as well as those in the village.

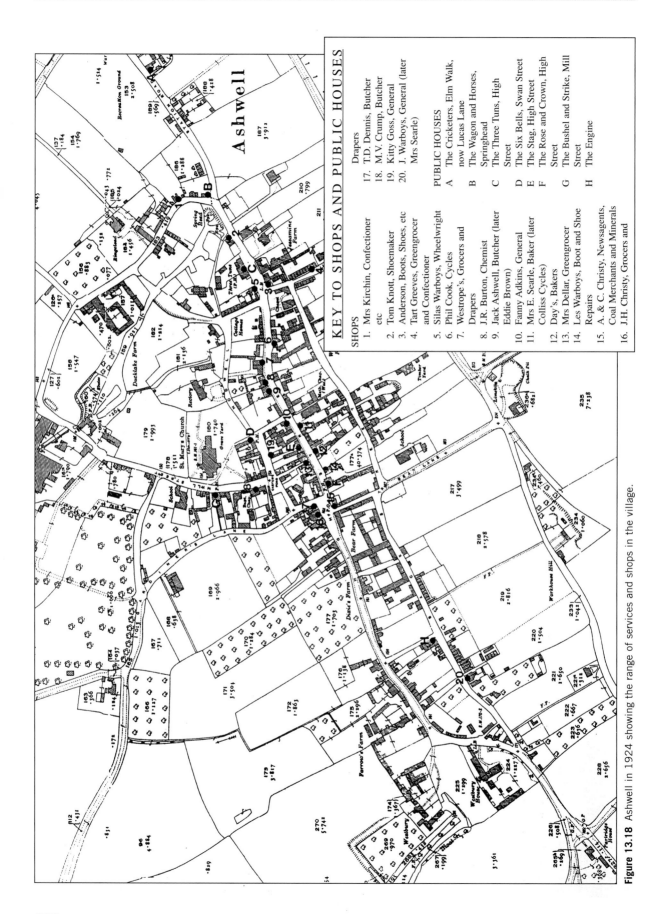

KEY TO SHOPS AND PUBLIC HOUSES

SHOPS
1. Mrs Kirchin, Confectioner etc
2. Tom Knott, Shoemaker
3. Anderson, Boots, Shoes, etc
4. Tart Greeves, Greengrocer and Confectioner
5. Silas Warboys, Wheelwright
6. Phil Cook, Cycles
7. Westrope's, Grocers and Drapers
8. J.R. Burton, Chemist
9. Jack Ashwell, Butcher (later Eddie Brown)
10. Fanny Adkins, General
11. Mrs E. Searle, Baker (later Colliss Cycles)
12. Day's, Bakers
13. Mrs Dellar, Greengrocer
14. Les Warboys, Boot and Shoe Repairs
15. A. & C. Christy, Newsagents, Coal Merchants and Minerals
16. J.H. Christy, Grocers and

Drapers
17. T.D. Dennis, Butcher
18. M.V. Crump, Butcher
19. Kitty Goss, General
20. J. Warboys, General (later Mrs Searle)

PUBLIC HOUSES
A   The Cricketers, Elm Walk, now Lucas Lane
B   The Wagon and Horses, Springhead
C   The Three 'Tuns, High Street
D   The Six Bells, Swan Street
E   The Stag, High Street
F   The Rose and Crown, High Street
G   The Bushel and Strike, Mill Street
H   The Engine

Figure 13.18 Ashwell in 1924 showing the range of services and shops in the village.

Service industries		
Acupuncture (1)	Electrician (2)	Power tools repairs (1)
Agricultural machine services (1)	Export consultants (1)	Professional associations (2)
Builders (4)	Financial services (1)	Publishing/Print preparation (2)
Business services (8)	Heating oils (1)	Reflexology (1)
Car repairs, haulage (2)	Home service centre (1)	Sports clothing (1)
Carpenter (2)	Industrial cleaning services (1)	Structural engineer (1)
Catering services (1)	Interior design (1)	Surveyor (1)
Child minders – Registered (2)	Leisure (1)	Trees surgeon/Logs (1)
Children's entertainer (1)	Naturopathy (1)	Voice training (1)
Communications, TV, etc. (1)	Osteopathy (2)	**Manufacturing**
Computer services (1)	Painting/Decorating (3)	
Chiropodist (1)	Patent agent (1)	Engineering (1)
Day nursery (1)	Petrol/Car repairs (1)	Trade suppliers (1)
	Photography (1)	[Source: Ashwell Year Book 1999]
	Physiotherapist (1)	

Figure 13.19 Table showing different employers and businesses in Ashwell, 1999.

Commercial.
Early closing day, Thursday.

Adkins Fanny (Miss), draper, High st
Ashwell Saddlery (The) (Mrs Phyllis Fordham, proprss.) High st
Bailey F. J. & Co. Ltd. bldrs. T N 27
Barclays Bank Ltd. (sub-branch to Royston) (hours, Thurs. 10.15 a.m. to 12.30 p.m.), High st.; head office, 54 Lombard st. London EC3
Barnes Jas. Hare & Hounds P. H. Slip end
Bayliss Wm. Heath F.G.S. sec. to Messrs. E. K. & H. Fordham Ltd
Beadle Wm. Simon, beer retlr. West end
Bean Jn. blacksmith, High st
Bennet Jas. Biscoe, fruit grower, The Poplars
Bentley Edwyn, rate collector & clerk to the Parish Council, High st
Bonnert Hy. tailor, High st
British Legion Club (s. Bryant, hon. sec)
Brown Edwd. Arth. & Tyrell, Christphr. Jn. butchers. T N 32
Bryant Emily (Miss), dressma. High st
Bullard Alfd. beer retlr
Burns Wallace Bruce, Rose & Crown P.H. High st
Carter Jn. beer retlr. Station rd
Christy A. & C. coal mers. High st T N 47
Christy J. H. & Son, grocers, High st. T N 20
Collis Chas. Herbt. cycle agt. High st
Crump Mark V. butcher, Mill st. TN 5
Day Chas. & Son, bakers, High st
Dennis Thos. D. butcher, High st T N 36
Dolly Mrs. shopkpr. Swan st
Fordahm E. K. & H. Ltd brewers, maltsers & bottlers of ale & stout, Ashwell brewery (T N 42 Ashwell); (William Heath Bayliss F.G.S. sec)
Fordham S. H. & Co. Ltd. coal mers. Ashwell station
Fordaham Phyliss (Miss), dairy
Gentle Frank, haulage contrctr
Hall Benj. cattle dlr. Park farm. T N 29
Holoway Hy. Geo. fruitr. High st
Keearney Dermot Thos. insur.agt
Ketteridge Alfred, motor propr
Kingsley Ernest William, cashier to Messrs. E. K. & H. Fordham Ltd
Kirchin Arth. shopkpr
Knott Thos. boot mkr. High st
Letchworth, Hitchin & District Cooperative Society Ltd. High st. T N 39
Merchant Taylors' School (G. C. Whitby M.R.S.T., M. Coll. H., head master)
Miller Fredk. Chas. agricltrl. engnr. Kingsland rd
North Metropolitan Electric Power Supply Co. High st. T N 22
Potton David, beer retailer
Prime Edwd. A. motor engnr. High st. T N 31
Rand Wm. Fredk. bldr. High st T N 57
Roper Geo. Alfd. Bushel & Strike P.H. Mill st
Russell Eardley Edwd., M.P.S. chemist, High st. T N 50
Saunderson Ebenzr. corn mer. High st
Three Tuns Hotel (Edwyn Bentley), High st
Thurgood Edwin Jn. cooper, The Green
Tyrrell Christphr. Jn. butcher, see Brown Edwd. Arth. & Tyrrell Christphr. Jn
Wallace Wm. Hamilton, farm bailiff to Mrs. W. A. Fordham J. P. Blue Gates
Westrope Chas. Hall, draper, High st. T N 34
Worboys Leslie, boot & shoe repr. High st

Figure 13.20 Extract from Kelly's Directory 1933, showing employers and services in Ashwell.

Figure 13.21 Dixie's Farm, a conversion of farm buildings into small employment premises in Ashwell. Most of these are small businesses which employ local people.

1 Study Figure 13.19. Which businesses seem to have been omitted from Ashwell's main list of services? Look at figure 13.17 to help you.
2 In pairs, group and classify Ashwell's main employers for 1999, as shown in Figure 13.19.
3 Now compare these with main employers and services in 1933, as shown in Figure 13.20.
4 Between 1900 and 2000 the UK changed from a mainly primary and manufacturing economy to a predominantly service economy. How far does Aswell reflect this?

Transport and traffic

By comparison with many rural communities in the UK, Ashwell fares well for employment, and the fact that it has a railway station enables people to look further afield for employment. However, many rural areas of the UK have poor bus services, both as a result of their remoteness from major population centres and because car ownership in rural areas is high. Although wages for locally employed people may be relatively low in rural areas, a car is perceived as a necessity for work, family and other reasons.

Accessibility to Ashwell by bus suffers as a result. Figure 13.22 shows the bus schedules for Ashwell in 1998. Its route is to Hitchin, via Baldock and Royston. Find these in Figure 13.1. Compared to city provision, public transport by bus is infrequent. However, support and subsidies from Hertfordshire County Council has enabled services to survive. Analysis of Figure 13.22 shows that bus transport is available during priority times such as the starting times of school and full-time work. Other service providers supplement these, such as

Mondays to Fridays Ashwell buses to Baldock, Letchworth and Hitchin at				
0642	0755	0810 (school days)	1009	1346 plus a service to Baldock at 1632
Saturdays Ashwell buses to Baldock, Letchworth and Hitchin at				
0800	1009			1346 plus a service to Baldock at 1632

The same number of services return from Hitchin, Letchworth and Baldock.

Figure 13.22 Bus schedules from Ashwell 1998.

1 Which people benefit from the schedules shown in Figure 13.22? Which people do not?
2 How might provision be built in for people who do not benefit? Who might be responsible for this?

the local taxibus which meets trains in the morning and evening.

In 1998, a Saturday evening bus service began, linking Ashwell with Letchworth, returning at 11.30 p.m. Designed mainly for young people returning home, the bus coincides with cinema and pub closing times in Letchworth. In 1999, it ceased; only one person ever used it. It was supported financially by a local trust. A new bus scheme has been introduced to Baldock, Letchworth and Stevenage, supported by a subsidy from Hertfordshire County Council. Other centres to which young people are likely to travel, such as Stevenage, are served by rail but the 2-mile distance to Ashwell's station along a quiet rural road is not conducive to its use, without support from parents to collect their children from the station late at night.

Landscape, wildlife and open spaces

Hedgerow destruction by farmers has not been a major issue in this part of North Hertfordshire. Attention is being given to conservation of green spaces and 'corridors' along which wildlife can move freely, away from the intensive cultivation of farmland, and over-treatment with herbicides and pesticides. Planting is encouraged in 'stands' of small woodlands that may act as wildlife sanctuaries, rather than lines of 'shelter-belts' between fields, which do little to encourage wildlife. Where hedge planting is encouraged, native species are preferred – ash, wild plum, blackthorn, hawthorn, for example – rather than 'garden' varieties. Native species are usually better producers of edible fruit, rather than garden plants which are usually sold for design and colour.

In the village itself, the issue is similar. Gradual in-filling of open land between houses in the centre of the village has led to the deterioration of green 'corridors' between houses. Gardens and parks are essential for wildlife movement. The Design Statement encourages further planting of trees and extension of trees among new housing developments.

Development beyond the village

Ashwell does not exist in a vacuum, but is affected by events and trends elsewhere. It is not alone in experiencing increased demand for housing. The UK government announced during October and November 1999 that new housing stock is required to provide accommodation for changing housing demand and increasing population in south-east England in the first 20 years of the twentieth century. 4400 new houses are expected to be developed in north Hertfordshire by 2010.

Proposed housing expansion in Stevenage

Stevenage, some 13 miles away from Ashwell, is planned to expand substantially in future, which is likely to impact on Ashwell. Hertfordshire County Council has been forced to adopt a plan to build 10 000 houses in the Green Belt, west of Stevenage. CASE – the Campaign Against Stevenage Expansion – has stated that it will be a car-dependent development away from existing rail and bus networks. The development is also likely to lead to net inward migration into Hertfordshire, as housing will be for new inhabitants, rather than replacing houses for the existing population. The issue has caused hostility between the main political parties on Hertfordshire County Council and the Conservative-run council (1999) opposing it.

Plans for new development are likely to increase as a result of demand for new housing. There is bound to be some 'ripple effect' from new development in Stevenage, as people look for 'character homes' away from large modern estates.

Decision-making: What kind of place should Ashwell be in future?

This exercise is designed to help you understand how decisions might be made in Ashwell concerning the appropriateness of any new development. You should consider some of the issues presented, such as:

- the appearance of new building styles and their suitability. Do new buildings compare well with conversion of traditional buildings? How should Ashwell look in the future?

Figure 13.23 Colbron Close, a recent development in Ashwell.

Ducklake Farm Barns
Ashwell

We are delighted to announce possibly the last chance in Hertfordshire to acquire a unique and substantial barn conversion with the highest specifications in a peaceful and sought after historic village. Three Grade II listed barns will be sympathetically restored and converted to offer stunning accommodation, two of which will enjoy river frontage with views over the village church and countryside beyond.

A rare opportunity indeed to acquire a tailor made house by choosing kitchens and floorings at an early stage. Due for completion in the Spring/Summer 2000.

Tenure: **Freehold**

Figure 13.24 Plans and designs for conversion of barns at Ducklake Farm, just outside the village.

- who the new residents will be. Consider the pressure on housing in Ashwell and the need for low-cost housing and for extra housing for the elderly.
- the impact new developments might have on landscape and wildlife, on the community, or on open spaces.

The Design Checklist shown in Figure 13.25 was drawn up by about 50 residents in Ashwell where they decided what they felt their village should be like in the future. While accepting that local councils would make decisions about what was or what was not appropriate development, they drew up the Design Checklist as guidance for local developers. Using these, you should consider the appropriateness of Colbron Close as a recent development and whether you think the proposal at Ducklake Farm to convert three barns into housing should go ahead.

Notice the housing style, and judge for yourself the extent to which builders and developers have attempted to keep within the guidelines given in Figure 13.30. Consider too whether or not housing styles complement those shown earlier in the chapter on village growth and development, in Figure 13.8.

Design checklist

If you are considering any sort of building work in Ashwell, please take a few moments to look through the following checklist before formulating your plans.

Developers

Will what you are planning:
- Blend in with the landscape?
- Make allowance for local wildlife?
- Enhance the social mix?
- Strengthen Ashwell as a working community?
- Respect the traditional settlement pattern of the village?
- Blend in with the surroundings in terms of scale, density, character and building numbers?
- Provide additional open space?
- Provide new or preserve existing vistas into, out of or within the village?
- Provide pedestrian access and link in with existing footways?
- Include good quality street furniture which is appropriate for its setting?

Architects and designers of new buildings

Please ask yourself the following questions about the building or extension which you are planning:
- Is its position and size in keeping with neighbouring buildings?
- Is the roof height and pitch appropriate for the area and the style of building?
- Are the construction materials in common use in Ashwell and is their colour appropriate?
- Are the types of windows to be used appropriate for the building and the area, and is their size and proportion in keeping with the historic norm?
- Are the parking arrangements sufficiently inconspicuous so that, for example, the garage does not dominate the frontage?
- Will there be sufficient storage to allow the garage to be used for its proper purpose and avoid on-street parking?
- Am I retaining, or improving, all existing hedgerows and native trees?
- Will what I am proposing harmonise with existing buildings and make a positive contribution to the character of the area?

If you are considering altering the exterior of your property, or changing any external detail of the building, its paintwork, signs, garden or surrounds, please undertake your own design assessment by studying each elevation or aspect of the building, and asking yourself the following:
- What are the distinctive features of the property?
- Are there any particular features which are out of character with the building itself or with neighbouring properties or with the Design Statement?
- Is what I am proposing in accordance with the Design Statement?
- Does what I am proposing enhance distinctive features or help remove uncharacteristic ones?
- Will it make a positive contribution to the character of the area?

Figure 13.25 Design checklists for new development in Ashwell. These are guidelines published in the Village Design Plan for Ashwell.

The future of Ashwell

Work in pairs on the following decisions. Present your findings to the rest of your group and compare your decisions with other groups.

1 Study the Design Checklist in Figure 13.25. To what extent do you agree or disagree with the criteria as shown?

2 Are there any other criteria that you think should be included in the Village Design Checklist? Justify your choices.

3 Use the village Design Checklist to decide:
   a) ways in which you feel Colbron Close is in keeping with the character of Ashwell.
   b) ways in which you feel Colbron Close is not in keeping with the character of Ashwell.
   c) whether or not you feel Colbron Close is an appropriate development for Ashwell. Give your reasons.

4 Use the same checklist to decide whether Ducklake Farm represents appropriate development for Ashwell.

5 As a group, discuss to what extent Ashwell is:
   a) *typical* of rural villages that you know of.
   b) *atypical* of villages that you know of.

6 As a class, discuss the following statement: 'If villages are to survive in the UK, we need to make sure that they remain communities for everyone, and not just for the wealthy few.'

Summary

You have learned that:

- Villages date back for many centuries; almost all settlements in Britain existed by the time of the Domesday Book in 1086.
- Clues such as architectural style and maps help to trace the origins and growth patterns of villages.
- Mixtures of architectural styles and building features help to define the 'character' of a place; so too do people and housing, the local economy and employment and features of transport, landscape, wildlife and open spaces.
- Different pressures may threaten to alter the character of a place; these include changes in population, caused by ageing and inward migration.
- The movement of people away from large urban centres to live in smaller rural communities is part of the process of counter-urbanisation.
- Housing demand and rising prices may affect different people in different ways; some benefit while others are driven away.
- Economic change both within and outside a village can affect its people and employment. Villages which were predominantly agricultural 30 years ago have altered, with new employers and sources of work available.
- Development in housing and the use of space may take different forms; some may be regarded as more appropriate than others.
- Political decisions to build new housing may threaten the nature of a village.

Ideas for further study

- Identify local villages known to you and compare to what extent the issues faced by these places are **a)** similar to, and **b)** different from those facing Ashwell.
- Compare Ashwell or another village known to you with a village in a LEDC. Use Chapter 14 to help you. What are the key issues facing the village in the LEDC? To what extent are the issues similar or different to those facing Ashwell?

References for further reading

Owen, Andy (1998) *Managing Rural Environments*, Heinemann (especially Chapters 1 and 2)
Ashwell Education Services web page HYPERLINK http://www.ashwell-education-services.co.uk

14 Rural issues in Malawi

Malawi is one of the ten poorest countries in the world and is less than half the size of the UK. The country is mostly rural; only 14 per cent of people live in towns or cities. It relies on agriculture for its economic growth; 85 per cent of the economically-active population are involved in it. Its output accounts for 90 per cent of Malawi's overseas earnings, most coming from Malawi's 22 000 estates that grow tobacco, tea, sugar and coffee. Agriculture's share of the total value of production each year – known as Gross Domestic Product (GDP) – is about 40 per cent. It is the main source of livelihood for well over two-thirds of Malawi's 10 million people.

This chapter will investigate rural issues facing Malawi. It includes fascinating diary extracts and photographs from Ruth Totterdell, a Sheffield teacher who visited Malawi in 1999. Her observations tell the human stories behind some of the 'official' data published about Malawi. Her rural study area lies near the city of Blantyre (Figure 14.1).

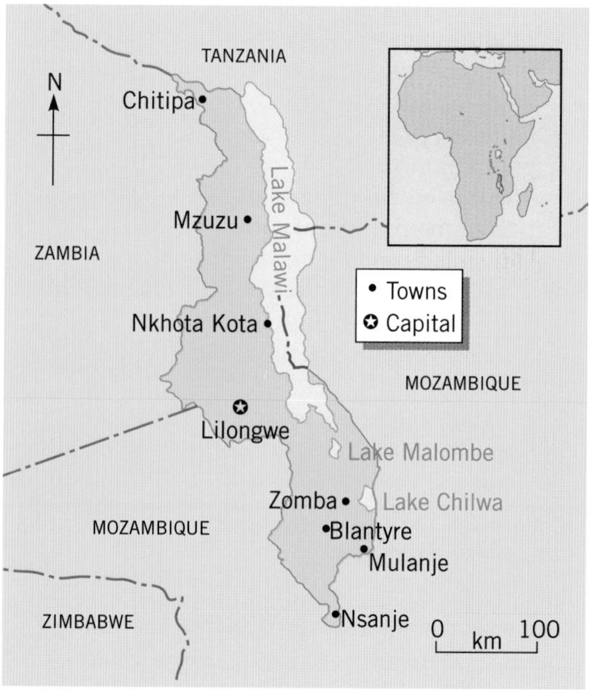

Figure 14.1 Malawi's key features.

Economic crisis and the rural economy

Malawi is among the world's poorest countries, with an annual income per person of only US$170. Over one third of this comes from overseas aid. It owes US$2.1 billion in debts, but earned only a quarter of this (US$549 million) in exports in 1998. One quarter of its export earnings is soaked up by debt repayment each year, and it has little chance of repaying its debts. It also suffers from a skewed distribution of income, where the poorest are the greatest number of people. In 1997 it ranked 161st out of 175 on the United Nations Human Development Index. Between 1990 and 1995, the percentage of the population without access to health services was estimated at being 65 per cent. Figure 14.3 shows some basic socio-economic indicators for Malawi.

---

1 On an outline map of Africa, locate Malawi and its neighbouring countries, using Figure 14.1. Write a paragraph describing its location (include latitude and longitude).
2 Use an atlas to identify each location named in this chapter.
3 Use Figure 14.2 to describe features of the climate of Malawi, e.g. seasons, temperature patterns, rainfall distribution, the best growing season, and any problems likely to be faced by farmers.

Figure 14.2 Climate in Malawi.

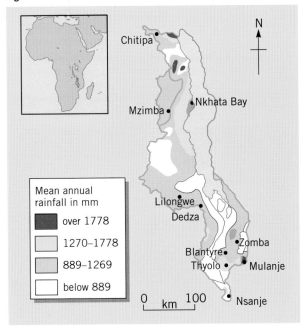

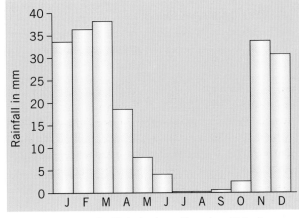

b) Rainfall pattern for Mulanje, in south-eastern Malawi.

Location	Altitude (m)	January mean temperature °C	July mean temperature °C
Nsanje	210	27.2	25
Chitipa	4612	21	17

c) Temperature variations in Malawi.

a) Rainfall distribution map for Malawi.

Indicator	Malawi	UK
Population data		
Population 1998	10 million	59 million
Birth rate 1996	39/1000	13/1000
Death rate 1996	23/1000	11/1000
Infant mortality per 1000 live births 1999	132/1000	6/1000
Total fertility rate 1999	5.5 births per woman	1.7 births per woman
% of population under 15 years	48%	19%
% of population over 65 years	3%	16%
Urban population 1996	14%	90%
Women using contraception 1997	22%	72%
Human development indicators		
Life expectancy 1999	36.3 years	77 years
Reasonable access to safe drinking water 1996	37%	100%
Malnutrition: % of children under 5 showing stunted growth 1997	48%	0%
Maternal mortality per 100 000 births 1990	560	9
Number of doctors per 100 000 population 1998	2	164
Male adult literacy rate 1995	72%	99%
Female adult literacy rate 1995	42%	99%
Economy		
Gross National Product (income) per person 1995	$170	$19 600
PPP$ (or real GNP) 1995	$773	$19 302

Figure 14.3 Quality of Life Indicators for Malawi (with UK comparison) in the late 1990s.

Measuring development

Measuring development is difficult. Most measurements rely on data based on incomes in the formal economy, and ignore informal and unpaid work. They are also average figures and ignore how wealth is distributed between people or regions. Some data are economic, e.g. Gross National Product (GNP), number of telephones per 1000 people, whereas others are social, e.g. number of people per doctor, life expectation, percentage of children enrolled in school.

Economic indicators

In order to compare countries, money has to be converted into a common currency, usually US dollars. However, these figures do not say how much different currencies will buy in their own country. While GNP (average income) and GDP (total production) give an idea of average wealth and economic development, they are very blunt instruments. To improve on this, the United Nations have introduced:

- Purchasing Power Parity $ (PPP$), sometimes referred to as 'real' GDP per person
- The Human Development Index (HDI).

Purchasing Power Parity $ (PPP$) and 'real GDP'

PPP$ assesses the value of a currency in its own country. For example, in 1993 the cost of living in the USA was more than in Malawi, according to the exchange rate of the US$, so you could buy more for one US$ in Malawi than in the USA. On this basis, Malawi had a GDP per person of US$170 in 1995, but based on PPP$ had a 'real' GDP per person of US$773 – about 5.5 times greater. This means that US$1 will buy 5.5 times more in Malawi than it will in the USA. However, even US$773 is nowhere near enough to ensure a good standard of living for most people.

The Human Development Index

The United Nations introduced the Human Development Index (HDI) to measure development for each country as a whole, and not just economically. The HDI uses selected indicators to show human well-being or deprivation. Three key indicators used are Life expectancy, Education and Standard of living. The index averages these data for each country, and combines them into a single figure – the HDI – whose maximum possible value is 1. It is calculated as follows:

Life expectancy – The longest global life expectancy in the world is 85 years and the shortest, 25 years. In the HDI, values are converted to a number between 0 (worst) and 1 (best); 25 years therefore becomes 0, and 85 years becomes 1. A country whose life expectancy is 55 therefore 0.5, half-way between the lowest and highest.

Education is measured by averaging adult literacy rates to get one figure, and the number of years of schooling to get another. Lowest and highest values are expressed as a figure between 0 and 1; 75 per cent adult literacy would be 0.75 on the HDI. The same is done for schooling. If a pupil spends six years out of 12 in school, then this is expressed as 0.5 in calculating the index – i.e. 50 per cent of a 12-year maximum.

Standard of living is measured using a scale from the lowest PPP value (assessed by the United Nations in 1995 as US$200) and maximum (US$34 000). These are used as a scale from 0 to 1, as before, so that US$17 000 would be approximately 0.5.

The four indices – Life Expectation, Adult literacy rates, Number of years schooling, and Standard of Living – are added and averaged to give one figure, with '1' the maximum score. In 1995 Canada (the top) had an HDI of 0.96, and Sierra Leone (the bottom) 0.185. Malawi was 161st out of 174 with 0.334.

Quality of life indicators

1 Study the data in Figure 14.3. In pairs, select five sets of data which you think best show the level of development in Malawi.
   a) Justify your selection.
   b) Explain how these may be related, e.g. infant mortality and the number of doctors.
   c) Say how these each affect the quality of life of people in Malawi.
2 Use the diary of the visit to the health centre at Namitambo (Figure 14.9) and the village at Chiradzulu (Figure 14.17) to illustrate the data. Do the stories confirm the patterns shown in the data?

Distribution of rural wealth

Wealth in rural areas is not equally distributed. Malawi's colonial past has given rise to a few large estates (Figure 14.5). Most are owned by private landowners, rather than big corporations. These exist alongside small-holder agriculture. The estates produce most of the country's exports, such as tea, sugar and tobacco, cropped by estate farmers who can afford irrigation, transport and storage services, and global marketing. However, they also use a lot of fertiliser compared to small-holders who use natural or no fertiliser. Plantation labour is hired and paid, usually landless labourers, or those supplementing incomes from other rural areas.

In spite of the large estates, most rural areas are dominated by small-holders (Figure 14.6). Around 1.8 million farming families occupy 1.8 million hectares of land and produce 80 per cent of the country's food. The main crop is maize, the staple food; rice and groundnuts are also grown. Farms are small; over half are less than 1 hectare. Most people are locked into subsistence agriculture. The dominance of subsistence farming means that there is little money in circulation; people eat what they grow. Most food never finds its way into markets or export and few have money to invest.

Unlike large estates, small-holders rely on rain, and cultivate maize, sorghum, millet, cassava and horticultural plants. Small holdings produce little surplus beyond that needed for each household and are the poorest sectors of the population. Especially poor are seasonal workers, female-headed households and tenant farmers who work on tobacco estates and who rely on these for income and food.

Figure 14.4 The Nchima Tea Estate.

We were up early and on the bumpy track south to Thyolo to visit the Nchima Tea Estate. The estate produces tea, coffee and macadamia nuts. At this time of year (July) most of the 7000 workers are involved in maintenance, spraying and pruning. The workers get 25 kwacha (45p) for 44kg of tea leaves or 25 kwacha for coffee cherries. This is very little. But they also get a house with water, firewood for cooking and a cooked lunch of maize and beans; there are several cookhouses around the estate. Another 'perk' is a free coffin. The demand for coffins is increasing as AIDS spreads. This year so far – 300 coffins.

Also on the estate is a coffee processing plant and a tea processing plant, each fuelled by wood. There are large areas of eucalyptus on the estate which are coppiced every eight years. There is some mechanisation of tea processing, but most sorting of the coffee beans is by hand. There was a shed of about 40 women sorting beans from a conveyor belt. Sacks of coffee beans and tea leaves leave the estate and compete on a world market. It is not good quality and used only in blended products. The only way it can compete is because it is cheaper than Kenyan or Brazilian, as wages in Malawi are so low. The land can be used for little else but tea and coffee as rains are so intense - 120mm can fall in an hour. Bushes provide protection from soil erosion that maize and other food crops cannot. The soil is very poor quality and crop yields soon decline without fertiliser.

Figure 14.5 From Ruth's diary - the Nchima Tea Estate.

Figure 14.7 Unpaved roads in rural Malawi prevent more rapid economic development.

Figure 14.6 A small-holding of land in rural Maiawi.

1 Draw a table to compare plantations and small-holdings in terms of **a)** ownership, **b)** crops produced, **c)** who works there, **d)** what happens to the produce, **e)** investment.
2 Summarise the advantages and disadvantages of large estates and small-holdings for Malawi's **a)** people, **b)** economy, **c)** environment.

Why do rural areas in Malawi have low levels of economic development?

Malawi has several problems in developing economically. Much stems from a lack of infrastructure, such as weak communication links and poor telecommunications both within the country and outside. Population issues and rural deprivation are made worse by health issues, including the impact of AIDS. Each of these is considered in turn.

Poor communication infrastructure

Of the 28 400km of roads in Malawi, 5254km are paved and 23 146km are unpaved. Many links are difficult to negotiate (Figure 14.7), especially on dirt roads during the wet season. These often make rural areas inaccessible, and journey times are long.

Missing out on the Internet revolution

Rural areas are sidelined in electronic development. Increasingly, telecommunications are an essential part of economic development. The internet was introduced in Malawi in 1994, but there is only one service provider – Malawinet. Malawinet had 2500 subscribers in 1999, but the number of users was small compared to MEDCs; it charged US$50 as an initial set-up fee and a monthly fee of US$20 – high for a population whose per capita income was only US$170. It suffers slow service and congested telephone lines. Malawi has 65 000 telephone lines of which only 36 000 work – about 0.3 telephone lines per 100 people. Of these, 28 000 are accessible by 2.4 million people in urban areas (1 line per 86 people), while the remaining 8000 are shared by over 7.5 million users in rural areas (1 per 9375 people).

Although better than the African average of 1 user per 5000 people, Malawi had 1 internet user per 4000 people in 1999, compared to the global average of 1 internet user for every 40 people, and 1 internet user per every 3-6 people in North America and Western Europe.

3 Construct graphs to illustrate Malawi's level of development as illustrated by the data in telephones and internet users.
4 Why should the lack of internet service be so important?
5 Explain the disparity between rural and urban internet telephone lines, and use of the internet.

Population issues in Malawi

Figure 14.8 is a population age-sex diagram of Malawi. It shows a typical profile of a population in a LEDC. Pyramidal in shape, it shows the following pattern:

- a broad base, where 45 per cent of the population is aged 0-14 years. The male:female ratio is broadly equal – there are 2 265 526 men as opposed to 2 246 135 women. This reflects the high birth rate compared to many MEDCs, shown in Figure 14.3.
- gradual reduction in each older age band. 52 per cent of the population is aged 15-64 years; the male:female ratio begins to develop a skew, as more men die at an earlier age than women. There are 2 580 125 men and 2 637 464 women.
- a small proportion of people aged 65 years and over: only 3 per cent of the total population. Compare this with the UK's 15.5 per cent in 1999. At this stage, women far outnumber men (158 353 against 112 813).

Only a small proportion of its people are able to generate wealth. 64 per cent of the UK's population are of working age, compared to Malawi's 52 per cent. The Malawi government is unable to raise sufficient taxes for a welfare state, so there is no support for children or the elderly. Families have to support both groups. Like medical care, schooling has to be paid for.

Rural deprivation and health

On almost all indicators, rural areas in Malawi are worse off than urban. A study in 1998 showed that a rural African farmer typically spends 43 minutes a day collecting firewood, 48 minutes in walking to the farm, 28 minutes to reach the grinding mill, and 128 minutes to walk to market – a total of nearly five hours! There is little time for improving care of children or the aged, or tending crops more

Namitambo is a rural health centre serving about 70 villages, about 1 hour by dirt road from Blantyre. One of the research projects run by Welcome is researching into anaemia in pregnancy and they are doing a trial with 700 pregnant women to see if taking vitamin A along with iron tablets will improve the absorption of Haemoglobin – weak anaemic mothers produce weak anaemic babies. The midwives organise the taking of blood samples for the research as well as normal ante-natal care.

The centre has an out-patients department with a medical assistant and a nurse. It deals mainly with chest problems and malaria. There is an under-5s centre, where children are weighed and checked, and a malnutrition centre. Some children who are severely malnourished are kept there and their mothers shown how to cook nutritious meals. The staple food is nsima (ground maize) – it provides little nutritional value unless eaten with beans or meat. The causes of malnutrition are ... lack of food, especially in the dry season when the new crops are just planted and the previous year's are gone. Many farms cannot feed a family for a year as they are too small and the land is under-utilised.

The ante-natal clinic gives the usual check-ups. It starts with the midwives leading some singing of songs about health education. There is also a labour ward at the back. Very basic, but all the women are encouraged to have their babies there in an attempt to reduce maternal mortality. If there are any complications, an ambulance ride of half an hour over bumpy roads is the alternative. Babies' weights are much lower than in the UK.

Family planning is a crucial part of the service. I chatted for a long time to the woman in charge of that. She had some free time as she had no syringes and had to send her ladies away. The most popular form of contraception is a depo-provera injection every three months. There is a lot of counselling involved. I saw the pictures they talked through, showing a worried man with lots of poor threadbare children (I have seen plenty of those), contrasting with a happy family of four children off to school and nice clothes and food (I have seen fewer of those). The aim is to get men convinced that contraception is a good thing, and away from the idea that the number of children you have is a measure of your manhood. The village women ... have had very little education and are well in to the grind of motherhood at 19.

Figure 14.9 From Ruth's diary – Health issues in rural Malawi.

Figure 14.8 Population age-sex diagram of Malawi. This is typical of a LEDC population structure. In 1997, the birth rate in Malawi of 39 per 1000 people was three times greater than that of the UK. Yet its death rate has fallen sharply in recent decades, and is now 23 per 1000 – only twice that of the UK. The causes of early death – such as disease (e.g. malaria) and infected water – have been reduced by improved domestic water supply. Now, there are 16 extra people each year for every 1000 people.

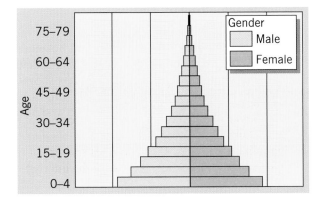

Name	Doctors		Dentists		Nurses/Midwives	
	Number	People per physician	Number	People per dentist	Number	People per nurse / midwife
UK	94 955	620	23 004	2564	N/A	N/A
Malawi	257	38 910	29	345 000	638	15 670

Figure 14.10 Health indicators for Malawi.

effectively. In this way, many families throughout rural Africa are trapped in poverty. Many families are under-fed, poorly educated, and suffer worse health than urban populations. Ruth's diary (Figure 14.9) illustrates this, but also shows that progress is being made.

Health indicators for Malawi (Figure 14.10) are poor. Rural areas have fewer health care facilities. 46 per cent of the population are not expected to live until the age of 40. About 42 per cent of people earn less than US$1 per day; 94 per cent of people have no access to sanitation. Disease is still likely to spread, and the AIDS epidemic, described in the theory box, AIDS in Africa, has had enormous effects on the population in both urban and rural areas. However, some data are encouraging and offer a better outlook, such as vaccinations (Figure 14.11), are carried out by teams of people in mobile health clinics. The challenge is to produce permanent health care that covers a range of health conditions.

Disease	Vaccination coverage %	No. of cases
Tuberculosis	100	not available
Polio	94	2
Diptheria	95	0
Tetanus	95	8
Measles	87	10 845

Figure 14.11 Vaccinations in Malawi, 1997.

Disease	Number of cases per 100 000 people
AIDS	36.6
Tuberculosis	198
Malaria	49 410

Figure 14.12 The extent of different infections in Malawi.

1 Identify ways in which indicators of poverty described in this section form a 'vicious circle'?
2 Why should rural areas be worse off for health care than urban?
3 Use the data and Ruth's account (Figures 14.5 and 14.9) to identify three priorities that need to be dealt with in the next two years. Justify your priorities.
4 What evidence is there that health care is improving?

In recent years, HIV – the virus that causes AIDS – has spread far more widely throughout Africa. The majority of doctors and hospitals are in urban areas – so, again, the rural poor suffer most. This has had devastating implications for those with AIDS, as the theory box shows. AIDS has put further pressure on resources. However, as Figure 14.12 shows, there are other diseases that are far more widespread and also likely to result in death – though at a later stage.

AIDS in Africa

AIDS in Africa exists on a scale quite different to that of European countries and other MEDCs. In 1999, 34 countries had over 1 million HIV-infected people; 29 of these were in Africa south of the Sahara, Malawi included. Of the 30 million people in the world infected by HIV, 26 million (85 per cent) live in these 34 countries and they account for 91 per cent of all AIDS deaths in the world. Figure 14.13 shows countries worst-affected by AIDS deaths in 1996. The scale of the extent of AIDS has led the World Health Organisation (WHO) to label the spread of AIDS as a **pandemic**, that is, greater than an epidemic in its extent and scale.

The nature of the epidemic is different to that of the MEDCs, where the majority of cases of AIDS have been among homosexual men. Figure 14.14 shows that the African pandemic is largely heterosexual in nature, but that, like the pandemic in MEDCs, it affects largely young and sexually active young people.

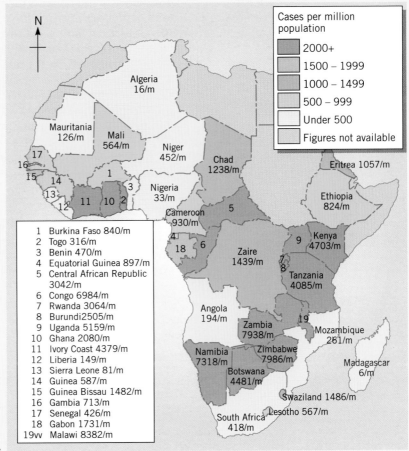

Cases per million population

- 2000+
- 1500 – 1999
- 1000 – 1499
- 500 – 999
- Under 500
- Figures not available

Algeria 16/m

Mauritania 126/m

Mali 564/m

Niger 452/m

Chad 1238/m

Eritrea 1057/m

Nigeria 33/m

Ethiopia 824/m

Cameroon 930/m

Kenya 4703/m

Zaire 1439/m

Tanzania 4085/m

Angola 194/m

Zambia 7938/m

Mozambique 261/m

Namibia 7318/m

Zimbabwe 7986/m

Madagascar 6/m

Botswana 4481/m

Swaziland 1486/m

South Africa 418/m

Lesotho 567/m

1 Burkina Faso 840/m
2 Togo 316/m
3 Benin 470/m
4 Equatorial Guinea 897/m
5 Central African Republic 3042/m
6 Congo 6984/m
7 Rwanda 3064/m
8 Burundi 2505/m
9 Uganda 5159/m
10 Ghana 2080/m
11 Ivory Coast 4379/m
12 Liberia 149/m
13 Sierra Leone 81/m
14 Guinea 587/m
15 Guinea Bissau 1482/m
16 Gambia 713/m
17 Senegal 426/m
18 Gabon 1731/m
19vv Malawi 8382/m

Figure 14.13 Countries most affected by HIV and AIDS in Africa 1996.

The toll from AIDS deaths in Africa is devastating. In the 29 worst-affected African countries, life expectancy at birth is already estimated at 47 years, seven years less than what could have been expected in the absence of AIDS (Figure 14.15). In the nine worst-affected African countries, the average life expectancy at birth is 48 years. Without AIDS, it would have reached 58 years. This group includes Botswana, Kenya, Malawi, Mozambique, Namibia, Rwanda, South Africa, Zambia and Zimbabwe. The impacts of HIV/AIDS are likely to get worse.

The disease is largely urban but has spread rapidly to some rural areas of Malawi where it has had major effects (Figure 14.16). Data from 17 rural health centres in Malawi between 1994 and 1996 showed that between 2 and 30 per cent of antenatal women were HIV-positive; 13 per cent of women under 20 years of age were infected with the virus, against 20 per cent in urban areas.

a) By mode of transmission, 1996

Transmission group	Numbers in Malawi 1996
Homosexual	0
Heterosexual	4729
Blood transfusions and from blood products	100
Transmission during pregnancy, birth or breast-feeding	577

b) By age group affected, 1996

Age group	Numbers in Malawi 1996
0–14	617
15–24	781
25–49	3674
Aged 50+	298

Figure 14.14 AIDS cases in Malawi notified to WHO. These are a small proportion of the total number of cases, whose extent is unknown in a country where rural medical care is unequal.

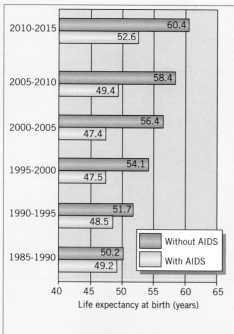

The impact on death rates is great. In sub-Saharan Africa, death rates are dramatically higher as a result of AIDS than would have been expected without it. In Cameroon and Nigeria, where HIV had infected 5 per cent of adults by 1995, death rates in 1998 were 20 to 30 per cent higher. By the year 2010, the crude death rate will be nearly twice as high in Cameroon and over twice as high in Nigeria. In Zimbabwe the crude death rate in 1998 is over three times as high as it would have been without AIDS and it will be more than four times as high by the year 2010.

1 Explain the impacts of AIDS shown in Figure 14.16. Why should AIDS have such an economic impact?
2 Use Figure 14.16 to say why rural families might be worse-affected by HIV and AIDS than urban.

Figure 14.15 How life expectation has reduced as a result of AIDS in Africa's worst-affected countries.

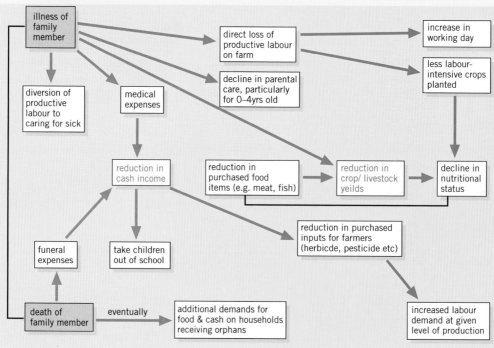

Figure 14.16 The economic impacts of HIV and AIDS upon rural families.

Population and migration within rural Malawi

For many, economic opportunities in rural Malawi are limited by small land-holdings, low wages on plantations, and the need to pay school fees or medical care. Many families are divided when some move away and work. Figure 14.17 describes Ruth's observations of one man, Philip, and his family;

Figure 14.19 describes the lifestyle of Liena, a grandmother living in a township outside Blantyre. Both Philip and Liena have common lifestyles – they are each from rural areas and they each return regularly. Their family links are still strong but their families are also dependent upon them for income. Their lives are the result of rural-urban migration, described in the theory box.

1 Why, when Philip has land, should he work for a family in Blantyre? What does this tell you about rural families?
2 In pairs, consider reasons why:
  a) there are far more men than women in his village
  b) Philip's family are economically better off than some
  c) subsistence farming is so labour-intensive.

I got another truck ride … to Philip's village, Chiradzulu. The village is very spread out but each extended family lives in close proximity. Philip works as a cook for an English family in Blantyre. His house … has a corrugated iron roof, walls of bricks, windows, curtains (patterned side so they can be seen from the outside!) a cooking hut, an area for washing, a pit latrine, and furniture. A family Philip worked for gave him some furniture so he has a bed, table and chairs. That is very unusual. I was then taken to see Philip's family in the surrounding houses, these all had mud walls and straw roofs. There were lots of women and children but no men.

Philip has some land, 3 plots all in different places. He grows mainly maize but only one crop a year during the rainy season. He also grows some nuts. It went dark at about 5.30 p.m., so by paraffin light I was served nsima and chicken. I began to realise exactly what subsistence farming was all about. To produce the nsima, the maize had been grown in the rainy season, harvested, then dried on straw mats, stored in sacks, pounded into flour, mixed with water fetched from the village pump, cooked in a pot heated by firewood that had been collected.

Figure 14.17 From Ruth's diary – Philip's house in Chiradzulu (see Figure 14.18).

Figure 14.18 Philip's house.

I have had a day out with Liena, a nanny for an English family in Blantyre. She is a Malawian Grandma, one of 12 children. She lives in her quarters behind her employer's house in Blantyre during the week but at the weekends goes to her home in outside Limbe about 10 km away (Figure 14.19b). About 5 years ago she bought a plot of land from the local chief and on it there are now 3 houses; one for each of her sons and one for her. They have each built their own houses and Liena's now has a corrugated iron roof, thanks to a loan from her employers. These areas of fairly densely-populated self-built houses outside the town are called townships.

Limbe Township was amazing - houses at all stages of development; some painted, with flower borders, others with mud walls and straw roofs. Some are rented, while others are occupied by the people who own the land, like Liena. I met Liena's two sons. One daughter-in-law is expecting her sixth child; the other has just one child as the other four died. Her sons have treadle sewing machines in their houses - no electricity - and make clothes to sell at the market in Limbe. I also met a one-day old baby born at the hospital in Limbe. The baby belonged to Liena's nephew, whose family has just moved to the township from his village and is renting a one-room hut. He is learning to sew with the help of Liena's sons. He hasn't got his own sewing machine yet.

Throughout the township are numerous little markets, some shops and lots of barbers. Few have electricity, I saw one 'Entertainment centre' with electricity, showing videos. I have seen countless women carrying babies on their backs and carrying water or other goods on their heads - as the township is only about 5 years old there are few supplies of water. There was a very long queue at the one tap I saw. The children all go to primary school in the township but either in the morning or afternoon. Security is a problem and all the houses had locks and some had intimidating walls with broken glass or barbed wire. There is no official policing of the township.

Figure 14.19a From Ruth's diary – Liena's lifestyle.

1 Identify **a)** common and **b)** different features of Philip's and Liena's lifestyles.
2 Would you describe Philip and Liena as relatively poor or wealthy? Justify your answer.
3 How far do you think their work in the city is beneficial to: **a)** them, **b)** their families, **c)** rural areas in Malawi generally?

Figure 14.19b Liena and her son at Limbe market, selling the clothes they make.

Theory

Rural-urban migration

Rural-urban migration is the movement of people from rural to urban areas. This causes population growth in the urban area, and decline or depopulation of the rural area. Urbanisation has occurred in most countries of the world to some extent. However, few countries have urban populations as low as in Malawi, where it is only 20 per cent. Compared to this, MEDCs have high levels of urbanisation, e.g. 90 per cent in the UK.

Urbanisation in MEDCs has occurred since the industrial revolution in the 1750s. In LEDCs, it is more recent. Until 1940, over 90 per cent of people in LEDCs lived in rural settlements or operate in nomadic or semi-nomadic systems such as the Fulani. Since then, there has been a rapid movement of people from rural areas to urban areas (Figure 14.20) either as a result of economic forces, rural poverty, or civil war. Small administrative centres developed in the colonial late nineteenth century have grown into megacities.

Causes of rural-urban migration

Rural-urban migration occurs as a result of two sets of reasons: the disadvantages of rural areas – known as 'push' factors – and attractions of urban areas – known as 'pull' factors. These are summarised in Figure 14.21.

	1960	1980	1990	2000 (est.)	2020 (est.)
World total	34.2	39.6	42.6	46.6	57.4
MEDCs	60.5	70.2	72.5	74.4	77.2
LEDCs	22.2	29.2	33.6	39.3	53.1

Figure 14.20 Percentage of urban population, 1960-2020.

Who migrates?

People most likely to migrate are the economically active, able to find work and earn money in the city. Migrants are often young, male and better-educated. They usually go alone, sending money back to their rural families, or, like Philip, visit at weekends or for family occasions. Sometimes whole families may follow the initial migrant – at first, immediate family, but later the extended family as the initial migrant becomes established with a home and employment. Migrants rarely lose contact with their villages and bonds are strong. Funerals, weddings and other family events continue to be based there. Contact weakens with second and subsequent generation of migrants.

Consequences of rural-urban migration

Rural-urban migration leads to strains on cities and urban resources. Land and houses are in short supply, formal jobs are difficult to find, and most jobs are informal; small businesses and trading are the usual ways of earning money. The loss of the young, skilled males from villages worsens the problems of rural areas. The elderly, women, and children are left behind and,

Push factors-reasons for leaving the rural area	Pull factors-attractions of the city
• Poverty • Inefficient farming methods on small farms. Often land is split up • Inadequate food for healthy growth • Poor education, only primary schools • Poor heath care, in remote rural clinics • External factors; natural disasters such as drought, flood, and human disasters such as war	• Opportunities for work in business and trading • Secondary schools • Hospitals • Opportunity to earn money to send back to the village • Exciting and busy • Hope for a better future

Figure 14.21 Push and pull factors.

with fewer hands to work, land is farmed with less efficiency. Much rural land is under-used and is less likely to use newer farming techniques or seed which give higher yields.

Rural-urban migration

1 Construct a graph to show urban population growth patterns in Figure 14.20. Annotate it to identify and explain the trends shown.
2 What problems are evident in rural areas in Malawi as a result of rural-urban migration? Use all the information in this chapter so far.

Rural-urban contrasts in Malawi

Why do people move to cities, and to what extent do their lives improve, having moved? Ruth Totterdell's diary extracts in Figures 14.22 and 14.24 show aspects of health care and education in Blantyre, the largest town in southern Malawi. Blantyre has a hospital and offers secondary schooling at the secondary school. Neither of these are generally available to rural people.

Melita took me to see the hospital this afternoon. I looked at wards 3B and 4A. They are the medical wards where money from the church has been spent. Each patient in Malawi hospitals has to have a guardian with him. The guardians have to look after the patients by cooking for them, washing their clothes, and any other tasks. All the patients are very poorly. Almost all have AIDS and 20% die on the wards. It was very shocking to see. There are also end wards where diarrhoea patients are kept. There didn't seem to be any beds in those.

I was there to look at the improvements. But it was difficult to see beyond the immediate horror of it all. Apparently, before the improvements I would not have been able to stay in the ward for more than a few minutes for the smell. It is incredible to see what money has been spent on: drains for surface water and underground ones for the sewage, paths over what was once muddy smelly ground, washing lines and seats for the guardians, toilets and showers that work, 4 sinks for 60 patients where there were none, curtains and painted walls, a drugs trolley where there was none, screens to give patients some privacy. Also impressive is that Melita has managed to find Malawian workers to do the work. It has been a boost for their businesses.

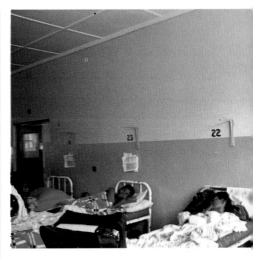

Figure 14.23 The Hospital in Blantyre.

Figure 14.22 From Ruth's diary – a visit to a hospital in Blantyre

Today I went to Blantyre Secondary school for a morning's visit. The school is a government school and has 600 students. At the end of Primary school (Standard 8) students take a national exam and the best go on to secondary school. Children at Primary school continue to take exams each year until they pass, so they can be of any age when they get to Standard 8. Primary schools can have children up to 18 years old!

I was taken into the staffroom – a very bare room with 20 staff discussing the disco which was to follow Graduation (for school leavers) on Friday. The staff want the disco in the day but the students want it in the evening. The staff were concerned about behaviour and security! As some of the students are 20 years old I did think a daytime disco was a bit unfair!

Blantyre Secondary School has four years. The first two years have a broad curriculum from which they sit a national exam, Junior Certificate, and have to pass this before going on to Forms 3 and 4, where they specialise in 7-8 subjects to gain the Malawi School Certificate of Education. From there students go on to jobs, University or Polytechnic. There are very low school fees for boys, and all girls' education is free in government schools. About half the students are boarders, but this is being phased out as the government are trying to put secondary schools in rural areas.

Unemployment is a big problem among the educated. Most graduates go on to be teachers as there are few white-collar jobs. The government are trying to promote rural areas and are encouraging the educated and skilled to stay there in a hope that conditions in rural areas can be improved, and to stem rural-urban migration. This tends to leave the elderly and inactive in the village, thus furthering the inefficient agricultural system.

Figure 14.24 From Ruth's diary – Education in Blantyre.

1 Read the diary extracts (Figures 14.22 and 14.24) and list the pull factors that would attract people to move to a township like Liena's from Philip's village.

2 Might life in the township be worse than, better than, or the same as life in the village?

3 Why should the government want people to return to rural areas? How might it benefit both areas?

4 Why, in spite of the fact that people in towns feel very attached to their village, do you think that people in towns rarely want to return there permanently?

5 Form groups of 2–3. What possible solutions could be used to improve life in rural Malawi? How could these benefit the development of the whole country?

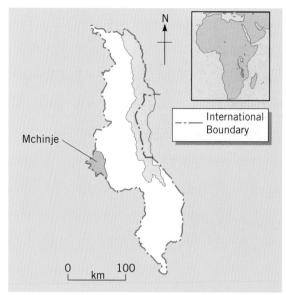

Figure 14.25 Location of Mchinji, western Malawi.

Resolving rural poverty in Malawi

Mchinji (Figure 14.25) is a province of Malawi in which several rural development programmes are being run by the United Nations Development Program (UNDP). Each focuses upon household food production and is designed to support a variety of activities identified by villagers, and aimed at improving food production sustainably. Four villages have been studied in Mchinji, in western Malawi, and programmes were devised, based on need. UNDP workers identified four areas for improvement with villages – food and nutrition, fuelwood, water supply and investment.

Improving nutrition in rural Mchinji

In Mchinji, rain-fed agriculture is the main livelihood with tobacco, groundnuts, cassava and soya beans grown as cash crops, and maize, yams, velvet beans and pumpkin as food crops. Secondary activities, especially during the dry season, include brick-making, beer brewing, bicycle repair, carpentry and casual labour on tobacco estates. Cash crops such as tobacco and maize earn additional income, but they require inputs such as pesticide, costs of which have risen. Neither is resistant to drought, in a region where there is low rainfall.

Until recently, farmers had to sell maize at fixed prices to the government. This gave all farmers – rich and poor – a chance to sell at a fixed income. However, the government has deregulated these, leaving many small-holders with no official way of marketing crops. The collapse of Malawi's currency in the late 1990s has made imported oil expensive, and transport costs to nearest markets have increased sharply. There is now little incentive to travel any distance to sell maize for an uncertain price. Traders and middlemen have taken over as dealers, who purchase maize from small-holders, then transport it to sell in Lilongwe or Blantyre. This results in food shortages in Mchinji, with the worst effects on casual, landless labourers. In one village, over a third of households experience food shortages between December and February, at the start of the rainy season, when men and women are busy preparing fields and have only one meal a day.

Attempts are being made to improve farm production. Figure 14.26 shows a UNDP project, designed to help farmers boost maize production by promoting compost from animal manure or crop residues as a fertiliser. This prevents the need for expensive chemical fertiliser. There are moves to:

- grow crops that require little or no fertiliser such as millet, sorghum, or sweet potatoes
- encourage Malawi's National Agricultural Research System to develop drought-resistant maize varieties.

Efforts are also being made to use 'dambos', land depressions close to riverbanks which are naturally irrigated and fertile. They are now used for horticultural crops such as tomatoes, rice, turnips, onions and green maize. Crops are either

Figure 14.26 Improved land designed to enhance maize yields, by using compost instead of fertiliser.

consumed or sold in rural markets. It also offers people food during the drought season (Figure 14.2) where maize crops have failed.

Linked to these projects, a village health centre is being established to monitor weight of under-5 children.

The problem of fuelwood

Each village experiences fuelwood shortages, increasing levels of soil erosion and scarcity of water. Two villages - Ironzi and Nyamawende – have tried to reverse this by adopting methods of contour ridging, planting of grasses along riverbanks and afforestation with the assistance of the Malawian Forestry Department. Both villages have planted wooded areas with bluegum trees (Eucalyptus), which can later be sold for making poles - therefore providing income - and, in extreme cases, used as firewood. Afforestation is underway to conserve soils and limit run-off, using trees, shrubs and grasses which can improve soil fertility or ones that help to fix nutrients.

Providing safe drinking water

Over 60 per cent of people in Malawi lack access to safe drinking water. UNDP is working with the Malawi government to provide simple water pumps for safe drinking water in villages. The target of the government has been to provide clean water to over 75 per cent of the population by 2000. Thousands of boreholes have been drilled, most of which are equipped with hand-pumps. The project was set up to establish 1000 boreholes in the Southern Region of Malawi, equip them with locally manufactured hand-pumps, and organise villagers to operate and maintain them and to adopt better sanitation practices.

Figure 14.27 A water pump.

Figure 14.28 A UNDP micro-credit loan made it possible for this woman in Malawi to start her own bee-keeping business.

Small-Scale enterprises

Over 80 per cent of rural households and small-scale rural enterprises do not have access to formal financial services, such as banks. One UNCDF project has been set up to provide credit to help the poor. Tax incentives and loans have been targeted towards the formation of small and medium-sized businesses, especially for existing trades of carpentry, brick-making, or car and bicycle repairs. Income from these helps families to invest in education, health, clothing and to save. Figure 14.28 shows one project which has resulted from this.

Summary

You have learned that:

- Malawi is among the world's poorest countries, as defined by a range of socio-economic data.
- Rural areas of Malawi suffer greater deprivation than urban.
- Agriculture is by far the most important sector of Malawi's economy. Its output and wealth are distributed unequally, between the large cash-crop estates and small land-holdings of poorer farmers.
- Economic development in rural areas of Malawi is held back by many factors, e.g. economic infrastructure, population structure, rural deprivation and health, and by rural-urban migration.
- Rural-urban migration affects many parts of Malawi. However, it mainly involves single family members who move away to earn extra income; most people keep contact with their rural home.
- Several strategies are being implemented by the United Nations to improve rural development, e.g. improving provision of food, firewood, clean water, and rural investment.

Ideas for further study

1 Compare the issues faced by rural areas of Malawi with those facing rural areas of the UK, described in Chapter 13. To what extent are the issues **a)** similar, **b)** different? Can differences simply be explained by economic factors?

2 Research rural development projects carried out elsewhere by national governments (e.g. by finding out how the UK government helps countries such as Malawi) or by non-government organisations (charities and voluntary organisations such as ActionAid, Oneworld, FarmAfrica) in other countries of Africa.

References for further reading

FarmAfrica web page
http://www.charitynet.org/~farmafrica/index.html
Owen, Andy *Managing Rural environments*, Heinemann, 1998, especially Chapter 3
UNAIDS page
http://www.unaids.org/hivaidsinfo/statistics/june 98/fact_sheets/pdfs/malawi.pdf
United Nations Development Program, web page
http://www.undp.org

The impact of tourism on rural areas

Tourism is the world's largest and fastest-growing industry. Not only is it a major industry in MEDCs, but many LEDCs perceive it as a main route by which to attract investment and achieve economic growth. However, it has its effects on people, the economy and environment. This chapter will explore some of the effects of tourism in VietNam, the UK, and northern Queensland, Australia.

Tourism and the rural economy in VietNam

VietNam stretches along a 3200km section of the South China Sea, from China in the north to the Gulf of Thailand in the south, and bordering Laos and Cambodia to the west. Most inland border areas are mountainous and sparsely populated. Almost 80 per cent of its population, estimated at 78.5 million in 1998, lives in rural areas. Ethnic Vietnamese – known as Kinh – account for about 87 per cent of the total population, and live mainly in the major delta areas of the Mekong and Red Rivers and their coastal plains. The rest of the population consists of 53 ethnic minority groups, generally located in the mountainous areas. The two largest cities, Ho Chi Minh City (4.5 million people), and Hanoi (2.5 million people), account for about half of the urban population. Average income per capita (GNP) is around US$310 per year, although this figure masks a gap between rich and poor. On the basis of this alone, VietNam could be considered one of the world's poorest countries.

However, VietNam has achieved a relatively high level of social development. Its progress is mapped out in Figure 15.2. It ranks 122nd out of 174 countries on the United Nations' Human Development Index (HDI) based on life expectancy, education-al attainment and income. VietNam's average life expectancy of 66 years (1995) is well above the average of 51 years for LEDCs. The adult literacy rate was 93 per cent in 1995, compared to 49 per cent for LEDCs. Infant mortality is estimated at 33 deaths per 1000 live births (compared to 65 for all LEDCs), in spite of under-developed health care facilities and limited access to safe water. However, only 21 per cent of the population has access to adequate sanitation and 43 per cent to safe water; 45 per cent of children under five are still malnourished.

Figure 15.1 Location of VietNam.

Factor	Progress	Deprivation
Income	Poverty reduced from 70% of people in the mid 1980s to about 50% in 1992	37 million people still suffer from poverty
Food	Between 1978 and 1994 food output rose from 238kg to 361kg per person	52% of children suffer from stunting a result of chronic malnutrition
Education	Literacy is more than 90% for both men and women under 40 years of age	Only 57% of children complete a primary education
Children	The infant mortality rate is about 36 per 1000 live births	In mountainous areas the infant mortality rate can reach 75 per 1000 live births
Health	Real government per capita on health tripled from 1986 to 1995	70% of the population has no access to safe water.

[Source: UN report: Poverty elimination in VietNam (October 1995)]

Figure 15.2 VietNam: progress and poverty.

VietNam has developed considerably under communist rule since 1975, the end of the VietNam war. In 1986, its government committed itself to a policy of *Doi Moi*, designed to reduce state control on the economy, and to encourage private investment. It made economic growth and the attraction of overseas investment a major objective. Tourism is fundamental to this, and the country has attracted increasing numbers of tourists since.

Rural tourism in VietNam

VietNam has excellent potential for the development of tourism with its long coastline and mountain scenery. Its scenery and climate are ideal for year-round tourism. Along the Vietnamese coast, from north to south, there are many beaches, especially that at Ha Long Bay (Figure 15.3) with more than 3000 islets and recognised by the UNESCO as World Heritage coastline. VietNam also has a range of virgin rainforests with rare animals and plants. Mountains and highlands occupy three quarters of the country's area. Mountain systems stretch from the north-west border to the east of the South VietNam, with the total length of 1400km; the highest peak – Phanxipan – is 3143 metres.

Figure 15.3 Ha Long Bay.

Since 1991 there has been an annual growth in tourism of over 40 per cent. In 1997, 1.7 million overseas visitors came to VietNam, which is likely to grow to 4 million by 2000 and 9 million by 2010. Of these, 40 per cent were paying tourists, i.e. staying in hotels or resorts, and 15 per cent were visiting relatives. Another 400 000 or so came to VietNam on business visits. In 1997 tourism earned the country US$538 million, 14 per cent more than

for 1996. The contribution of tourism to GDP is expected to be 15.4 per cent by 2010, almost four times that in 1984.

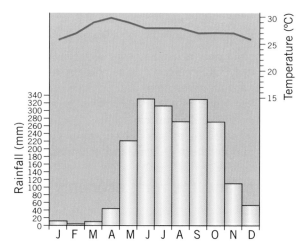

Figure 15.4 The climate of Ho Chi Minh City.

The VietNamese government have devised a 15-year plan for tourism development 1995 to 2010, shown in Figure 15.5. This plan contained various objectives:

- Economic objectives – to exploit the tourism potential and maximise the contribution of the tourist industry to VietNam's GNP
- Environmental objectives – to preserve and protect the natural environment
- Social and cultural objectives – the development of tourism should preserve and promote the culture and tradition of the VietNamese people.

VIETNAM TO SPEND US$3.3 MILLION ON RAISING ITS PROFILE

An estimated 46 billion VietNamese dong (over US$3.3 million) will be spent on the VietNam: Destination for 2000 promotion to encourage tourists to visit the country. The money will be spent on a variety of events to raise the country's international profile. A meeting to examine the campaign drew up a six-point plan of action which included the improvement of several of the country's top tourist resorts. The action plan also called for the promotion to emphasise VietNam's traditional culture.

Figure 15.5 From *VietNam News*, 19 August 1999.

Its policy is to develop tourism in ways that will preserve and promote rural cultural traditions. VietNam is rich in historical interest and cultural architecture, and tourism is seen as a means of enhancing the economy. The government has identified tourist development 'hotspots' where investment will be focused. While some of these are urban, (e.g. Ha Noi, the capital), rural areas have a part to play. Specific locations, shown in Figure 15.1, are:

- Ha Noi (the capital)
- Ha Long, shown in Figure 15.3
- The Hue – Da Nang – Lao Bao region, an area of cultural history and architecture
- The Nha Trang – Ninh Tru – Da Lat area
- Marine recreation at a number of coastal resorts
- The Ha Tien – Phu Quoc (Kien Giang) area, an area with potential for the development of eco-tourism.

Tourism and rural poverty

Government policy now is to promote tourism inland, in order to improve living standards among rural areas and ethnic minority groups. What impact will tourism have on rural VietNam?

Most rural areas in VietNam are poor; 90 per cent of its poorest people live in rural areas. UNICEF reported in 1996 that 53 per cent of VietNam's urban population had access to safe water, compared to 32 per cent in rural areas. The proportion of population with access to adequate sanitation was 47 per cent in urban areas and 16 per cent in rural. Average annual per capita income is US$310 for the country as a whole, but less than US$200 in rural areas. One in five people eats less than the minimum 2100 calories a day required for adequate nutrition. VietNam's 53 ethnic minorities live mainly in remote rural areas; 38.5 per cent of the majority Kinh VietNamese live in poverty while the figure for ethnic minorities is 66 per cent.

Now, *doi moi* reforms have improved the lives of rural VietNamese people. Land was returned to families from the state, and they have greater choice on how best to farm and manage agricultural land. Farming families now grow cash crops such as coffee, tea and rubber. The development of farm co-operatives has helped to produce more food for cash sale. As a result, VietNam has been transformed from a significant rice importer in the mid-1980s to the world's second largest rice exporter.

Traditional lifestyles abound as shown in Figure 15.6, and it is these that tourists wish to see. While the central and southern highlands are attractive to tourists, it is often the rural lifestyle which is most photogenic. Traditional ways of life, such as

Figure 15.6 Working in the rice fields.

farming and wood gathering, are perceived as more 'real' than many urban lifestyles. The highlands are home to many of VietNam's ethnic minorities (such as the Muong, Koho, Hmong and Ede), most of whom make a living by growing avocados (Figure 15.7), rice, coffee, black beans and sweet potatoes. Some is grown in plots within the village and others on the lower slopes of surrounding hills, where land is terraced to allow cultivation. Although tractors are replacing water buffalo in many rice-producing communities around VietNam, in this village the work is done by hand using basic implements. Many families rely on collecting wood to provide fuel for cooking, though some have electricity.

Figure 15.7 Family loading bag of avocados on to motorcycle to sell at market.

Social and economic effects of tourism in rural communities

While tourism has economic impacts as the fastest growing industry within VietNam, it can have effects on people and culture. Customs may be

Figure 15.8 More modern housing in the 'Klong' village.

Figure 15.9 Tourist bus at the 'Klong' or 'Chicken village', near Da Lat.

One village affected in this way is in the village known as the 'Klong' village, just outside the town of Da Lat (Figure 15.1). To English-speaking tourists it is known as the Chicken village, for reasons shown in Figure 15.9. It has become a popular stopping point for tourists wanting to visit an ethnic minority community. Tourists arrive expecting to see traditional housing, basic wooden structures with thatched or corrugated iron roofs. While some of these exist, most people live in more modern housing (Figure 15.8).

The first point on arrival in the village is a large thatched hut, where textiles woven locally are sold. The tribes people who sell tapestries do not live there, and income generated from sales does not contribute to the village economy. The loom, on view in a small hut is meant to represent traditional crafts, but, textiles are made in an urban factory, not here. There is a feeling of 'staged authenticity' about this village; everything is put on for tourists, rather than reflecting genuine rural life. Tourists are invited to walk around the village, and residents become like museum pieces for tourists to peer at.

Environmental impacts of tourism in rural areas

The development of the long-haul market in tourism in western Europe and the USA has been expanded in recent years by the emergence of a wealthy socio-economic class in south-east Asia. Nearly half of overseas visitors to VietNam come from Taiwan, China, Thailand, Singapore, and Hong Kong; all are within a flight of under four hours. For VietNam, as elsewhere, tourism has several benefits and problems. Figure 15.12 summarises these, and seeks to provide the balance of benefits and issues presented by tourism.

Part of the expansion of tourism in south-east Asia is fuelled by demand for golf courses. Financial returns are large. A golf course has been created in Da Lat, in the Southern Highlands of VietNam, near the 5-star Sofitel Palace hotel. Course fees are about US$100 a round, in a country whose per capita income is only three times that

revived and monuments restored. New markets are created for indigenous arts and crafts and presentations for song and dance.

However, the culture of indigenous people can be exploited as a tourist attraction to provide tourists with packaged amusement, and their beliefs and customs ignored or subverted in the name of tourism. Much-visited communities may see their traditions and lifestyles adapted to suit visitors' rather than their own needs. This is known as 'staged authenticity', where culture exists more as an economic activity than a reflection of people's beliefs and customs.

per year! The tourist industry had also taken a knock in the late 1990s with the collapse of the south-east Asian currencies. Golf has become a serious industry in south-east Asian countries; it relies heavily upon capital investment, and economic collapse in the region – although temporary – has major effects.

According to Tourism Concern, the boom in golf tourism can have devastating effects on people's health. An estimated 350 new golf courses are built world-wide each year, maintained using massive amounts of fertilisers, pesticides, herbicides and water. In Thailand the health of many people living near golf courses has been affected by drinking water or eating fish contaminated by toxins, which have mixed with rainwater and flowed into rivers and reservoirs. Fears are growing that rural VietNam could face the same problems (Figure 15.11).

GOVERNMENT SEIZES PEASANTS' RICE FIELDS FOR GOLF COURSE

Thousands of peasants in Kim No, to the north of Ha Noi, fighting to preserve their rice fields, clashed with hundreds of Public Security cadres and army troops dressed in riot gear. At the beginning of the year, the Ha Noi government announced its decision to sell the Kim No food planting fields to Korean developers, Daeha, to build a luxury golf course to accommodate foreign vacationers. Given the general food shortage in northern VietNam and the lack of any offer of replacement fields, the residents of Kim No petitioned the government to change its mind.

On May 13th the Public Security cadres and army troops came equipped with tear gas, riot shields and electric prongs. They started to uproot the rice plants. A village woman tried to stop the perpetrators but was beaten unconscious and was later found dead at the site. Thousands of villagers awakened to what was happening, vented their anger at the attackers with rocks, forks, sickles and poles. The melee lasted for hours.

Figure 15.11 From *VietNam News*, 13 May 1997.

TOURISM NEEDS AN ECOLOGICAL OUTLOOK

The tourism industry has been recognised as one of the fastest growing industries with great economic potential. It has created jobs, generated revenue, improved infrastructure, restored historical and cultural relics, and given the whole national economy a boost.

However, many tourism activities have had negative impacts on the environment and disrupted traditional cultural and social values due to the push for short-term benefits.

While there has been ample warning about these impacts, not enough has been done to minimise them.

In September 1999 the VietNam National Administration of Tourism stated that eco-tourism is an unavoidable option for VietNam, which requires concerted efforts from the State and the different players in the tourism industry.

Figure 15.12 Adapted from *VietNam News*, 10 September 1999.

Consider the social, economic, and environmental effects of tourism in rural VietNam. Do you think that the benefits outweigh the problems caused? Write a report for Tourism Concern, in which you:
• highlight the need for growth in the tourist industry in VietNam
• outline the impacts of tourism under these headings – social, economic and environmental
• outline the problems created by tourism under these headings – social, economic and environmental
• reach a verdict: is tourism worth the problems caused?

Kinder Scout in the Peak District

Kinder Scout is one of the highest peaks of the southern Pennines in Britain and a popular destination for many visitors to the Peak District National Park. It lies on the first stage of the Pennine Way from Edale to Kirk Yetholm on the Scottish Border. This part, known as The Dark Peak, is wild open moorland, lying at about 600 metres above sea level. It is wild and featureless; at 636 metres its summit is broad and flat, and could certainly not be called a peak. Gritstone Edges are found around the plateau such as The Edge, shown in Figure 15.14. Its moorland environment is increasingly threatened by its proximity to large cities such as Sheffield and Manchester, and by increasing car ownership.

In summer, the purple slopes of the heather moorland are stunning, but in winter much of the vegetation dies down, leaving a barren landscape. Poor weather makes this a cold, bleak, desolate and dangerous place. Annual precipitation is about 1550mm. In winter this may fall as snow, and snow may lie for more than 70 days. Average temperatures are only 11°C in July and fall to 0°C or below in winter. Strong winds make it feel even colder.

Kinder Scout is part of the Dark Peak Site of Special Scientific Interest (SSSI), owned by The National Trust who protect and manage it. The Dark Peak is an SSSI because it is an extensive area of semi-natural upland vegetation, supports rare species of bird, and it is a location of geological interest. The geology and features of Kinder Scout are shown in Figure 16.13.

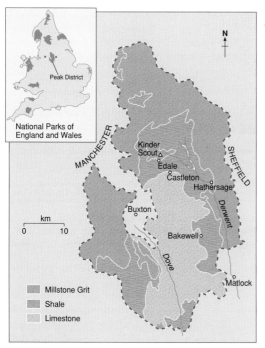

Figure 15.13 The Peak District showing Kinder Scout, underlying geology and the SSSI.

Figure 15.14 The Edge / Gritstone Edge.

1 Using an atlas and information from this page, draw a map to show the location of Kinder Scout. Annotate it to show rock type, soils and climate.

2 Suggest reasons why Kinder Scout is popular with visitors, and what kinds of visitors.

National Parks and Areas of Outstanding Natural Beauty

National parks are large areas of beautiful countryside, specially protected so that they remain beautiful and can be enjoyed now and in the future. Between 1951 and 1957 ten national parks were set up in England and Wales, following the National Parks and Access to the Countryside Act in 1949 (Figure 15.15). In 1989 The Broads became the eleventh National Park, and in September 1999 the government announced that The New Forest and The South Downs were to be given the extra protection of National Park status.

The two main aims of the National Parks are:

- to conserve the natural beauty, wildlife and cultural heritage of the area
- to provide opportunities for people to understand and enjoy of the areas.

These two aims inevitably bring conflicts, and it is the job of National Park Authorities to balance varied demands on the areas. The Planning Authorities have to devise policies that balance needs of conservation with need for houses, jobs and services in the local communities, the needs of visitors and the needs of local industries, such as mineral extraction, water storage, farming and forestry. It is the policy of all National Parks to aim for sustainable development. About half the money needed to fund the National Parks comes from national government. The rest comes from local residents, through Council Tax, and from visitors themselves.

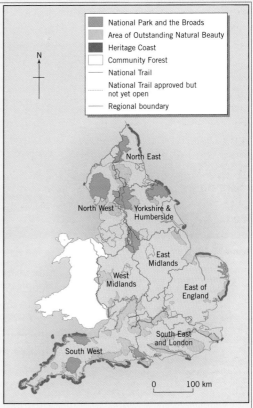

Figure 15.15 The National Parks, Areas of Outstanding Natural Beauty and Long distance footpaths in England and Wales.

The Peak District was the first National Park created in 1951. Its central location in the country and proximity to large cities such as Sheffield and Manchester mean that it remains the most visited of all the National Parks. Up to 30 million visits are made to the Park each year, mostly by car. Most are day visitors but there are around a million overnight stays each year. Visitors in these numbers inevitably cause problems but they also bring business to local traders. It is estimated that about £137 million is spent in the Peak Park by visitors, each year.

The map in Figure 15.15 also shows thirteen National Trails, or long-distance footpaths. Kinder Scout is at the southern end of The Pennine Way, Britain's first long distance footpath. Until the 1930s, people were not allowed to use large tracts of moorland and open countryside such as the Peak District. Water companies owned much of the land and were worried that reservoirs would become polluted, and landed gentry did not want walkers disturbing grouse reared for the shooting season. Kinder Scout became a focus for protests against these restrictions and in 1932 there was a famous Mass Trespass. Despite campaigning, it was not until the National Parks and Access to the Countryside Bill in 1949 that agreements with landowners began; even now, landowners have the right to close grouse moors on shooting days.

In 1999 the government announced their intention of opening up huge areas of countryside in private ownership. The 'right to roam' legislation, when passed (probably by early 2001), is likely to offer a legal right of access to much land in the wilder upland regions of the UK: mountains, moor, downland heath and commons.

Figure 15.16 Blanket Bog with cotton grass on the Kinder plateau.

Plant and animal communities on Kinder Scout

Blanket peat covers the whole of this area, formed as a result of clearance of natural oak woodland which impoverished the soil. Peat formation has been more substantial in this area than anywhere else in Britain because deforestation took place earlier here than elsewhere. The moorland divides into two, the plateau blanket bogs, where the peat is deepest, and surrounding hillsides. Cotton grass, crowberry and bilberry dominate the blanket bog, although heather is also found. In some places, mosses such as Sphagnum occur but high levels of atmospheric pollution have drastically reduced the numbers and variety of bog mosses.

On drier slopes, heather dominates with more grasses here than on the plateau. Many birds and animals living on Kinder Scout find their food from these two plant communities. Moorland vegetation must adapt to live in acid, wet peat and survive in wet, cold conditions at relatively high altitudes. There are limited numbers of soil fauna and an absence of earthworms because wet soils contain little oxygen. Butterflies too are rare, although moths, flies and mosquitoes occur in substantial numbers.

Peat accumulation

High rainfall and poor drainage on Kinder Scout mean the process of decay is slowed down because of reduced supplies of oxygen, high soil acidity and low temperatures. As plants die, the remains only partly rot and begin to accumulate, forming peat. Sphagnum moss is the most important plant in the formation of peat because it absorbs a lot of water and nutrients.

How is heather moorland managed?

Heather (Figure 15.17) survives on acidic, nutrient-poor soils and produces acidic leaf litter. Soil acidity reduces availability of nutrients for other plants. Few soil animals survive in organic matter produced beneath heather, and there are no earthworms. As a result, decomposition of organic matter is slow, leading to layers of acid peat on the surface. Heather is semi-natural vegetation maintained by grazing and burning. This allows it to regenerate rapidly with little competition from other plants. If it is left without management for 15 years or more, it degenerates, plants become woody and the branches fall outwards, allowing other species to take over. Without management, the area would revert to scrub and at lower altitudes, oak woodland would establish itself through plant succession.

Managing Kinder Scout

Kinder Scout's location as a feature along one of the most used long-distance footpaths in the UK's most used National Park creates pressure. This section explores the influence of human activities on Kinder Scout, especially tourism.

Evidence of an environment under stress

There has long been concern about peat erosion in The Peak District. This led The Moorland Erosion Study set up by the Peak District Planning Board (PPPB) in 1979 to examine the increase in moorland erosion in the National Park, and to try to find the causes of it. The results showed how serious the problem had become, particularly on Kinder Scout.

The extent of peat erosion can be seen clearly on Figure 15.18. The Point was erected in the 1920s and the concrete base was level with the peat surface. By 1981, over a metre of peat had been removed. Peat erosion is caused by two processes: the destruction of vegetation, and the removal of exposed peat by running water, wind or gravity.

The blanket peats of Kinder, over 4m deep in places, are now severely eroded. Up to 60mm of peat a year is being lost, and gullies over 2m deep are found on the plateau. Where vegetation and peat are lost, the balance of the ecosystem is upset. The survival of birds and animals is at risk from the reduction in number and variety of primary producers, and the loss of plants and soils upsets nutrient cycling.

Causes of peat erosion

The rapid rate of erosion of the peat on Kinder Scout is due to a number of factors: pollution, sheep grazing, tourism and climate.

Pollution

Kinder Scout lies between Manchester and Sheffield and suffers severe atmospheric pollution. Sphagnum moss and other bog plants do not tolerate high levels of sulphur dioxide nor nitrogen oxides, yet these are brought to the moorlands as acid rain. Sphagnum moss acts like a sponge and protects the peat surface. Pollution levels are too high to allow it to re-establish.

Building Phase 7–15 years
Heather grows fast, developing a dense canopy, up to 90% cover. Most other species are excluded by the lack of light and moisture as rainfall is intercepted.

Pioneer Phase Up to 7 years
Young plants grow from seed, or new shoots sprout from the base of burnt stems. About 10% cover, and little effect on the microclimate.

Heather should be burnt before reaching the mature stage.

Mature Phase 15–25 years
Plants become woody and heavy branches fall outwards. About 75% cover. Moss and lichens may become established, on stems and in the gap. Other species begin to re-invade.

Degenerate Phase Over 25 years
Heather begins to die, and now contributes only 40–50% cover. The gap in the centre of the bush allows other plants such as bilberry, wavy-hair grass and bracken to grow. New heather seedlings may also establish themselves, beginning a new cycle.

Once heather has degenerated, other plants, particularly bracken, take over. If conditions are right, bracken can take over the whole site as it is able to reproduce using spores and rhizomes. It is poisonous and unpalatable to grazing animals, and it suppresses other plants by producing large amounts of leaf litter.

Further succession could lead to the establishment of birch trees and scrub, and eventually to oak woodland on lower slopes where climatic and other factors allow. Bracken is unable to tolerate too much shade, so will die out. Grasses and billberry will emerge beneath the canopy.

Figure 15.17 Life cycle of heather.

Figure 15.18 Triangulation Point on Kinder Plateau.

Sheep grazing

The number of sheep on Kinder and Bleaklow increased from 17 000 in 1914 to 60 000 in the 1970s, when government subsidies paid farmers according to how many sheep they kept. When the National Trust bought Kinder Scout in 1983, there were often 2000 sheep grazing there. Lack of fences and low levels of shepherding meant that sheep could wander freely. Most were breeding ewes which are selective in their grazing, choosing heather and bilberry rather than other plants. This has increased the spread of moorland grasses at the expense of more valuable shrubs. The 30 per cent loss of heather in the Peak District last century could be attributed to overgrazing by sheep. In places, grazing has led to extensive areas of bare ground broken only by coarse grass. The loss of food and cover has led to a decline in bird species.

Walkers

About 10 000 people annually walk the length of the Pennine Way. More than 150 000 visit Kinder Scout. This exceeds the capacity of every major and most minor paths in the Kinder area. Between the mid-1970s and mid-1980s, the average bare width of peat on the Pennine Way increased from 1.45m to 3.53m. Plants such as cotton grass, mosses, bilberry and heather are easily damaged by trampling,

Technique

Measuring and recording footpath erosion

Field work can easily be carried out on a footpath to measure its width and depth, and to assess damage caused by trampling boots or by other forms of erosion.

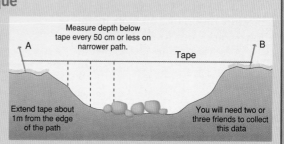

Figure 15.19 Measuring footpath erosion.

1 Stretch the tape across the footpath.
2 Extend the tape at least one metre from the edge of the path at each side in order to record untrampled vegetation.
3 Record the width of the path (A to B).
4 Record the depth of the path at point A, then at regular intervals, for example every 0.5 metres, and at point B. Present information as cross-sections drawn to scale.
5 Repeat the measurements at other places on the footpath. You could select sites systematically, for example, every 10 metres, or randomly using Random Numbers.
6 Present your data as an isoline map to show depth of the footpath, shown in Figure 15.20.

If the footpath is worn with patches of bare soil, but not badly eroded you can use a quadrat as shown in Figure 15.21. You should record data about bare soil for both an eroded and an undamaged area away from the main area of trampling.

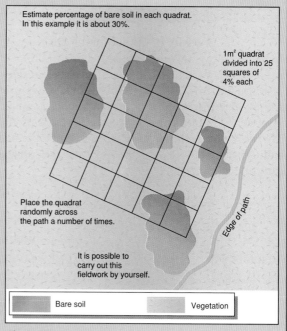

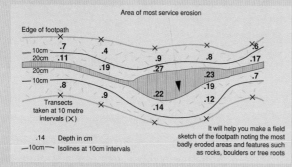

Figure 15.20 Isoline map of an eroded footpath.

Figure 15.21 Measuring damage on a worn footpath.

particularly on slopes and in wet areas. Once plants are removed, peat is churned up by boots and becomes waterlogged. People avoid boggy areas, but widen the footpath by damaging vegetation at the edges. A sponsored walk can cause as much damage to a path in one day as a year's normal use. Walkers go on to Kinder at all times of the year, giving vegetation no time to recover.

Moorland fires and tourism

Heather moorland fires are beneficial, removing old stands of heather and returning nutrients to the soil. Accidental fires in summer, however, can have devastating effects, especially on dry peat. They burn longer and deeper, destroying all vegetation, dormant seeds in the peat and sometimes the peat itself. Insects, birds and animals may also be killed. Once vegetation cover has gone, peat is easily removed by rain and wind; some areas never recover. High rainfall, high intensity rain storms, and strong winds cause great damage. Summer fires are almost always associated with visitors; 60 per cent start on or near a footpath or road, and most start at the weekend.

Management strategies

In 1982 Kinder Scout was bought by the National Trust. It was suffering from severe erosion. In consultation with others, the Trust drew up a plan which had three main objectives:
- to preserve and enhance the natural beauty of the area
- to halt, if possible, to reverse the moorland erosion
- to continue to provide access for the public, while acknowledging the fragility of the area.

The main management strategies are outlined below.

Figure 15.22 Walkers on the Pennine Way, Kinder Scout.

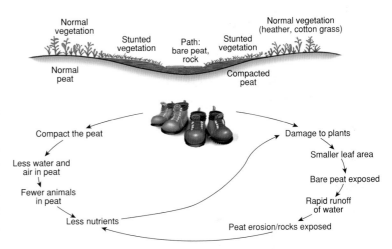

Figure 15.23 Damage done by walkers to footpaths.

1 What evidence is there of severe peat erosion on Kinder Scout? What was the average annual rate of removal of peat on Kinder? How does this compare with average surface lowering of bare peat which is thought to be 5–10 mm per year?

2 The first stage in peat erosion is removal of vegetation. Discuss the causes of vegetation loss.

3 Complete Figure 15.25 which summarises peat erosion. Use colours to highlight aspects that could be tackled in attempts to reduce peat erosion on Kinder, and those that would be difficult to tackle.

4 Form groups of 3–4. Decide how you would tackle each problem shown in your diagram. Present your plan to the group. Discuss difficulties you might face.

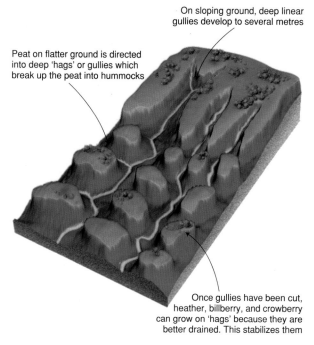

On sloping ground, deep linear gullies develop to several metres

Peat on flatter ground is directed into deep 'hags' or gullies which break up the peat into hummocks

Once gullies have been cut, heather, billberry, and crowberry can grow on 'hags' because they are better drained. This stabilizes them

Figure 15.24 Bower's classification of peat erosion. Peat erodes in two ways.

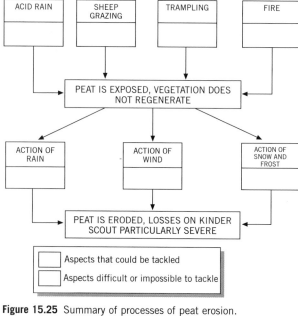

Figure 15.25 Summary of processes of peat erosion.

Sheep grazing controls

There is no value in trying to manage tourist impacts if farming undoes all the good work. The National Trust's first action in 1983 was to ban all sheep grazing on Kinder moors. Fences and drystone walls help to keep the sheep out. The ban is still in place in 2000 and will stay for the forseeable future. In 1995 cattle grazing was introduced on the lower slopes. This helps to break up matt grass and allow heath to spread.

Revegetation

The National Trust and the University of Sheffield have been investigating how bare peat can be re-vegetated. As a result of their work applications of lime and fertiliser were applied using helicopters. This reduces soil acidity and adds nutrients, aiding growth and germination. Lime is slow to break down and the benefits can still be seen years later without requiring further applications.

Revegetation is also carried out by spreading seeds and cuttings of native species such as cotton grass and heather. In 1999 seeding started to be carried out by helicopter instead of hand, using specially developed seeds that have a higher chance of germination. The seed can be spread over a much larger area, more quickly and this outweighs the high costs. Seeds are usually spread on existing vegetation which provides shelter and prevents the seed being washed away by rain and surface run-off.

The bare peat on Kinder plateau presents more serious problems and reseeding is not effective here. In places, small dams made from local stones have been built across the groughs. Cotton

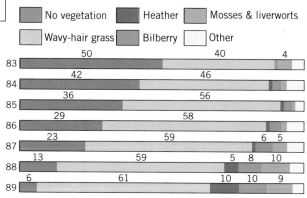

Figure 15.26 Graph showing changes in ground cover, 1983 to 1990.

grass is planted behind the dams and is trapped with particles of peat, enabling it to become established. Elsewhere the best solution is to cover the peat with large pieces of turf consisting of crowberry, bilberry and cotton grass. The underground runners enable the turf to become established. This has been a successful method and is particularly used close to footpaths where bare peat is exposed.

The application of lime and fertiliser encourages the growth of wavy hair grass which soon seeds itself on surrounding bare peat. As the grass develops it protects the peat and allows other plants to develop, such as heather.

Footpath repair

Paths across Kinder are particularly affected by trampling. The National Trust are anxious that as far as possible this should not threaten the natural landscape. In some places drainage channels or the relocation of boulders and turf is enough to protect a path from erosion but elsewhere it has been necessary to lay new surfaces. Several types have been tried:

- two major routes have been 'stone pitched'. Large blocks of local Gritstone have been embedded into the ground, creating a hard wearing path.
- floating paths have been laid in badly drained places. These involve laying fibre matting and covering it with either aggregate or boulders.
- Gritstone flagstones have been laid on gentle slopes to provide paths across deep or waterlogged peat.

The footpath repair work is still being done. In 1999, 1000 tonnes of flagstones were used to repair about 25km of footpath. Funding from the National Trust and the Peak District National Park has been boosted by national lottery funds. The results are encouraging and the majority of the walkers now keep to the paths and avoid the open plateau where the less damaged paths are recovering. Furthermore, the improved surface has meant that older walkers are now able to walk up Kinder Scout where in the past the rough terrain made it too difficult.

Fire lighting

Damage caused by accidental fires has been tackled by the National Trust by:

- providing specialised fire fighting equipment in strategic locations on Kinder, which Wardens are trained to use
- working closely with the Fire Service and developing a Fire Plan to deal with fires
- developing a fire warning system with the Meteorological Office, which allows the Met Office to provide information about conditions likely to lead to fires; the National Trust can then put out signs to warn the public to be particularly careful
- better education of the public such as posters, information leaflets and the 'No Moor Fires' campaign.

Figure 15.27 Footpath repair on Kinder Scout. Footpath repair and maintenance is extremely costly particularly as paths may be a long way from roads. Materials may have to be air lifted in and people may have to walk for over an hour to the site.

How successful are the National Trust's management strategies?

1 Study the map showing management strategies employed by the National Trust on Kinder Scout, and the information in this chapter. Summarise the benefits and drawbacks of each strategy: Sheep grazing ban; Revegetation; Footpath repair; and Fire fighting.

2 Form groups of 3–4. Select one of the management strategies below. Write a paper to justify a strategy for **one** the following, where you will meet:
  • the local Ramblers Group, upset about the expense of footpath repair, and the number of surfaced paths which they consider to be unnatural. They wish you to look at alternative strategies.
  • local farmers concerned about sheep grazing bans, who want to graze sheep on Kinder.
  • the Finance Committee of The National Trust who wish you to justify the costs of managing Kinder Scout. It suggests that you reduce costs by concentrating on a smaller number of projects.

3 In your opinion, is the National Trust justified in the policies it has chosen? Which of the projects, if any, might be abandoned? How else might erosion be reduced on Kinder Scout?

Figure 15.28 The location of Port Douglas in far northern Queensland.

The challenge of eco-tourism in Queensland

Tourism is one of Australia's fastest growing industries. It is a major source of foreign exchange, GDP, income and employment. In 1993–94, the tourist industry contributed an estimated 6.6 per cent to Gross Domestic Product (GDP). In 1995 international tourism to Australia generated export earnings of $13.1 billion (an increase of 17.2 per cent on the previous year). This was 12.6 per cent of Australia's total export earnings and 62.2 per cent of services exports. It directly accounts for employment of around 500 000 people (7 per cent of the workforce). While Japan, south-east Asia and New Zealand are presently the largest single markets, tourists from other Asian countries are increasingly important.

The effects of rural tourism on Port Douglas

Port Douglas is 70km north of Cairns on the far northern Queensland coast of Australia. In recent years it has grown rapidly as a result of tourism. Domestic tourism is helped by the climate of this part of northern Queensland, shown in Figure 15.28. Its tropical location – Cairns is 16°S of the equator – brings year-round warmth, and the Australian winter from June to August coincides with the dry season here. While Melbourne may have winter temperatures of about 12–13°C, Cairns has daily winter temperatures of 28°C.

The growth in overseas tourism is affecting Australia generally. At the time of winning the bid for the 2000 Olympics in 1993, Australia received about 1.2 million overseas visitors. The Olympics are expected to generate about 9 million overseas visitors to Australia in 2000. Because of its distant location, most visitors are unable to visit frequently and therefore combine several locations within their 'big trip' to Australia. Port Douglas is increasingly one of those locations.

The attractions of Port Douglas include its 'four mile beach', shown in Figure 15.30. Its village 'feel' is boosted by the influx of tourists, who maintain an increasing number of services in and around the village. Recent development has increased, so that For Sale signs on building plots are common as land prices rise (Figure 15.31). Such is the restriction on land space that holiday apartment blocks are being developed on the site of a previous, older but much smaller blocks. Plots of vacant land are increasingly difficult to find.

The map in Figure 15.32 shows its confined geographical location on a sand spit. Developments such as the Sheraton Mirage hotel have gradually increased the size of the village southwards. Such is the speed of development that people are increasingly concerned about the changing character of Port Douglas and the whole northern Queensland coastline. Fears about increasing development include concern for the Great Barrier Reef, which is closer to the coastline of northern Queensland than any other part of Australia.

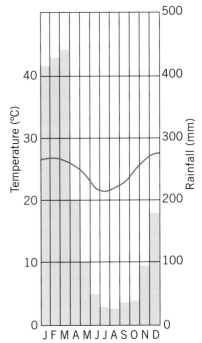

Annual average temperature: 24.8°C
Rainfall total: 2025 mm

Figure 15.29 The climate of Cairns.

Figure 15.30 Port Douglas and its four mile beach.

Figure 15.31 Signs such as this are increasingly common as people sell land for development.

With this in mind, the whole northern Queensland coast has become designated a World Heritage Site. The combination of tropical rainforests and palm-fringed coast around the Daintree river make for a coastline which is regarded as sufficiently special to need protection. Already, much forest has been sectioned into plots of land for sale. Many people fear that Port Douglas has already become too developed. Now a road to the north of Port Douglas into previously inaccessible rainforest has been tarmaced as far as Cape Tribulation, 70km north (Figure 15.33).

Global prosperity among professional and managerial income groups has led to an increase in the proportion of well-educated people with high disposable incomes. Many have a high degree of environmental awareness and believe that natural environments are being permanently altered or destroyed, or that traditional tourist destinations are over-crowded.

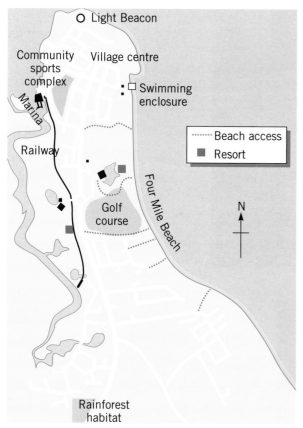

Figure 15.32 Port Douglas, confined on a sand spit. This has forced many developers to seek land further north on the World Heritage Coast.

Figure 15.33 Cape Tribulation. In 1992, this was accessible only by 4wd vehicles; now the road is sealed or tarmaced.

There is a global trend towards nature-based eco-tourism. Particular visitors wish to encourage the tourist industry to be ecologically sustainable and for tourist development to take place in ways that enhance and complement the environment and blend with rather than destroy it. Eco-tourism has been defined as 'tourism that involves travelling to relatively undisturbed natural areas with the objective of admiring, studying and enjoying the scenery and its wild plants and animals, as well as any cultural features that may be found there.' (Ceballos-Lascurain 1991).

The Ecotourism Association of Australia seeks to ensure that eco-tourism:

- is ecologically sustainable
- contributes to conservation of places and of biodiversity ('bio-diversity' refers to the wide range of plant and animal species)
- benefits local communities
- offers a range of opportunities to visitors of different socio-economic backgrounds.

It follows the principles of sustainability (see Chapter 4). The case study of Bloomfield Lodge at Cape Tribulation shows how these principles may or may not work in practice.

Bloomfield Lodge, Queensland

Located within Queensland's World Heritage Daintree Rainforest, Bloomfield Rainforest Lodge is a recent development set in tropical rainforests beside Cape Tribulation. The Daintree Rainforest is acknowledged as a World Heritage Site. One fifth of Australia's bird species resides in the area, including cockatoos, parrots, honeyeaters and lori-keets; some of the world's largest butterflies live amongst the branches. Bloomfield Lodge along lies along a stretch of coast on the Barrier Reef, (Figure 15.34).

Figure 15.34 Bloomfield Lodge amid the Daintree forest, overlooking the Barrier Reef coast of northern Queensland.

It is not easy to get to Bloomfield Lodge. The Lodge is two hours away from Cairns International Airport via a half-hour scenic flight to an airstrip close to the rainforest. From there, a short trip on a dirt road to the banks of the Bloomfield River leads to a boat cruise down-river and across Weary Bay (Figure 15.35) to the Lodge, nestling in the rainforest. The boat is pulled ashore. No access exists by road.

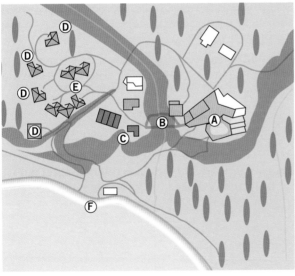

Key

A Main lodge, dining area, bar and pool

B Boardwalk and Jungle Bridge

C Queenslander cabins

D Honeymoon cottage

E Deluxe cabins

F Weary Bay and jetty

Figure 15.35 Plan of Bloomfield Lodge.

The Lodge is planned as an eco-tourist resort. Designed to blend in with, and take advantage of the views from, the rainforest, it consists of amenities that tourists expect – bar, dining terrace, pool, games room are housed close to accommodation in timber cabins. All timber used was from local fallen trees; no new timber was cut. No plastics were used in construction and there is no air-conditioning; all air movement is directed by ceiling fans. No materials were used that require high amounts of energy to manufacture. Fresh drinking water is taken from local sources and the Lodge generates its own electricity on a 240V system. Chemicals used in cleaning and hygiene are bio-degradable. Waste water contains only such chemicals and is either disposed of biologically, or is recycled for garden sprinklers. Tourists are asked to comply with guidelines shown in Figure 15.36.

Staying at Bloomfield Lodge is not cheap. Figure 15.37 shows prices per person for stays of between four and seven nights for 2000. These prices include flight and water transfers from Cairns, as well as accommodation, meals, and some activities. A seven-night stay in a 'Rainforest' cabin – the lowest price band – costs over £630 per person. For anyone travelling to Cairns from elsewhere, flight costs must be added on; in December 1999, economy return air fares from Melbourne were over AU$80 0 (£315). Part of the problem is that the resort can take a maximum of only 34 guests. To maintain a profit margin means higher prices than a mass tourist resort might be able to achieve.

Compare the prices in Figure 15.37 with mainstream accommodation in Port Douglas. At the expensive end of the range, the Sheraton Mirage resort charges AU$350 (£137) per night. However, this is charged per room, so two people staying would be half of this. Lower down the scale of charge, there is a large number of apartments, which offer 2-bedroom accommodation at AU$110 (£43) per night for a group of up to four people. Almost all apartments offer a pool plus en-suite accommodation. A three-course meal in a restaurant in Port Douglas costs in the range £16–£40, depending upon restaurant position and quality.

Other eco-tourist facilities are developing in Australia, including one in the Crystal Creek rainforest near Murwillumbah on the Queensland/New South Wales border. Like Bloomfield Lodge, it is small-scale; its scale is part of the reason why it tends to be more expensive. Low environmental impact tourism is likely to be a feature of many new developments early in the twenty-first century. However, the debate remains whether eco-tourism will just be a feature among high-income groups, while lower income groups are forced into mass tourism in crowded resorts.

While you are away …
Minimise the negative impacts of your visit by:

Social impacts
- Be culturally sensitive and respect local customs
- Remember you are a guest
- Try to allow enough time in each place

Environmental impacts
- Leave an area cleaner than when you found it
- Be efficient with natural resources
- Travel by your own muscle power where possible
- Stay on the trail
- Take only pictures to remember the places visited
- Be careful not to introduce exotic plants and animals
- Do not exploit an area when food gathering
- Do not disturb wildlife or wildlife habitats
- Familiarise yourself with the local regulations
- Do not use soap or detergents in natural water bodies
- Consider the implications of buying plant and animal products. Find our whether they are rare or endangered species, or taken from the wild, and if the trade is approved of by wildlife authorities.

Economic impacts
- When travelling, spend money on local enterprises
- Do not encourage illegal trade by buying products made from endangered species

Figure 15.36 Guidelines for tourists published by the Eco-tourism Association of Australia.

Figure 15.37 Prices per person for stays at the eco-tourist Bloomfield Lodge. Prices are in AU$ – in April 2000, there were 2.5 Australian dollars to £1. These prices include meals.

Accommodation	4 nights	7 nights	Extra night
Rainforest	998	1621	107
Queenslander	1135	1789	120
Deluxe	1295	2027	139
Honeymoon Cottage	1465	2278	151

1 Compare the cost of staying for seven days in Port Douglas at different types of accommodation with that at Bloomfield Lodge. Use the data in this study to copy and complete the following table:

Type of accommodation	Cost for 7 nights per person (£)	Plus meals per person x 7 days	Total cost £
Apartment in Port Douglas Sheraton Mirage Bloomfield Lodge – Rainforest cabin			

2 How do the different types compare What does this say about eco-tourism at Bloomfield Lodge?
3 Why should Bloomfield Lodge costs be so different? Are they worth paying?
4 What does this say about the market for eco-tourism?
5 Is there any way that mass tourism can ever be environmentally sustainable? Discuss.
6 To what extent does Bloomfield Lodge seem to be environmentally sustainable in its design and materials?

Summary

You have learned that:

- Tourism is a rapidly growing industry at all scales; almost all countries are affected.
- Tourism has a number of social, economic and environmental impacts both in LEDCs and MEDCs. These may be beneficial or present problems.
- National parks such as the Peak District are under threat from increasing numbers of visitors.
- Kinder Scout has a fragile and sensitive ecosystem which has been damaged by human activities, particularly agriculture and recreation.
- Management strategies need to be carefully planned and implemented in order to protect fragile environments.
- Strategies such as eco-tourism may have several environmental advantages, but many at present are expensive.

Ideas for further study

1 Research tourist development in a LEDC and compare it with VietNam. What are the key features of tourism? How are they similar to, or different from, tourism in VietNam?

2 There are many opportunities for fieldwork in moorland areas. Recording changes in vegetation, soils and microclimate can provide you with a focus for fieldwork studies. Studies could focus on:
- The characteristics of the ecosystem and vegetation along a transect
- The impact of humans on the ecosystem and any damage being caused
- An evaluation of management strategies to manage these impacts that are already in place, or are planned
- A presentation of your solutions for resolving current problems.

3 Other eco-tourist projects exist, both in Australia and elsewhere. Use a search engine on the Internet (Altavista is especially good for geographical research) to search for locations and characteristics of eco-tourist projects. How do their costs compare with Bloomfield Lodge?

References for further reading

The Eco-tourism Game – an interactive simulation on the internet, reference
http://www.eduweb.com/ecotourism/eco1.html
O' Hare, Greg (1988 reprinted 1994) *Soils, Vegetation and Ecosystems*, Oliver and Boyd.
The National Trust Kinder Scout Ten years On, 1994
RSPB (1994) *Ecosystems and Human Activity*, Collins Educational

Challenges facing rural areas

This chapter is about the challenges that face rural areas. Villages are places into which urban commuters and retired people are increasingly moving, away from city lifestyles. In MEDCs, agriculture has ceased to be the single most important economic activity in rural areas, but it remains one of the most significant. During the late twentieth century, farmers in the UK found it increasingly difficult to make a living, as Figure 16.1 shows, resulting in considerable deprivation in rural areas. This chapter shows how rural deprivation has arisen and how it might be managed. It focuses upon the UK generally, and Cornwall specifically, to see how challenges might be met.

Farming hit by worst crisis since the Thirties

■ Millions of animals now worthless

■ Produce price plummets to all-time low

■ Minister: No cash for sheep farmers

Figure 16.1 From *The Independent*, 28 August 1999.

The changing UK rural economy

Agriculture is no longer the main source of employment in many rural areas, and employment is as varied there as elsewhere (Figure 16.2). People in rural areas are as likely to work in mining or service industries as if they lived anywhere else in the UK. However, there are some distinctions that make rural areas different:

- a greater proportion of people work in farming (4.4 per cent) than in England as a whole (1.7 per cent)
- even where farming is the most important activity, only 15 per cent of the population work in agriculture; others work elsewhere
- a higher percentage of people there are in part-time employment (26.4 per cent) compared to the UK as a whole (25.1 per cent)
- more people in rural areas are self-employed (15 per cent) than in the UK as a whole (12 per cent)
- more people work from home in rural areas (7.2 per cent) than in the UK as a whole (4.9 per cent)

- small businesses make up the great majority of employers; in Rural Development Areas in England and Wales, over 91 per cent of employers had between one and nine employees in 1999.

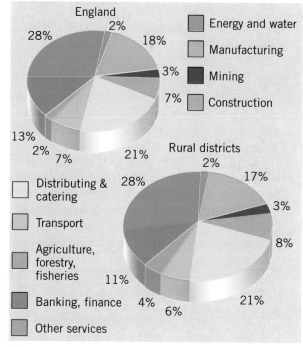

Figure 16.2 Employment diversity in rural areas of the UK.

In spite of this, unemployment in rural areas in August 1998 (4.2 per cent) was lower than for the country as a whole (6.1 per cent). However, some rural counties of England have much lower wage rates than average, as shown in Figure 16.3. Cornwall's average weekly wage is £88 lower than the national average. Rural areas such as these are often geographically, and economically, peripheral, as explained in the theory box, *Core and periphery*.

---

1 Study Figure 16.2. Suggest reasons why each employment sector is as varied in rural areas of the UK as for the country as a whole.
2 Why should wage rates (Figure 16.3) be lower in some parts of England and Wales than others? What impact are such rates likely to have on those parts, and why?

Figure 16.3 Wage rates in rural counties compared to England as a whole.

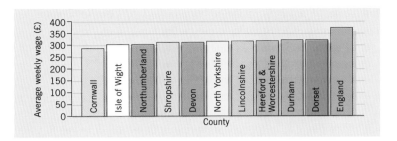

Theory

Core and periphery

'Core and Periphery' is a useful concept in geography because it helps to show how some parts of the world or a country become more developed and wealthier than others. The idea is a useful aid to understanding economic patterns. The relationships between different places and changes that affect them happen at all scales, local to global. Figure 16.4 shows how one set of processes cannot exist without the other. The formation of capital and wealth with the core processes is made possible by the labour and resources in the periphery.

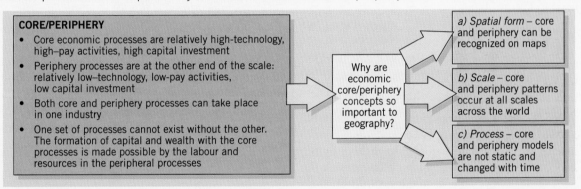

Figure 16.4 The concept of 'core and periphery' in geography.

The south-east of England is the wealthiest part of the UK; it is usually accepted as the core region of the UK. In this region, employment develops which uses relatively high technology, offers higher pay than elsewhere and promotes high capital investment. Wealth of the region attracts high densities of service industries designed to provide for the population. Most decisions involving investment and government are made there. The relative wealth of the region means that house prices are higher there than elsewhere.

In contrast, periphery processes occur at the other end of the scale. Regions on the periphery attract employment that uses relatively lower technology, offers lower pay and attracts lower capital investment. However, there are benefits, as such regions are often cheaper for people to live in. Traditionally, Cornwall, Northern Ireland and Scotland have been quoted as examples of peripheral parts of the UK. However, this should not detract from the economic growth that does occur there.

Why economic core-periphery concepts are important in geography

- They present different patterns over space. It is usually easy to recognise core and peripheral regions using simple economic data. Core processes become concentrated in areas that become affluent and in turn, gain political and economic control. Peripheral processes occur in poorer areas with higher unemployment. Often, both sets of processes occur at the same place, so that economic growth in one part of the economy (e.g. high-tech industries in central Scotland) may balance decline in others (e.g. coal and engineering).
- Globally, core and periphery patterns occur at all scales. Global-scale core regions such as North America and Western Europe contain many smaller-scale core regions, e.g. Southern California, M4 corridor. Other areas within global cores are smaller-scale peripheral areas, such as Western Ireland or Labrador. In the same way, global scale peripheral areas such as Latin America and Africa contain smaller-scale core regions like Accra or São Paulo.
- Core and periphery models are not static. The location and nature of core and periphery regions are constantly changing. The world's Newly Industrialising Countries (NICs) such as Thailand and Malaysia were far less economically developed 30 years ago. Now they are core areas in their own right.

The changing pattern of UK farming

Most farms in England are small. About 43 per cent of all farms, shown in Figure 16.5, are less than 20 hectares in size. However, together these cover only about 4 per cent of agricultural land in the UK. The largest proportion of land – 43 per cent – is held by less than 3 per cent of farms which are over 300 hectares in size. These latter farms are usually run as agri-businesses, using machinery and high levels of capital investment. There is an increasing gap between the prosperous larger units and the deprivation of areas where small farms are the norm. Competition from EU countries (e.g. milk from France, apples from the Netherlands), limits on production imposed by the EU (e.g. on milk) together with cheaper foodstuffs from countries outside the EU have led to falling farm incomes, as shown in Figure 16.6.

As a result, agricultural employment has declined. Between 1987 and 1997, 14 per cent of jobs in agriculture were lost, mostly among employees rather than farm owners. Mechanisation has resulted in fewer employee vacancies. The collapse in beef and dairying during the late 1990s has resulted in transfer of some farms from cattle farming to arable land; this in turn requires fewer employees.

Many farmers are now unable to make a living. Newspapers reported in 1999 how farmers in the UK were no longer able to make a profit on their products (Figure 16.7). Falling prices mean that farmers have to look elsewhere for income, or sell off farm buildings and land. Farms sell their milk quotas, equipment, and then seek planning permission to have outbuildings or barns converted for residential use or holiday accommodation.

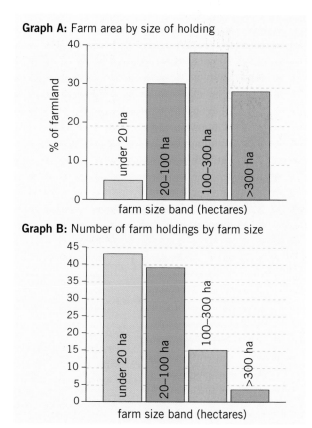

Figure 16.5 Farm area by size of holding (Graph A) and number of farms within categories of farm size (Graph B).

Housing and services in rural areas

While there is extensive rural poverty, some rural areas enjoy a high quality of life. Owner-occupation is greater in rural areas (74 per cent) than nationally (67 per cent), and there is less social housing (15 per cent in rural areas against 23 per cent nationally).

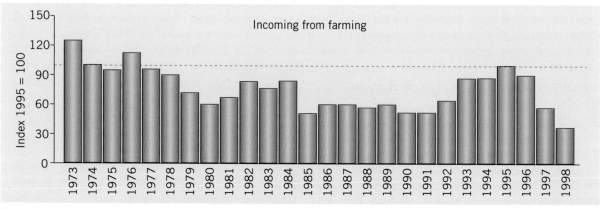

Figure 16.6 Variations in farm incomes since 1973, when the UK joined the European Union. The graph is not actual figures, but an index, where 1995 production is used as a base line of 100. By reading the graph compared to 1995, it is possible to tell whether farm incomes were better or worse from year to year.

Product	Cost of production	British farm gate price	Supermarket cost
Beef	95p (43p per lb) to £1 per kg	94p per kg (105p to 110p per kg before BSE crisis)	Mince: £5.58 per kg; Sirloin: £11.99 per kg
Milk	About 18.5 per litre	17.8 per litre	99p per litre
Lamb	90p to 95p per kg	74.3p per kg (down from £1 per kg in 1998)	Cutlets: £7.78 per kg; Loin fillet: £14.49 per kg
Pork	95p per kg	84p per kg	Loin steak: £6.09 per kg; Fillet: £7.99 per kg
Eggs	45.3p per dozen	27p per dozen	Fresh: £1.45 per dozen; Free-range: £1.95 per dozen
Wheat	£95 per ton	–	£115 per ton (including subsidy) for bread wheat; £100 per ton for poorer wheat
Potatoes	£77 per ton	£50 to £60 per ton	27p per kg (£270 per ton)
Apples (Discovery)	55p per kg	Top grades: 66p per kg; Less than 55p per kg overall	86p per kg
Savoy cabbages	13p per head	10p to 12p per head	47p per head
Iceberg lettuce	25p to 35p per head	30p per head	47p per head

Figure 16.7 The collapse in farm incomes 1999. The data show how much it costs a farmer to produce particular items, compared to the price they actually get. In 1999 supermarkets were accused of not passing on these prices to their consumers; farmers became increasingly angry at how others were making a profit while they could not.

However, there is increasing need for social housing as more and more people are unable to afford homes to purchase. Figure 16.3 has shown how wage rates are lower in peripheral rural counties. Two-thirds of households could not afford to purchase an appropriately-sized home, according to one study in 1998 by the Department of the Environment and Transport. The introduction of the 'right to buy' scheme, whereby people in social housing could purchase their property, resulted in a reduction of 91 000 houses from the social sector in rural districts between 1985 and 1990. Only 17 700 new properties have been built to replace this loss.

Homelessness is a feature normally associated with urban rather than rural life. However, in the early 1990s, about 16 000 families in rural districts were accepted as priority homeless each year, about 12 per cent of the national total. Between 1987 and 1991, homelessness rose more rapidly in rural than in urban areas.

Services and functions in rural areas

Figure 16.8 shows how badly rural areas fared for shops and services in 1997. The past 20 years has seen the huge growth of superstores, many of which are accessible to shoppers with cars from rural areas. As a result, over 4000 food shops closed in rural England between 1991 and 1997, as well as banks and police stations. The pattern depends upon population; 59 per cent of parishes with fewer than 1000 people lacked a shop in 1997, compared to only 1 per cent with populations between 3000 and 10 000. Similarly, two-thirds of

- 42% of rural parishes had no shop
- 70% did not have a general store
- 43% had no post office
- 28% had no village hall or community centre
- 75% had no daily bus service
- 49% had no school for any age
- 29% had no public house
- 83% had no GP based in the parish.

Figure 16.8 Availability of services and functions within rural parishes of England and Wales, 1997.

parishes with fewer than 1000 people had no school, compared to 1 per cent of larger parishes.

The old and the young

Figure 16.9 shows how the young and the elderly fare within rural parishes. These groups depend upon government services, such as schools, medical and care facilities. Government policies towards rural areas and spending have affected them (Figure 16.10). Governments assume that urban areas need more funding than rural, perhaps because deprivation is more concentrated. Deprivation in rural England is more dispersed and less easy to identify.

Transport in rural areas

The absence of services in rural areas is often compounded by lack of public transport. In 1997, 75 per cent of rural parishes had no daily bus service and 65 per cent had no service six days per week. Bus services, where they exist, run during school times or at the start and end of a full working day. About 93 per cent of rural parishes

Older people
- 91% of rural parishes have no day care for older people
- 80% have no residential care for older people
- 61% have no recreational clubs for older people

People with disabilities
- 96% of rural parishes have no day care for people with disabilities
- 91% have no residential care for people with learning difficulties or physical disabilities

Younger people
- 50% of rural parishes have no school for 6-year olds
- 90% have no school for 12-year olds
- 96% have no school for 18-year olds
- 68% have no youth club or other young people's club

Young children
- 89% of rural parishes have no post-natal clinic
- 93% have no public nursery and 86% no private nursery
- 61% have no parent-and-toddler group
- 59% have no pre-school playgroup
- 92% have no provision for out-of-school care

Figure 16.9 Services for the young and elderly in rural areas.

Spending per resident (1997/98)	
Shire (mostly rural) areas	£670
Unitary authorities	£717
Metropolitan districts	£778
Outer London boroughs	£832
Inner London boroughs	£1 160

Figure 16.10 Government spending in rural and urban areas of the UK.

have no rail service; 77 per cent have no bus service after 7 p.m. Community schemes are on the increase but often attract few people.

With few services, people are forced into car ownership: 84 per cent of rural households have a car, compared to 69 per cent nationally. More than 80 per cent of rural journeys to work are by car, compared to 66 per cent nationally. The result is faster growth in traffic in rural areas than in urban. Vehicle flows increased by 23 per cent in built-up areas between 1981 and 1997, compared to 75 per cent outside these areas. (Built-up areas are defined as those areas with a 40mph speed limit or less, including some rural villages.) The majority of road deaths are on rural roads, as Figure 16.11 shows. Some people are left more isolated than others. While 38 per cent of rural families have more than one car, many elderly people have no means of travel. Where public transport does exist, there are fewer concessionary fare schemes for them (Figure 16.12). Only 41 per cent of pensioners in rural areas take advantage of concessionary schemes, compared to 82 per cent in urban metropolitan areas, often because of the lack of services.

Casualty types	Motorway	Urban	Rural
Killed	4	41	55
Seriously injured	3	62	35
Slightly injured	4	70	26

Figure 16.11 Percentage of road casualties in areas of Great Britain, 1997.

Percentage of pensioners for whom:	Rural	Metropolitan urban areas
Concessionary fares are available	93	100
Travel is free	5	67
Their travel pass must be bought	38	4

Figure 16.12 Concessionary fare schemes in rural and metropolitan urban areas.

Work in groups of two or three people.
1 Make a large copy of the table below on to a sheet of paper. Complete it, using the material in this chapter so far.

Strengths of living in rural areas	Weaknesses of living in rural areas	Opportunities for rural areas	Threats to rural areas

2 Do the positive aspects of rural life outweigh the negative? Give reasons for your answer.
3 Identify three priorities for rural areas in the UK as a whole that you believe should be tackled in the next two to three years.
   a) Say what these priorities are.
   b) Justify why you believe these are priorities.
   c) Outline actions that you believe should be taken, and why these should be taken.
   d) Say who you believe should be responsible for these actions.

What kind of future should Cornwall have?

Cornwall, shown in Figure 16.13, is a county of contradictions. It is a prime holiday destination in the UK, for which it is probably best known. Yet it is a poor county, with wage rates lower than anywhere else in England and Wales. Its unemployment is above the national average. Two of its six council districts figure among England's most deprived areas. Only tourism is increasing as an economic activity; most other employment is declining. Yet Cornwall's other characteristics do not seem to make it poor. Unlike many urban areas of deprivation, car ownership and home ownership are high. Its population is increasing at a rate well above that of the UK. It has fewer lone parents than the country as a whole. Its schools perform better than the national average.

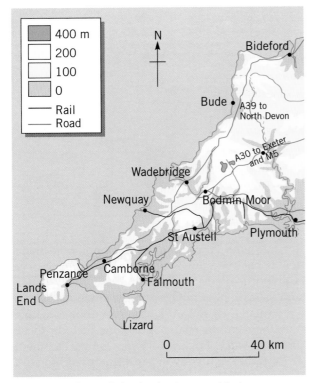

Figure 16.13 Cornwall showing key towns and features.

Figure 16.15 The coast of southern Cornwall, near Gorran Haven. This is an Area of Outstanding Natural Beauty, owned in this case by the National Trust.

Cornwall is a distinctive geographical area. It is a rural, maritime county with over 570km of coastline, the longest of any English county. It has over 300 public beaches and many secluded coves. To the east, Cornwall's border with Devon is formed almost entirely by the River Tamar, which forms a physical and cultural divide with the rest of Great Britain for all but 18km of its length. Some 26 per cent of Cornwall is designated as Areas of Outstanding Natural Beauty (Figure 16.15). The county has 17 per cent of the defined Heritage Coast in England and Wales. There are over 1700 recorded Ancient Monuments, and a further 3000 currently awaiting designation. Cornwall had a population of 485 600 in 1997

Figure 16.14 The character of Cornwall, as described by Cornwall County Council.

Population change in Cornwall

Cornwall is part of the South West Standard Economic Region, is the second largest county in the region, but has the lowest population density. Its population has grown by 27 per cent since 1983

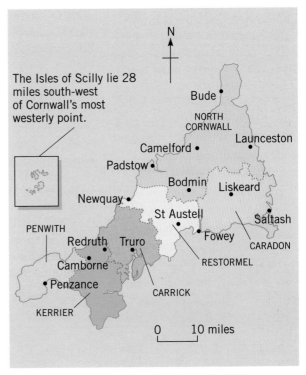

Figure 16.16 The council districts in Cornwall, 1999.

and its working population by 24 per cent. The comparable rates for the UK and the European Union are 15 per cent and 6 per cent. Figure 16.16 shows the six council districts in Cornwall. While the county as a whole has had substantial population increase since 1971, it varies geographically (Figure 16.17).

Districts	1971	1996	1971-96 actual number increase (+) or decrease (-)	1971-96 % increase (+) or decrease (-)	Most recent change 1991-1996
Penwith	35 475	39 700	4 225	11.9%	-2.4%
Kerrier	33 525	39 320	5 795	17.3%	0.5%
Carrick	43 140	49 060	5 920	13.7%	0.9%
Restormel	42 415	55 415	13 000	30.6%	1.8%
North Cornwall	27 070	38 730	11 660	43.1%	5.0%
Caradon	27 940	42 950	15 010	53.7%	4.8%

Figure 16.17 Population change within Cornwall, 1971 - 1996.

1 Make two copies of Figure 16.16. Shade in three categories of population change as shown in Figure 16.17 for:
a) 1971–1996
b) 1991–1996.

2 Where does population increase most rapidly? Refer to Figure 16.13 and suggest reasons why this might be.

3 In recent years, the main A30 trunk road that links Penzance, Bodmin and Exeter has been substantially improved. How might this affect population growth in Cornwall, and why?

Town	Population 1996	Population of catchment 1996	% difference
Penzance	18 305	36 860	101%
Camborne & Redruth	27 480	59 130	115%
Falmouth	19 785	35 380	79%
Truro	16 705	52 395	214%
St Austell	21 185	61 725	191%
Newquay	17 955	32 890	83%

Figure 16.18 Shopping catchment sizes of towns in Cornwall, whose catchment exceeds 25 000 people.

Most of Cornwall is rural, with 70 per cent of its population living outside urban centres. It has nine towns with a population of over 10 000 – Penzance, Camborne, Redruth, Falmouth, Truro, St Austell, Bodmin, Liskeard, and Saltash. By comparison with towns elsewhere in the UK, its towns are small (Figure 16.18). However, they contain a range of functions, as their commercial populations include those who travel in to shop or to work. Figure 16.18 shows the catchment area of each town. Truro is the county town, has a larger workforce, a greater proportion of professional and managerial people with higher spending power, and therefore supports a larger range of shops and wider range of functions than its population might suggest.

Cornwall's population is older than most other parts of the UK, as shown in Figure 16.19. For many years, it has been a favourite place for retirement. Most people who retire to Cornwall do so because they can afford to sell up and move, often paying less for a house in Cornwall than the one they left. This has implications for the county's social services. Generally, incomes of the elderly are lower and they are more demanding upon health resources.

	% people aged 0-15	% people of working age 16-64	% Retired people, aged 64+
1991			
Cornwall	19.2	61.2	19.6
UK	20.2	63.9	15.9
1997			
Cornwall	18.9	61.1	20
UK	20.5	63.7	15.8

Figure 16.19 Population structure of Cornwall, 1991-1997.

%	1981	1997	Change 1981–1991
Primary	13.6%	10%	-3.6%
Secondary	13.1%	10.5%	-2.6%
Tertiary	73.3%	79.5%	+6.2%

Figure 16.20 Changing employment sectors, 1981-1995.

4 What trends are shown by the population data in Figure 16.19?

5 How would you use these trends to assess:
a) the needs of the county now?
b) the needs of the county in 20 years time?

The changing economy of Cornwall

Cornwall has undergone considerable economic change in recent years (Figure 16.20). Traditional industries such as farming, fishing and china clay extraction (Figure 16.21) have dominated the economy of the county in the past, but are now weakening. There is a reasonable manufacturing base, but by far the dominant sector of the economy is in tertiary – or service – industries.

Every year, the population of Cornwall swells considerably in summer. Over 4 million people visit Cornwall each year, spending some £930 million; because of its remote location, most visitors stay over for periods of a week or more. This adds considerably to the economic potential of the county. In 1996 Cornwall's Gross Domestic Product was estimated to be £3680 million. Of the 587 jobs created in Cornwall in the first three months of 1999, 185 were in Leisure and tourism, and another 106 in Food and drink. Tourism is the only industry which is currently expanding. The industry accounts for around 30 000 jobs with many more at the peak of the season. However, much investment in tourism is from outside the county, so that only about a third of tourist spending is retained within Cornwall.

A summary of employment found in Cornwall is shown in Figure 16.22. Small firms dominate the economy. In the late 1990s, 87 per cent of Cornish employees worked in small firms with fewer than 10 employees, compared to 83.6 per cent for the UK. 21 per cent of the workforce were self-employed, compared to the UK average of 11.6 per cent. Most work is, predictably, found in towns. For the 70 per cent of people who live in rural parts of Cornwall, this means a sometimes lengthy journey to work.

Why is Cornwall so poor?

With a prosperous tourist industry and a broad economic base, Cornwall is still the poorest county where people are worse-paid than anywhere in England and Wales. Figure 16.24 shows wage levels for Cornwall up to 1998. Not only are wages low, they are falling further behind. Much employment in the county is low-wage, such as those jobs in tourism and in food and catering, which are semi- or un-skilled.

Figure 16.21 China Clay extraction in mid-Cornwall, north of St Austell. This has given mid-Cornwall a characteristic appearance as a result of the 'white hills' consisting of quartz waste. It is in this landscape that the Eden project is to be developed.

Employment type	Examples and locations	Current trends
Primary		
Agriculture and forestry (9670 people in 1991)	County-wide; mainly dairying and early vegetables, as a result of the mild but wet climate	Lost large numbers of workers during the 1970s and 1980s; milk quotas imposed by the EU have had drastic effects on milk production; farm incomes are collapsing.
Fishing (1120 people in 1991)	Several fishing ports, e.g. St Ives, Mevagissey, Fowey, Padstow, Polperro	Currently losing many traditional fishing grounds to EU competition; price of fishing boats has collapsed; overfishing by EU fishing fleets generally are depleting fish stocks.
Mining – surface (3250 people in 1991)	The country's largest producer of china clay, shown in Figure 18.24; mainly on St Austell Moor in mid-Cornwall between St Austell and Newquay;	Until recently, Cornwall's biggest booming based at St Austell industry; world glut has caused a price collapse and the main producer, English China Clays, has been bought out by a French company; falling job prospects.
Mining – underground (260 people in 1991)	One of Cornwall's oldest industries, found mainly on granite moors near Redruth and Camborne	The last tin mine in Cornwall, South Crofty, closed in 1997
Secondary		
Manufacturing (over 19 000 people in 1991)	Mainly located in urban areas, especially Camborne, Penzance Falmouth, Newquay and Bodmin	Several companies in engineering, IT equipment, timber and building, and food processing; strong push towards developing 'Cornish' food products, such as ice cream, pasties, clotted cream, etc.
Tertiary		
Retail (21 000 in 1991)	Mainly urban; several main superstores in the main towns shown in Figure 18.19; recent out-of-town shopping complexes at St Austell, Truro, Penzance, Falmouth	As elsewhere in the UK, booming superstores and deteriorating town centres. However, Truro attracts several 'quality' shops, but other towns such as St Austell are declining; most new development focused on Truro, especially as tourism develops all year round; most employment is low-wage.
Hotels and catering (14 000 in 1991)	Mainly coastal; mixed between mass resort tourism (Newquay), shown in Figure 18.25, heritage tourism using Cornish culture and history (county-wide), and environmental quality (especially gardens, such as Heligan Gardens, shown in Figure 18.26, and trails, such as the River Camel trail)	Variable; most employment is low-wage; hotels in Newquay are suffering at the lower end of the market; the sun's total eclipse in August 1999 brought an extra £51 million into the county; the industry is expanding, with White Acre Country Park in Newquay created 70 new jobs, and new Chef and Brewer hotel complex at Hayle another 60; the self-catering, holiday home market is growing, especially out-of-season; environmental and heritage tourism are booming, with Heligan Gardens attracting 10 000 people weekly in summer; television series such as Rick Stein's cookery programme have made Padstow (where the series is based) more popular
Banking and finance	Mainly at the county town of Truro	Growth industry, and likely to be more so as home finance via the internet enables banks to locate anywhere
Education	County-wide	Primary school rolls falling as the county ages; most 18 year-olds move away in search of higher education; the nearest universities at Plymouth and Exeter still force students to live away from home, making it expensive for many; plans for a new University of the South West in Cornwall are still not finalised; the main problem is a suitable site in a county which is 100 miles long and has slow communication links.
Medical	Several small 'cottage' hospitals have closed; most hospital care centred at Treliske Hospital, Truro; no teaching hospital exists in Cornwall	Increasingly, people in Cornwall have to travel to Plymouth for major surgical care (including neurology) and some cancer treatment.
Public administration	Mainly at Truro, the location of Cornwall County Council	Faced by cash cutbacks; however, the civil service Benefits Agency has created new jobs in Truro, Penzance, and St Austell.

Figure 16.22 Employment in Cornwall – a summary of the trends in 1999.

Seasonal unemployment

The tourist industry is seasonal, and employment opportunities increase substantially in summer. Once school terms begin in September, the industry slows until March. Figure 16.25 shows how unemployment rates vary according to time of year.

Men	Cornwall	Great Britain	% by which Cornwall is below GB average
1981	118.1	140.5	15.9%
1986	169.3	207.5	18.4%
1991	246.3	318.9	22.8%
1996	303.4	391.6	22.5%
1998	319.2	427.1	25.2%
Women	Cornwall	Great Britain	% by which Cornwall is below GB average
1981	82.3	91.4	10.0%
1986	119.4	137.2	13.0%
1991	178.2	222.4	19.9%
1996	224.6	283.0	20.6%
1998	250.6	309.6	19.0%

Figure 16.24 Average gross weekly earnings of full-time adult employees, 1981-1998 (£ per week, figures for April each year).

Figure 16.23 Heligan Gardens in south Cornwall. The gardens began in 1990 as a project to unearth and restore gardens left abandoned in 1914 on a former large estate. Now known throughout the world, it has been televised and has caught a surge of interest in gardens generally among older tourists.

During times of economic growth, Cornwall is slower to pick up than the rest of the country. Its remoteness from major cities and urban markets makes it less desirable for companies to locate there. Figure 16.26 shows how its unemployment lags behind the UK as a whole, and is also worse with increasing distance into the county. Four towns in Cornwall featured in the 20 worst locations in England and Wales for unemployment by percentage unemployed, in January 1999.

	January 1996	July 1996	Jan 1997	July 1997	Jan 1998
Cornwall	10.3	8.4	9.2	6.0	7.3
South-West region	7.4	6.2	5.7	4.2	4.1
Great Britain	8.2	7.6	6.7	5.6	5.2

Figure 16.25 Unemployment in Cornwall compared to the rest of the UK. The data show rates of unemployment for Cornwall January 1996 to January 1998. The figures given are registered unemployed as a percentage of the total workforce (employees in employment, the unemployed, self employed, HM forces, and participants on work-related government training schemes).

Town	Unemployed people March 1999
Penzance	2223
Redruth/Camborne	1518
Truro	1496
Falmouth	1040
Newquay	1295
St Austell	1444
Bodmin area	804

Figure 16.26 Unemployment in and around Cornwall's main towns March 1999. Recorded unemployment fell dramatically in Cornwall during the late 1990s, but at a slower rate than the rest of the UK.

1 Imagine a haulage company locating in Penzance. Use a computer journey programmer such as 'AutoRoute' to plan lorry journeys between Penzance and London. Set average journey speeds to 50mph for journeys on major trunk roads, and 60 mph for motorways. How long does the journey take?
2 Compare this for a haulage company in Exeter and travelling to London. How significant is the difference?
3 In what ways is geographical isolation a problem for developing businesses in Cornwall? How could this addressed?
4 Using a copy of Figure 16.16, map the data shown in Figure 16.26. How far do these data support the idea that Cornwall is a peripheral rather than a core region?

Geographical isolation and cost

Cornwall is one of the most remote counties of England and Wales. Its sheer physical size is a factor; though a narrow peninsula, it is long. From Plymouth to Penzance is over 80 miles. East-west communication through the county is served by a number of main roads, the chief of which is the A30 from Penzance to Exeter, the M5 to Bristol, London, the Midlands and the north. For most of its length, it is dual carriageway. West of Bodmin, the road becomes intermittent single- and dual-carriageway to Penzance. Traffic queues and bottlenecks in summer are legendary, especially at weekends.

Rail links are good but slow, and expensive at peak times. Several attractive branch lines to coastal and inland destinations have survived against all odds, thanks to transport subsidies from Cornwall County Council. They are being marketed towards an expanding tourist market. However, a business traveller from Truro to London has to leave at 5.45 a.m. to be in London for 10.00 a.m. and the last train leaves London at 6 p.m. In 1999, the fares were £120 Standard and £160 First class.

Air links exist between Newquay and London's Gatwick airport. However, there are only four flights per day and a full return fare cost about £270 in 1999.

Does quality of life compensate?

Some people imagine Cornwall as an ideal place to live. Its beaches and countryside and a slower lifestyle are undoubted attractions for many people – and not just for those who are retired. Village life appeals to many, where community spirit is strong. The attraction for many young people is strong when 'surf's up', even in mid-January! This is referred to as 'quality of life' and has to be balanced against economic standards of living, where Cornwall figures less well.

However, for many, there is another side. Working hours are long (Figure 16.27), and nearly a quarter of the population works over 40 hours per week. This pattern is an annual average, and seasonal variations may make it worse at certain times of the year. In August 1999, a report entitled *Inequalities and Health in the South West Region* highlighted several areas of deprivation. In Cornwall and the Scilly Isles, 3716 households did not have a toilet, bath or shower, or were forced to share one. 70 per cent of council housing tenants received housing benefit.

Yet home ownership is higher than elsewhere in the UK. Three-quarters of people own their own home in Cornwall, as many as in some of London's most affluent areas. Again, 75.5 per cent of households in Cornwall owned at least one car in 1991, compared to the UK average of 67.6 per cent. Cornwall also has a higher proportion of older cars than elsewhere in the UK.

% of people working	More than 40 hours/week	31-40 hours	30 hours or less
Cornwall	22.7	53.6	23.7
England and Wales	17.7	61.1	21.1

Figure 16.27 Hours worked by people in Cornwall, 1998.

Tenure	Cornwall	England and Wales
Owner-occupied	74.4%	67.8%
Privately rented	11.6%	9.3%
Housing Association and Local Authority rented	13.0%	19.8

Figure 16.28 Housing tenure and car ownership in Cornwall, 1991.

Attracting development into Cornwall

'It is economic growth which will provide the opportunity for individuals to achieve their own routes out of poverty.' These words by Cornwall's Director of Health, Dr David Miles, in 1999 show the significance of acting to help the county. A report in August 1999 showed that many people in Cornwall suffered ill health as a result of poor housing and low incomes. Increasingly, Cornwall County Council, the UK government and the European Union (EU) have sought to address rural deprivation by pursuing economic growth. Cornwall has some natural advantages for economic activity, as Figure 16.29 shows.

Location	Industrial land, cost per acre (purchase)	Office space cost per sq. foot to rent
Devon	£40 000 – £250 000	£6.50 – £14.00
Cornwall	£40 000 – £65 000	£5 – £10
Plymouth	£60 000 – £70 000	£7.50 – £10

Figure 16.29 Average costs of sites and premises in Cornwall, 1997.

In attempts to attract industry and employment to Cornwall, a number of strategies are being developed. These include the identification of Rural Development Areas by the UK government and those areas targeted for Objective Funding by the EU. Both are intended to promote economic growth. Figure 16.30 shows how Cornwall features in both lists.

Rural Development Areas

Areas of England and Wales where rural poverty is greatest have been nominated Rural Development Areas (RDA), shown in Figure 16.30. RDA are those parts of England that suffer greatest economic and social need. There are 31 RDA in England, together covering over one-third of the country by area. These areas are diverse in nature and vary in their causes of need; RDA include Cleveland and Redcar, with unemployment rates in 1999 of 8.1 per cent, Doncaster RDA 7.9 per cent, and Cornwall and Isles of Scilly RDA 5.6 per cent. However, the Rural Development programme for RDA is targeted using money from central government; the total sum of money available in 1999 was about £1.3 million.

1 Which people in which jobs in Cornwall are likely to work the longest hours? At what time of the year and why?
2 Why should:
   a) car ownership
   b) house ownership
   be so much higher than the average for England and Wales?
3 Why is the fact that Cornwall also has a high proportion of older cars, significant?
4 What other data would be helpful in deciding whether most people in Cornwall had:
   a) a high or low economic standard of living
   b) a high or low quality of life?
5 How might the data in Figure 16.29 be used to attract companies to locate in Cornwall? Which other data in this chapter might also be helpful in persuading them?
6 Why do you think that, in spite of these data, it is difficult to attract companies to the county?

7 Use a map of the UK to name each of the RDA areas or counties shown in Figure 16.30. What patterns do you notice?
8 Why should areas such as Doncaster and Cleveland and Redcar be classified as 'rural' areas in need of assistance? Use your atlas to suggest why this is so.
9 How might the problems faced by Doncaster and Cleveland and Redcar differ from those in Cornwall? In what ways might they be similar?
10 Read the theory text 'Core and periphery areas'. How far do you think this idea helps to explain the location of RDA in England and Wales?

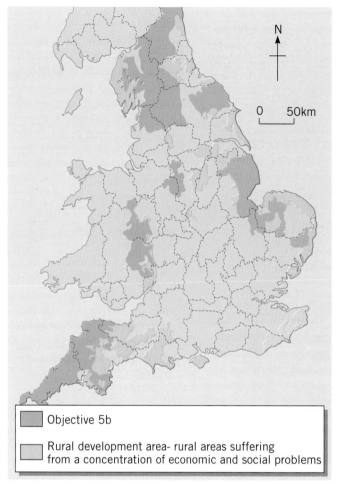

Objective 5b

Rural development area- rural areas suffering
from a concentration of economic and social problems

Figure 16.30 Rural Development Areas (RDA) in England and Wales,
April 1999.

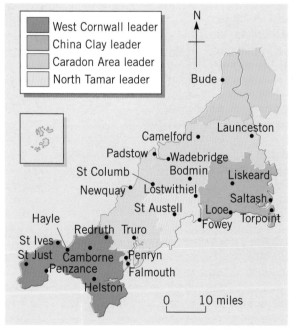

West Cornwall leader
China Clay leader
Caradon Area leader
North Tamar leader

Figure 16.31 Leader development areas in Cornwall.

EU Objective funding

The EU makes two kinds of funding
available – Leader and Objective Funds.

a) Leader programmes

This is a programme designed to promote
local small-scale development in rural areas.
There are five main areas of activity: rural
tourism, agriculture, small businesses, voca-
tional training, and community development.

In particular, the programme is keen to
develop and link rural tourism with farming,
to help farmers diversify and to develop a
range of styles of rural tourism. Cornwall has
four leader areas, shown in Figure 16.31. The
EU provides grants for projects in these areas,
provided that other sources at least match
grants awarded. The maximum amount
payable is therefore half of the cost of a
project. It depends upon the ability of those
proposing development to seek sources of
investment from elsewhere.

Identify the Leader areas shown in Figure
16.31 and, using data in Figures 16.14–
16.28 to help you, suggest:
   a) why these four areas should be
      chosen
   b) suggest why areas outside the
      scheme might have been left out.

b) Objective funding

Like Leader funds, these are targeted funds
aimed at increasing economic development.
However, in this case, the whole of Cornwall
and the Isles of Scilly is a European Objective
funding area. There are different funds within
the programme; in 1999, Objectives 1, 3 and
5(b) funds were available. These change
almost annually and up-to-date information
can be found on Cornwall County Council's
web site, referred to at the end of this chapter.

Objective funds are directed at the
following:
   • small- and medium-scale businesses and
     their needs, including creation of
     employment and training within the region
   • tourism, including training (e.g. in
     customer care) and creation of oppor-
     tunities for agri-tourism
   • agriculture, by encouraging diver-
     sification of produce, including higher-

value products (e.g. cheese, wine) and environmentally friendly approaches to managing the environment (e.g. planting of mixed woodland)

- community regeneration, by developing facilities such as training that will benefit local communities
- environment, in order to provide an alternative environment into which people from outside can enjoy leisure time

- long-term unemployment, training of young people and the promotion of equal opportunities between men and women. Funding is available for training, guidance and counselling, and job creation to improve employability of young people.

Four examples of Objective-funded projects are shown in Figure 16.32. They are mainly short-term and small in scale, but they are significant.

Name of project	Cost	Location	Purpose
CHEERS (5 years, ends 2002)	£1.2m	St Austell and china clay moors northwards to Newquay	Upgrading of council housing in six of Cornwall's most deprived estates; demolition of block of flats; provision of a multi-activity centre which will include community facilities and advice about employment and from the Benefits Agency
Cornwall Developing In Partnership (7 years, ends March 2003)	£7.5m	Town centres	Town centre improvements in St Austell, Saltash, Bodmin, Liskeard, Newquay, Camborne/ Redruth, Falmouth, St Ives and Hayle.
West Cornwall Employment Programme (4 years, ends March 2002)	£0.8m	West Cornwall	An employment programme which provides subsidies towards wages paid by employers and transport costs to encourage recruitment, capital grants for businesses setting up, free business advice (Business Consultan ts can otherwise charge up to £1500/day)
Engage (5 years, ends March 2004)	£2 million	County-wide	Targeted at youth programmes, including training

Figure 16.32 Four examples of Objective-funded projects in Cornwall in late 1999.

Winning Objective 1

Cornwall had been awarded Objective 1 status by the EU, which would provide £300 million in grants over six years, starting June 2000. Again, funding must at least be matched before any grants are awarded. The UK government will match the amount, thus bringing the total to £600 million. Each project will be funded to 75 per cent of its cost, the rest coming from private investment. A total investment of £1 billion could result.

Future projects

There are as many ideas on what to spend the money in Cornwall as there are people! However, the focus will be to improve infrastructure and develop larger-scale initiatives than other RDA or Objective funding. Some believe that the money should be used for large-scale projects that develop the infrastructure for Cornwall; they argue that this would encourage many more companies to locate

Figure 16.33 Objective 1 celebrations in March 1999.

there. Others believe that social issues are the most pressing; long-term deprivation has led to an ethos of low expectation among some of the young in Cornwall. More still believe that the route for Cornwall's future prosperity lies in enhancing and developing the quality of its environment, the very reason why people come to Cornwall.

Some of the proposals and projects currently alive in Cornwall are shown in Figure 16.35. Study these, and the feelings expressed in the letter in Figure 16.34. How should this opportunity best be developed?

I offer my services!

Now that we have European Grant Aid can we first consider what not to do with it.

Can we avoid giving large sums to companies which set up here only to disappear after three to four years? Can we avoid giving several hundred 'specialists' £40 000 a year for telling the rest of us how to find jobs at £3.60 an hour? Can we avoid throwing more money down ancient tin mines? Can the first five pounds of the money be spent on a plastic dustbin into which will go all the grant-spending suggestions? Then a panel of good, ordinary citizens can choose the best. Doing it that way we might do something sensible.

My suggestion will be to spend half of the millions on bargains at all the car-boot sales in Cornwall. These bargains can then be put for sale in those vacant warehouse premises we keep building. This wonderful scheme is environmentally friendly, benefits those most in need, will employ loads of people – at £3.60 an hour – will make a profit, and will only require one man on £40 000 a year to run it! I offer my services.

Malcolm Lindsay, Camborne

P.S. Can I apply for a grant to get this idea underway?

Figure 16.34 One person's feelings about Objective 1 funding in Cornwall.

Name of project or proposal	Location	Brief details
Actual project: Eden project, shown in Figure 16.36	Bodelva, 3 miles east of St. Austell, a 14-hectare former china clay pit, 60m deep with steep south-facing walls that catch the sun, even in the depths of winter. It had been worked for more than a century and by 1997, it was coming to the end of its economic life. Its dereliction is a 'challenge for regeneration' according to the project directors.	In a former china clay pit near St Austell, the Eden Project is building a living theatre to tell the stories of plants and people. Nearing completion, and the size of 30 football pitches, it consists of a visitor centre, on the lip the crater, two giant conservatories to house major 'biomes' of rainforest plants. The covered biomes will create a backdrop to a 'roofless' biome, home to temperate plants. Open all year. Local nurseries are growing many of the plants.
Proposed project: University of Cornwall	Various ideas for location, including Penzance, Truro, and another proposal that suggests splitting the site across the county, different courses being located on different sites	The demand for a University in Cornwall has been alive for decades. However, it came close in the mid-1990s when a shortlist of sites was drawn up. No further progress has been made. It is suggested that courses should be tailored to Cornwall's economy and culture, offering courses in tourism, engineering, ICT, etc.
Proposed project: Encouraging businesses such as call centres	Could be county-wide, as such businesses are footloose - their work is carried out by phone or electronically. Electronic banking is increasing rapidly. Cornwall has no such businesses at present.	Call centres deal with customer calls for anything from car finance, to mobile phones, to specialist ICT advice, or tele-sales. Training is usually intensive but fairly brief (in a few weeks rather than months)
Actual project: National Trust holidays	Lanhydrock House and estate near Bodmin	Consists of National Trust property on an estate which has one of the Trust's most popular attractions. As well as the main house, there are working farms and holiday cottages. The latest scheme is to offer working holidays on the Trust's farms.
Proposed project: Construct a new 6-mile dual carriageway stretch along the A30 west of Bodmin	Convert all single-carriageway stretches into dual carriageway, to provide a full dual route to Exeter and the M5 to London, Bristol, the Midlands and the North	The intermittent dual- to single-carriageway nature of the A30 west of Bodmin makes travel very slow especially in summer. It would be costly; the 6-mile stretch west of Bodmin alone would probably cost £10 million per mile.

Figure 16.35 Development projects for Cornwall – actual and proposed.

Figure 16.36 The Eden project, due for completion in Spring 2001.

Work in groups of two to three people. Evaluate the benefits to Cornwall of the projects shown in Figure 16.32 and 16.35.

1 Make a large copy of the criteria below.

Criteria for judging success	Project
How well does it add to Cornwall's employment base? How well does it address Cornwall's problem of seasonal unemployment? How well does it increase absolute prosperity? How well does it address Cornwall's geographical isolation? How well does it enhance the quality of environment in Cornwall? How well does it support communities faced with change? How well does it enhance Cornwall's 'distinctiveness'?	

You should be able to add to the list of criteria as you work.

2 Evaluate each of the projects in Figures 16.32 and 16.35.

3 Is it possible to evaluate which projects you feel are:
  a) most effective
  b) least effective
  in addressing Cornwall's problems?

4 Report your findings to the rest of your class. What is the general feeling about the most valuable projects? Why is this?

Summary

You have learned that:

- Rural deprivation exists, is serious, and can be defined by many criteria. These include low wage rates, higher than average unemployment and considerable seasonal employment.
- Many rural areas are geographically distant from economic core regions; they are regarded as peripheral.
- Some indicators of deprivation do not match with our expectations about poverty; many rural areas have high home and car ownership, yet are otherwise poor. Lack of transport infrastructure makes a car a necessity in rural areas
- Part of the solution to rural deprivation lies in maintaining economic diversity.
- Many rural areas are increasing in population, some of which results from migration from urban areas, including those who wish to retire there.
- Geographical isolation can be a hindrance to economic growth.
- Both local, national and European governments offer aid to rural areas, including RDA status and Objective Funding.
- Development aid money is often contentious in how it is spent.

Ideas for further study

1 Select two areas which are named RDA in the Midlands and northern England.
a) What circumstances have led to these areas being more deprived than other areas of the UK?
b) How are the circumstances there similar to or different from those in Cornwall?
2 Select one of the RDA in Figure 16.30. Identify the reasons why it is a RDA. Contact local councils to find out what is being done to address the issues there.

3 Identify a rural area close to you. How would you define whether or not it is a 'deprived' area? Which criteria identified in this chapter show that a) it is deprived, and b) it is not deprived?

References and further reading

Cornwall County Council at www.cornwall.gov.uk
Countryside Agency, *State of countryside report 2000*, available on the Countryside Agency website http://www.countryside.gov.uk
Owen, Andy (1998) *Managing Rural Environments*, Heinemann

Changing rural environments: summary

Enquiry questions	Key ideas and concepts	Chapters and case studies in this book
Introducing rural environments 2.1 What are rural and urban environments?	• The problems of definition within a rural-urban continuum. • Definition of rurality. • Perceptions of rural and urban living. • The changing relationships between rural and urban areas associated with development	• *Introduction to Rural Environments* • *Chapter 13 (Ashwell, and rural environments in MEDCs)* • *Chapter 14 (Malawi and rural environments in LEDCs)*
Process and change in rural environments 2.2 How and why do rural environments vary in landscape and character?	• Variety in rural environments results from basic differences in location and physical, socio-economic and cultural factors. • Contrasts exist between rural areas in MEDCs and LEDCs in terms of population density, settlement, and economic activities.	• *Chapter 13 (Ashwell)* • *Chapter 14 (Malawi)*
2.3a What are the processes of change which affect rural areas? 2.3b How are these leading to change and conflict in rural areas?	• Socio-economic processes have led to change and modification of rural environments and communities. • Changes in rural environments have led to changes in population - both decline and expansion, depending on the location and development potential. The effects of these changes. • Rural change and the changing demand for land can generate conflicts in rural environments and communities.	• *Chapter 13 (Ashwell) - changing rural environment, impact of counter-urbanisation on landscape and communities, problems of service provision* • *Chapter 14 (Malawi) - poverty in areas of outmigration and disease in LEDCs - e.g. AIDS in Malawi* • *Chapter 15 Development of rural areas for more varied uses such as tourism*
Rural planning issues 2.4a What are the challenges of managing rural environments? 2.4b Who should decide how rural areas should develop? Who are the main players? 2.4c How do planners and decision-makers attempt to resolve conflicts in a variety of locations? 2.4d How successful are the strategies of planners and decision-makers in managing change and conflict?	***Choose two*** of the following to illustrate the enquiry questions in 2.4, using examples from countries ***at differing states of development*** • Managing the countryside for recreational and tourist use. Issues and consequences. • Managing rural deprivation and rural poverty. Causes, consequences and solutions. • Managing the development of resources in the countryside. The conflicts generated by water, forestry and mineral use. • Managing rural environmental problems generated by agricultural land use.	• *Chapter 15 (VietNam, Kinder Scout, northern Queensland, including eco-tourism)* • *Chapter 16 - effects of tourism in Cornwall* • *Chapter 14 - rural deprivation in Malawi* • *Chapter 16 - rural deprivation in Cornwall* • *Chapter 6 (Murray-Darling River Basin)* • *Chapter 7 (Three Gorges Dam along the Yangtze River)* • *Chapter 6 (Salinity in the Murray-Darling River Basin)*
Rural futures 2.5 What is the future for rural areas?	• Sustainable development strategies are needed to improve the quality of rural environments and the lives of the rural poor. • Innovative solutions are required to ensure the survival of many rural areas.	• *Chapter 14 (Malawi and UNDP programmes)* • *Chapter 16 Rural development strategies in Cornwall*

Introducing urban environments

Figure 1 shows six images of cities. These photographs have been taken in different parts of the world and attempt to show the differences and similarities between cities. This section is about urban environments. It explores a number of themes, each focused on particular cities. It includes a detailed study of two cities – Manchester and Mumbai – and includes their origins, growth and development. Later chapters focus on urban issues, including studies of the environmental impact of transport and urban inequalities. Each asks whether urban living is 'sustainable', and whether cities can ever be sustainable.

Figure 1

A

B

C

D

E

F

1 Study Figure 1.
  a) In pairs, identify where each photograph might have been taken.
  b) What makes identifying some places easy, while others are more difficult?
  c) Which geographical ideas and issues can you identify in the photographs, e.g. transport, services, housing, quality of life?
2 On a large sheet of paper, write the letters A, B, C, D, E, F, spaced over the whole sheet.
  a) Using the photographs A to F, identify as many links as possible.
  b) Draw lines to show connections between the photographs and annotate each line with a possible link, e.g. E and F show inequalities.

How do people view urban environments?

For many people, their view about life in cities is based on their own life experiences. People who have lived traditional or rural lifestyles, such as Australian Aborigines or the American Indians, might have a different perception (view) to people who have grown up in a city. Chief Seattle, an American Indian who lived in the nineteenth century, had strong views about cities and was concerned with the apparent loss of environmental awareness amongst busy, noisy city dwellers.

What does 'urban' mean?

'Introducing Rural Environments' (p.159) has shown how 'rural' and 'urban' are best seen as a continuum. Urban can be defined as a 'built-up area', but how big does a 'built-up area' have to be before it becomes an urban settlement?

Three possibilities are:

- **Population density.** India for example, uses 386 people per square kilometre. However, the population density of some outer urban areas, such as that shown in Figure 3, might be greater than a rural settlement, shown in Figure 4.
- The **functions** of a settlement. In Pakistan, a town has a municipal corporation, a town committee or a permanent military station.
- The **number of dwellings.** In Peru, a settlement with more than 100 dwellings is a 'town'.

There is no quiet place in the white man's cities. No place to hear the unfurling of leaves in the spring, or the rustle of insects wings. And what is there to life if a man cannot hear the lonely cry of the whippoorwill [a bird] or the argument of the frogs around the pool at night? … Whatever befalls the Earth, befalls the sons of the earth. If men spit upon the ground, they spit upon themselves. This we know: the earth does not belong to man; man belongs to the earth. All things are connected like the blood which unites one family. Whatever befalls the Earth, befalls the sons of earth. Man did not weave the web of life; he is merely a strand in it. Whatever he does to the web, he does to himself.

Figure 2 From Chief Seattle, 1855.

1 Study Figure 2; what are the key messages that Chief Seattle highlights?
2 Compare the images of cities in Figure 1 to Chief Seattle's quote in Figure 2. How far do the photos support his view?
3 Form groups of two or three. Describe to what extent you a) like, and b) do not like cities.
4 Decide the advantages and disadvantages of living in a city for 16 to 19-year olds.

Figure 3 Outer urban housing on Long Island, on the edge of New York City. Here, each house is built on a 1-acre (0.4 hectare) plot.

Figure 4 Rural settlement in Arizona, USA. Houses here are built at greater density than urban housing in Figure 3.

There are many methods of defining 'urban' and many geographers use a combination. As population and settlements change, so do the words that are used to classify them, e.g. the word 'mega-cities' is used in the USA and elsewhere to describe cities with over 10 million people. Urban geography is therefore a constantly changing area of study.

What is urbanisation?

The term 'urbanisation' refers to an increase in the proportion of people who live in cities. A country becomes urbanised when the percentage of people who live in towns and cities increases above the percentage in rural areas. Urbanisation occurs in both More Economically Developed Countries (MEDCs) and Less Economically Developed Countries (LEDCs).

During the twentieth century, the actual number of people living in urban areas increased and a greater percentage of the total population now live in urban areas. The rate of change varies in different parts of the world. In MEDCs, urban populations were already high in 1900 because they increased during the industrial revolution of the nineteenth century. Most of the rapid growth in LEDCs has occurred since 1950. In 1980, most of the world's largest cities were still found in MEDCs – New York City, Tokyo, Paris, London. However, the populations of these cities have hardly changed since and in some cases, have even declined.

In 2000, the world's largest cities are referred to as 'mega-cities' because of their sheer size. Most are located in LEDCs. In 1994, 25 per cent of these cities were in MEDCs (Tokyo, New York, Los Angeles) and 75 per cent in LEDCs (São Paulo, Mexico City, Shanghai, Bombay, Beijing, Calcutta, Seoul, Buenos Aires, Jakarta). **Millionaire cities** have populations of over 1 million people. In 1997, there were 106 in MEDCs and 179 in LEDCs.

How does urbanisation take place?

Urbanisation is the process whereby the percentage of people living in towns and cities increases. Urbanisation takes place:

- where the movement of people (migration) from the rural to urban areas exceeds the migration of residents from urban to rural areas
- when the life expectancy is greater in urban than in rural areas, which results in more people living in the urban areas. Urban areas often have better access to medical care and other services than rural areas
- when natural population increase – the difference between birth rate and death rate – is greater in urban than in the rural areas, so that the urban population grows more rapidly.

---

1 Study and discuss Figure 5. Which continents:
   - are the most urban?
   - show the most change?
   - have the largest number of mega-cities?
2 Suggest reasons why this pattern has emerged.
3 How do you think this map will look in 25 years time? Explain your prediction.

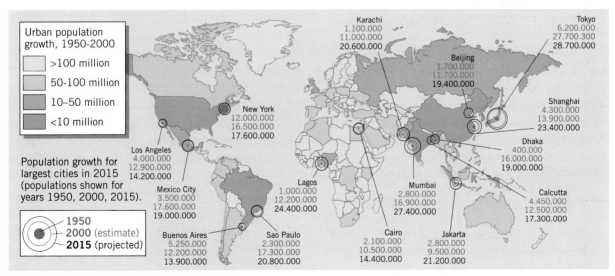

Figure 5 Urban population growth, 1950-2000.

How does urbanisation vary over time and space?

In 1995, 45 per cent of the world's population was classified as urban. However, the rate of urbanisation is uneven in the world. In the early 1990s, the United Nations estimated that over 70 per cent of the population in MEDCs lived in urban areas compared with less than 40 per cent in LEDCs. It predicts that by 2020 over 53 per cent of the world's population will live in cities in LEDCs.

Continent	Percentage of population in urban areas (1995)
North America	79.1%
Europe	75.5%
Latin America	75.0%
Oceania	73.5%
Asia	33.3%
Africa	33.2%

Figure 6 Urban residents of the world.

1 Choose an appropriate technique to display the urban percentage figures in Figure 6 on an outline world map. Justify your choice of technique.
2 Describe the distribution of urban residents in the world.
3 What do you think the distribution will be when you are aged 25? Describe and account for any changes.

Theory

The rank-size rule

Settlements can be ranked by population size, from largest (number 1), down to smallest. In investigating populations of settlements in countries, Zipf predicted that the second largest settlement would have a population half the size of the largest settlement, the third largest one-third the size, the fifth largest one-fifth the size, and so on. This formula is written as:

$P_n = P_1/n$

- P_n is the expected population of a settlement according to its rank
- P_1 is the population of the largest settlement,
- n is the ranked position of P_n

Figure 7 shows how some countries do not conform to this rule, because one city assumes dominance over the whole country. Such cities, of which London is an example, are known as **primate cities**. A primate city completely dominates a country through its size and attracts a disproportionate amount of investment, employment, political power and influence compared to the rest of the country. Primate cities are usually (but not always) the capital and are often several times the size of the second largest city.

Settlement	Population 1991	Rank	Expected Zipf population
London	7 567 000	1	7 567 000
Birmingham	1 014 000	2	3 783 500
Sheffield	471 000	3	2 522 333
Leeds	452 000	4	1 891 750
Manchester	440 000	5	1 513 400

Figure 7 Population sizes in England and Wales, 1991.

1 In this book, case studies are drawn from several countries, including Australia, Thailand, the USA, Malawi, China, India and Nigeria. Using atlas data, geography textbooks, or internet sources, research the populations of the five largest cities in each country.
a) Which do and which do not fit the rank-size rule?
b) Suggest reasons why this is the case.

This chapter studies Manchester to explore the growth and development of an MEDC city. Chapter 18 will provide a contrasting pattern in Mumbai, a LEDC city. Each has grown rapidly during its development and faces different issues.

The growth and development of Manchester

Figure 17.1 shows Manchester's location within the north-west of England, to the west of the Pennines. Many rivers, such as the Mersey, Irwell, Tame, Irk and Medlock, flow from the Pennines through Greater Manchester towards the Irish Sea. The city centre is approximately 30m above sea level and the altitude increases in the surrounding boroughs. To its west lie the Pennines, and a number of routes into the urban areas of West and South Yorkshire.

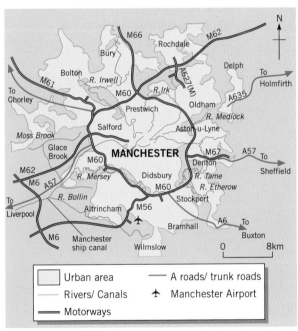

Figure 17.1 The location of Manchester.

Manchester has several images. Its image among young people is as a centre for music, sport, travel, nightlife and major venues such as GMex. It is a regional centre, with several theatres, concert halls, major shops and restaurants. It offers the opportunity to study all aspects of modern urban geography, including quality of life issues and conflicts of land use.

Defining Manchester

Within this book:
- **Manchester** means the city, with its population of 432 641 (1997/98)
- **Greater Manchester** refers to the county, which covers about 311 km² and has a population of 2 583 985 (1997/98), shown in Figure 17.2.

Greater Manchester contains the eight metropolitan boroughs shown in Figure 17.2, and the two cities of Salford and Manchester. The county was established in 1974, as one of the six new Metropolitan counties in England when local government was re-organised. There is no local government to control Greater Manchester; the Authority controlling the area was dissolved in 1986. Each borough and city has its own council. This produces problems in managing issues such as transport. The shape of Manchester as a city, (Figure 17.2), means that north-south transport links can be managed by Manchester City Council; east-west links rely on good relations with other local authorities. However, the county is still recognised as an administrative area for certain groups, such as the Greater Manchester Police Force.

How and why has Manchester developed?

Manchester has not always been a city. Pages 238 to 241 show a timeline of Manchester's development: pre-1800, in the nineteenth century, 1900-1970, and 1970 to 2000.

Figure 17.2 Greater Manchester showing the surrounding boroughs.

1 Make a copy of the Greater Manchester map (Figure 17.2). Label the ten areas and, using two colours, show which are metropolitan boroughs and which are cities. Highlight the boundary of the city centre.
2 What problems do you imagine exist in managing transport in the area?

Manchester before 1800

79 AD Romans founded a fort near the confluence of the River Irwell and the River Medlock. At this time woodlands, bogs and marshes surrounded the region. The population was 200 soldiers.

627 AD Saxon fort 'Manchester' built.

1550 Population estimated at 2500.

16th and 17th centuries Textile industries of wool and linen developed, followed by cotton.

18th century Dramatic changes established the town as the first major industrial centre of the world with canals and manufacturing using steam power. Local coal on the Lancashire coalfield helped to provide the steam power necessary to power the machinery in the mills. The population grew due to immigration.

1750 Population 43 000, establishing Manchester as a city.

1772 Duke of Bridgewater had a canal built, connecting Manchester to coal mines at Worsley.

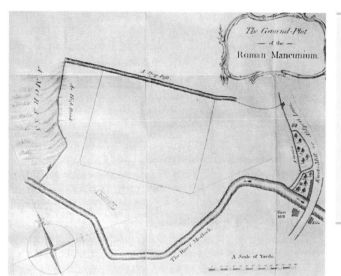

Figure 17.3 Plan of Roman Manchester.

1 Copy and complete the table below showing the population at different times.
2 Draw a line graph of the population of Manchester, with correct horizontal spacing between dates.

Date	Population of Manchester
79AD	
1550	
1750	
1801	
1850	
1877	
1901	
1921	
1991	
1995	
1999	

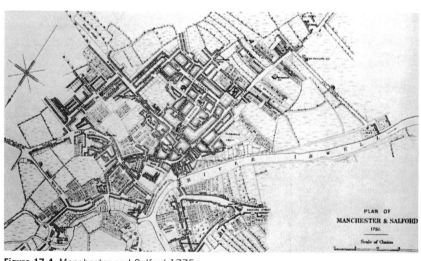

Figure 17.4 Manchester and Salford 1775.

"The conditions in the factory hit me really hard. Fifteen hours a day, non-stop, indoors, no air to breathe, steamy, hot and noisy."

Figure 17.5 Thomas Prince (born 1772) talking about factory work in Manchester in 1788.

"... in September 1793, I pinched a purse, and drank the findings. Next thing I knew! I was in the workhouse. Well you can't get any lower than that."

Figure 17.6 Thomas Prince, a migrant to Manchester from a Cheshire village in the early 1790s, talking about his experiences in 1793.

Prince took a job in the newly-opened McConnell Mills, attracted by the offer of a house to go with the job, No. 9, Loom Street. He became a skilled cotton spinner.

Figure 17.7 Thomas Prince in 1795.

Manchester in the nineteenth century

The nineteenth century was a period of immense civic pride, e.g. parks, squares and reservoirs. In the early stages, it was also a period of total urban squalor in some of the worst housing conditions known.

1801 Population 70 000.

1821 Manchester Guardian newspaper established.

1830 World's first city with a passenger railway linking Manchester and Liverpool.

1832 Major cholera epidemic highlighted the squalid conditions in the working class areas. It was thought to be due to 'bad air' – not infected water from open sewers, as proved to be the case.

1851 Population 316 000.

1868 Building regulations introduced. Previously there were no building regulations for housing. Builders could build as many houses on a site as they liked.

1877 Population 500 000.

1891 First public sector housing in Manchester was built on Oldham Road and Pollard Street; 5-storey flats were built around a central playing area for children.

1894 Manchester Ship Canal opened.

Street quality	House quality		
	FIRST: spacious, water laid on	SECOND: modest, no water	THIRD: inferior back to back
First class: wide, paved, sewers cleaned	19	22	27
Second class: Poorly paved, poor drainage, not cleaned	20	26	28
Third class: Unpaved, not drained, uncleaned, mainly courtyards	21	28	40

Figure 17.8 Death rates per street per 1000 population, Manchester, 1838. Environmental quality in Manchester was poor as a result of industrial growth that was unplanned.

In 1842, 23-year old Friedrich Engels came to work in Manchester at his father's cotton plant. The suffering he witnessed is recorded in his work, *Condition of the Working Class in England*.

Figure 17.9 Conditions in nineteenth century Manchester.

1 Figure 17.8 shows death rates per 1000 people in Manchester in 1838, according to the type of house and street quality. What does this tell you about Manchester at the time?

2 How might quality of environment have affected people's health in the nineteenth century?

3 In 50 words each, describe Manchester in the nineteenth century under the following headings: social, economic, environmental.

4 In pairs, discuss whether social conditions were a price worth paying for rapid economic growth.

…women from their door tossed household slops of every description into the gutter; they ran into the next pool, which overflowed and stagnated. Heaps of ashes were the stepping-stones, on which the passer-by, who cared in the least for cleanliness, took care not to put his foot. You went down one step even from the foul area into the cellar in which a family of human beings lived – the smell was so fetid as almost to knock the two men down – three or four little children rolling on the damp, nay wet, brick floor, through which the stagnant filthy moisture of the street oozed up.

Figure 17.10 From *Mary Barton* by Elizabeth Gaskell.

Figure 17.11 19th century Warehouses in Manchester.

Manchester 1900 to 1970

1901 Population of city 544 000. Within the south Lancashire conurbation it was 2 000 000.

1903 University of Manchester established.

1907 First garden village financed by the council, five miles south of the city in Burnage. Houses were built at a density of twelve per acre and each house was positioned to have the maximum amount of light and air. Places for recreation were provided, e.g. tennis courts and bowling greens. Quality was high and rents reflected this, making it inaccessible to working classes.

1911 Another garden village built at Chorltonville, South Manchester, by private developers. In the village, each pair of semi-detached homes differed externally and internally from other pairs. The houses had large bay windows to maximise the light received by each house.

1918 Manchester needed 17 000 'homes fit for heroes' at the end of WW1. Government funds were made available.

1921 Population 800 000

1922 Town Hall held a successful conference and exhibition on town planning. The need for land use plans became apparent and 15 town planning committees were set up, each responsible for its own area of Manchester.

1926 Wythenshawe was built, the largest housing estate in Western Europe. It aimed to re-house people from slum clearance programmes in Manchester. It aimed to build a complete town of 100 000 people, through developing shops and industry.

1927-29 Manchester Corporation built 2500 houses a year and was running out of suitable sites for future development.

1931 Population was 766 000 in the city of Manchester.

1930s Economic collapse created mass unemployment, especially in the textile mills of Manchester.

Before 1945, the rapid population growth in the city created a great number of slum areas with crowded living conditions and poor sanitation.

1940 – 29th August the first bomb was dropped. 22nd and 23rd December saw the Christmas Blitz. Manchester was heavily bombed and much of the city centre was devastated by bomb damage and fires, including 30 000 houses.

1945 The Manchester Plan offered the opportunity to create an international city. Extensive re-building was essential after the war. The Plan was the first development plan for housing, the environment, transport and the city centre. New housing density standards were established to reduce slum areas.

1947 Town and Country Planning Act

Figure 17.12 Garden village form the early twentieth century in Chorltonville.

1948 The world's first computer was developed and the first atom was split; each happened at Manchester University.

1961 Development Plan outlined future developments for 1961-71, e.g. public utilities, health, minerals, roads and open spaces. Compulsory Purchase Order powers allowed the council to re-develop slum areas. New housing density guidelines set an upper limit of 90 habitable rooms per acre. Many council housing estate projects were completed during this time, e.g. Hulme was created as modern, self-supporting communities.

1963 One third of the 80 000 houses in Manchester were unfit for habitation; this led to a massive slum clearance programme. These were replaced in many areas by high-rise flats.

Figure 17.13 Local Authority Housing in Hulme, Manchester 1960s.

Manchester 1970 to 2002

Since 1975 change in fortune with decline of many older industries leading to high unemployment, derelict land. Region fighting back, re-establishing itself as a major financial and service centre. Already, it is clear that attempts to plan the city in the 20th century have led to some successes (garden villages) and failures (high-rise flats, which become ghettoes of unemployment and social deprivation).

1974 Piccadilly Radio launched

1977 The Arndale Shopping Mall opens.

1981 Moss Side inner city riots make the headlines.

1991 Population 440 000

1991 M60 planned as a complete ring-road for the city.

1992 First British city to re-introduce trams.

1995 Population 450 000

1995 MEN (Manchester Evening News) Arena opened, Europe's largest indoor entertainment venue.

1995 New Manchester Plan to improve the city as a place to live, work and visit and to revitalise the local economy.

1996 15 June largest terrorist bomb ever exploded on mainland Britain in the city centre. An international competition to devise a master plan for rebuilding is launched.

Figure 17.14 The newly-opened Manchester Arndale Centre in 1999.

> The redesign and rebuilding of the city centre has been quick and the changes can only improve the already buzzing social and cultural scene, with its cutting-edge sports arenas, concert halls, theatres, clubs and leisure facilities, plus England's largest student population and a blossoming gay community whose spending-power has transformed a once derelict part of the city into an unrivalled Gay Village.

Figure 17.15 From the *Rough Guide to Manchester*.

Figure 17.16 Canal Street, centre of Manchester's 'gay village'. The area occupies what used to be tenements and warehouses; now it consists mainly of cafes, shops and clubs.

1999 4800 residents in city centre as warehouse conversions and 'cafe society' attract people back to the centre to live. City Council aims for 10 000 by 2002.

2000 Manchester re-defining itself as a city of leisure and tourism, and of media personalities – Oasis (Gallagher brothers born in Greater Manchester), Simply Red (Mick Hucknall owns Barca Pub in Manchester), Ryan Giggs and David Beckham (Manchester United), Coronation Street still in the Top 10 television listings after 40 years.

2002 Commonwealth Games host.

1 How do popular music, clubs, television programmes and sport help to:
   • generate economic activity in cities
   • provide other benefits for cities?
2 Copy the table below on to a large sheet of paper. Read pages 238 to 241 and classify the information about Manchester's development into the five headings.

	pre-1800	19th century	1900-1970	1970-2002
Population trends Changing employment Quality of environment Quality of life National/World reputation				

Analysing Manchester's growth using models

Geographers have developed models from their observations of patterns and processes to make sense of the world. Models are simplified views of the world. Models are applied to help focus on and analyse a situation or case study. The section on Rural Environments has shown how one cause of urbanisation is rural to urban migration. The Ravenstein model offers a method of studying patterns and processes of migration.

Theory

Ravenstein - Laws of Migration (1885)

Ravenstein stated that most migrants only travel short distances. He described a negative relationship between the distance travelled and the number of migrants. As the distance increases, the number of migrants decreases. Ravenstein also suggested that although the majority of migration is to an urban area, there will be some movement to rural areas from urban (a compensatory counter-movement). He suggested that migration occurs in stages or waves. Most people do not migrate directly to the largest settlement; instead, they move smaller distances to a minor town first, then a larger town and finally, a large city. Overall, people migrating are more likely to be from rural areas than urban. Ravenstein suggested that females were more likely to migrate within their country of birth whereas males were more likely to consider international migration.

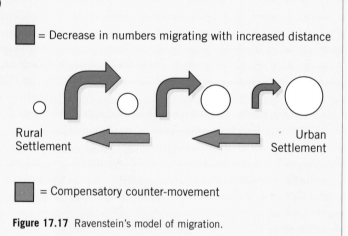

= Decrease in numbers migrating with increased distance

Rural Settlement

Urban Settlement

= Compensatory counter-movement

Figure 17.17 Ravenstein's model of migration.

1 Design a questionnaire to investigate migration. Some questions that you could ask includes:
   • Have you migrated?
   • Over what distances have you migrated?
   • What were your reasons for migrating?
   • How often have you migrated?
   • Would you consider migrating to these places: X, Y, Z?
2 Carry out a class survey on migration using your questionnaire. Do your results support Ravenstein's ideas?
3 What information would you need in order to establish whether Manchester fits Ravenstein's ideas?

Rostow's model (1960)

Another cause of urbanisation can be economic growth. This can attract migrants to the area looking for work. Economic growth can also lead to improvements in health and quality of life that would lead to an increase in urbanisation as life expectancy improves. In his model, Rostow outlined five stages of economic growth which an area may pass through in its economic development:

1 traditional society based on subsistence farming using little technology
2 pre-conditions for take-off with an increase in investment agriculture becomes more commercialised. A single industry dominates the economy, e.g. textiles
3 take-off manufacturing grows rapidly, supporting the infrastructure, e.g. transport is improved, growth is concentrated in one or two regions
4 drive to maturity – self-sustaining growth reaches all parts of the country as industry develops and there is rapid urbanisation
5 age of mass-consumption – there is rapid expansion of services and at the same time a decline in manufacturing.

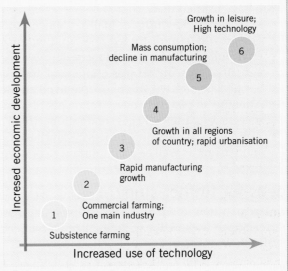

Figure 17.18 Rostow's model.

Recently a sixth stage has been added, called the post-industrial age and demonstrated by slow or zero growth. Manufacturing continues to decline and is replaced by high technology industries and services. People have increased leisure time. In Manchester, for example, attempts are being made to 're-define' the city and change it from a manufacturing environment of textile mills into a cleaner, more environmentally-friendly city. Manchester Science Park opened in 1984 as an attempt to generate high-income, hi-tech service employment. Retail and leisure industries have boomed, with the opening of the MEN Arena in 1995, and the Trafford Centre – an out-of-town retail and leisure complex – in October 1998. The Commonwealth Games are to be held in Manchester in 2002.

1 Make a copy of Figure 17.18. On it, select evidence from pages 238 to 241, showing the development of Manchester that illustrate the stages in the Rostow model.
2 To what extent do the following developments in Manchester refer to the sixth stage of the Rostow model?
   • Manchester Science Park (1984)
   • the MEN Arena (1995)
   • Trafford Centre (1998)
   • the Commonwealth Games (2002).
3 Using the information in this chapter, write an essay in response to the following statement that describes the changes in Manchester: 'Cities are a reflection of the times in which they develop.'
4 Which time period in Manchester's development has had the most influence in shaping the future of Manchester? Justify your answer.

Figure 17.19 Manchester Evening News Arena.

Summary

You have learned that:

- Manchester grew from a small settlement into a major city by a process of industrialisation.
- The growth of industry led to a rapid increase in jobs and therefore population.
- Environmental quality in Manchester was poor during the nineteenth century as a result of industrial growth that was unplanned and burned fossil fuels to create energy.
- Social class affected where people lived and the quality of environment in which they lived.
- Attempts to plan urban environments in Manchester in the twentieth century have led to some successes (garden villages) and failures (high-rise flats).
- Planning is an on-going process; Manchester now has plans to establish itself as a city of culture and leisure, as well as an important regional economic centre.

Ideas for further study

1 Research how town and country planning has developed in the UK e.g. the Acts of Parliament and their purpose.

2 Find out the contents of the Development Plan for your nearest urban area.

3 Investigate how Manchester is changing in the 21st century.

References for further study

Local Library

'Whatever happened to planning?' by Peter Ambrose

'Planning in Britain' by David Kirby and Richard Carrick (1985) University Tutorial Press ISBN 0 7231 0907 9

Manchester.

http://www.manchester.com/java/home.shtml

http://www.madforit.com

http://www.manchester.gov.uk

Manchester choosing our future by Chris Canham DEP ISBN: 1 869818 55 5

Manchester this good old town by Brooks and Howarth 1997 Carnegie Publishing Ltd ISBN 1 85936 034 3

Manchester 50 years of change 1995 HMSO ISBN: 0 11 702006 0

The growth and development of Mumbai

This chapter is about Mumbai, India's largest city. In 1996, the population of Mumbai was 13 532 000, the seventh most populated city in the world and nearly twice as big as Greater London. However, Mumbai ranked fourth in the world in terms of population density in 1996, higher in each case than London and most major cities in MEDCs.

Where is Mumbai?

Mumbai is the capital of the Maharashtra state, on the west coast of India. India is the second most highly populated country in the World, after China. In 1994, the World Bank estimated a population of 1 016 242 000 for India by 2000, compared with 1 255 054 000 for China. By comparison, the World Bank estimated a population of 58 882 000 for the UK in 2000.

The city of Mumbai was founded on seven islands. In the nineteenth century land reclamation schemes turned the area into a single island by draining wetlands. Now the single island makes up the landmass on which Mumbai is shown in Figure 18.1. The city now extends via roads and railway to the mainland.

1 Using an atlas, draw an annotated sketch map showing the location of Mumbai within India. Identify key relief features, and its location and distance from other cities.
2 Measure the extent of Mumbai north-south along the main built-up area. How does this compare with other cities that you know?

Mumbai or Bombay?

The names Mumbai and Bombay are interchangeable. In 1534, the city was taken over by the Portuguese, who named the area Bom Bahia (Good Bay) which evolved into the name Bombay. However, since May 1995, Mumbai has been the official name of Bombay, within India.

Mumbai is an interesting city. It is the centre of the Indian film industry – 'Bollywood' (from Bombay) is the world's largest, producing the greatest number of new feature films each year. Mumbai is also a city of great contrasts, with some of the most expensive property in the world, and some of Asia's worst slums. Of its population of over 13.5 million people, over half live on the streets or in slums, and many survive by begging.

The city is one of India's largest textile manufacturing centres. Over half of its factory workers are employed in cotton textiles, with others in silk, bleaching, dyeing and synthetics. Its industry also includes bicycles, printing, glassware and pharmaceuticals. The city is important for research and education; there is an atomic research centre, the University of Mumbai, the Institute of Technology, and the National Centre for Software Technology.

Study Figures 18.2 and 18.4, and the quotes from the Rough Guide to Bombay and the Lonely Planet website (Figures 18.3 and 18.5).
- To what extent do they fit your ideas of an LEDC city?
- In what ways are they different and why?
- In what ways are they similar and why?

Figure 18.1 Mumbai, India.

Figure 18.2 Dharavi, to the east of Mahim Bay, is known as Asia's largest slum. Here a quarter of a million people live in less than 1 km^2.

What is Mumbai like?

The Rough Guide series of books are a useful source of information for geographers and travellers alike (Figure 18.3). They offer a vibrant insight into life in the city. Mumbai is economically the most developed city in India, yet it has poverty.

The growth and development of Mumbai

Study the timeline of key events that have helped to shape Mumbai (on pages 247 to 250). This shows a pattern of when, how and why Mumbai developed as a city.

'Young, brash and oozing with the cocksure self-confidence of a maverick money-maker, Bombay, or Mumbai as it's now officially known, revels in its reputation as India's most dynamic and westernised city ... The roots of the population problem lie, paradoxically, in the city's enduring ability to create wealth. Bombay alone generates 35% of India's GNP, its port handles half the country's foreign trade, and its movie industry is the biggest in the world.'

Figure 18.3 From the *Rough Guide to Bombay* (Mumbai).

Figure 18.4 Afternoon tea in the exclusive Malabar district.

Mumbai is the glamour of Bollywood cinema, cricket on the maidans on weekends, bhelpuri on the beach at Chowpatty and red double-decker buses. It is also the infamous cages of the red-light district, Asia's largest slums, communalist politics and powerful Mafia dons. This tug-of-war for the city's soul is played out against a Victorian townscape more reminiscent of a prosperous nineteenth century English industrial city than anything you'd expect to find on the edge of the Arabian Sea.

Mumbai is the industrial hub of everything, from textiles to petrochemicals, and responsible for half of India's foreign trade. But while it aspires to be another Singapore, it is also a magnet for the rural poor. It's these new migrants who are continually re-shaping the city, making sure Mumbai keeps one foot in its hinterland and the other in the global marketplace.

Figure 18.5 From the *Lonely Planet* web site.

Bombay before 1800

Pre-history The Kolis settle on the islands of Bombay.

1100 Mahim was the first of the seven islands to support a significant population, when Hindus settled there; today Hindus account for 75% of Mumbai's population.

1300s Muslims invaded the islands; today they are the largest minority group, with 15% of the population.

1508 Portuguese arrived led by Francis Almeida.

1534 The Sultan of Gujarat handed over the main islands to the Portuguese.

1626 British and Dutch arrived in Bombay.

1640 The first Parsis arrived in Bombay.

1661 Portuguese passed the islands to the British as the dowry of Catherine on her marriage to Charles II.

1661 Population 10 000.

1668 British East India Company acquired the modern city, on lease from the crown for an annual sum of £10.

1672 Governor Gerald Aungier promised a society open to all racial and religious communities. He formed a judicial system, an army, opened a mint, constructed docks, encouraged companies and migrants.

1675 Population 60 000.

1687 East India Company moved their headquarters to Bombay, attracted by its sheltered harbour.

17th century A variety of migrants, especially skilled workers and traders e.g. Parsis, Armenians, Jews, etc., were attracted through incentives to set up business.

1720 Land reclamation began to join the seven islands.

1736 Ship building developed in Bombay. It was so successful it threatened the UK industry, providing frigates for the Royal Navy.

mid-18th century More migrants arrived, including slaves from Madagascar; the British segregated themselves from other groups.

1780 Population 13 726 as Bombay attracted fortune hunters.

Late-18th century Large-scale engineering works and land reclamation attracted construction workers.

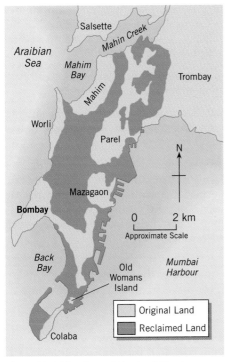

Figure 18.6 The Seven Islands of Bombay.

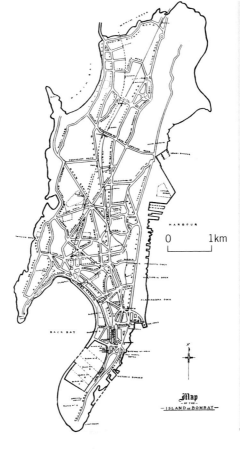

Figure 18.7 Early Bombay.

1 Study the timeline for Bombay's history before 1800. In groups, devise ten newspaper headlines to illustrate the events, taking your style of writing from the following: *The Sun; The Mirror; The Mail; The Guardian; The Times.*

2 What impression do you get of Bombay in 1800?

3 In pairs, select what you consider to be the four most significant events in Bombay's early development. Justify your selection, and compare it to that made by others.

Bombay in the nineteenth century

1803 The island of Salsette was joined to Bombay by the Sion causeway. A great fire in this year demolished large parts of the town.

1838–45 A number of roads were constructed, between Bombay and the hinterland. By 1845 all islands were joined.

1853 Construction began of the first railway in India between Thana and Bombay though it was not opened until 1869.

Figure 18.8 Victoria, Mumbai's rail terminus.

1854 The first cotton mill in Bombay opened, which attracted workers to the city. Previously the raw cotton was shipped to Manchester. Bombay became known as the 'Manchester of the East'.

1858 Following the First War of Independence, the East India Company was accused of mismanagement and the islands reverted to the British Crown.

1860 The Great Indian Peninsular and the Bombay Baroda and Central India Railway were started.

1860 First piped water supply in the city; a good drainage system was also constructed.

1860 40% of Bombay's exports was the drug opium, to China.

1861-5 The cotton trade boomed due to American Civil War. Personal fortunes were made during this period from the resulting increase in demand. Prices rose by 800%, leading to a property boom.

1869 Suez canal opened, increasing trade with England.

1870 The Bombay Port Trust was formed.

1871 Population 644 405.

Figure 18.9 Colonial-style buildings in Mumbai, surviving from the British rule of India from the mid-nineteenth century.

1872 Jamshedji Wadia, a master ship-builder constructed the Cornwalis, a frigate of 50 guns, for the East India Company, a success which led to several orders from the British Navy.

1875 The Stock Exchange at Bombay was established. Now it is the premier Stock Exchange in India.

1891 Population 821 000; the increase was due to investment in infrastructure, such as transport, power supplies, water and sewerage.

1899 75 000 people employed in the cotton mills of Bombay. Increasing urbanisation put pressure on the environment, especially sanitation. Bubonic plague spread, claiming nearly 3000 victims a week at its peak. Half the city's population fled to the countryside, which affected the 1901 census.

1 Study the timeline of the events in the nineteenth century. Create a flow chart of the improvements to the city's infrastructure.

2 In pairs, select which events were the most significant in Bombay in the nineteenth century. Now decide if they were political, cultural, social, economic or environmental events.

1900 – 1980

The city expanded onto the mainland. Bombay expanded northwards. Buildings such as the General Post Office and Town Hall were also built. 1880 acres of land were reclaimed.

1913 First Indian feature film released, *The Raja Harishchandra.*

1915 Gandhi returned from South Africa and reached Bombay to support the campaign for independence of India from Britain.

1927 The first electric locomotives manufactured by Metropolitan Vickers of England began on passenger trains to Poona and Igatpuri.

1933 First major film studio established.

1944 April, a mysterious fire started in one of the holds of the ship, *Fort Stikine*. It was thought that this might have been connected to a rumoured Japanese invasion, given the progress of World War Two.

1945 Bombay becomes Greater Bombay. India becomes a free country on 15 August.

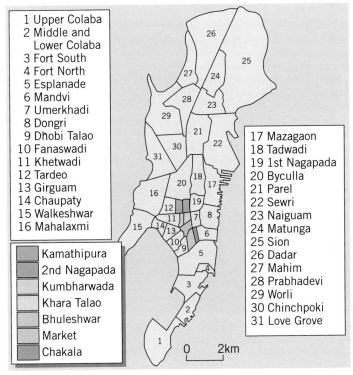

1 Upper Colaba
2 Middle and Lower Colaba
3 Fort South
4 Fort North
5 Esplanade
6 Mandvi
7 Umerkhadi
8 Dongri
9 Dhobi Talao
10 Fanaswadi
11 Khetwadi
12 Tardeo
13 Girguam
14 Chaupaty
15 Walkeshwar
16 Mahalaxmi

17 Mazagaon
18 Tadwadi
19 1st Nagapada
20 Byculla
21 Parel
22 Sewri
23 Naiguam
24 Matunga
25 Sion
26 Dadar
27 Mahim
28 Prabhadevi
29 Worli
30 Chinchpoki
31 Love Grove

Kamathipura
2nd Nagapada
Kumbharwada
Khara Talao
Bhuleshwar
Market
Chakala

Figure 18.10 The wards and sections of Mumbai.

The *Fort Stikine* was about to unload a lethal combination of cargo in the port. Dried fish, cotton bales, timber, gun powder, ammunition and gold bars were on board valued at £2 million. Nobody is certain how the fire started. Two explosions following the outbreak of fire were so loud that windows shattered eight miles away. Destruction in the docks was immense and several hundred dock workers were killed. Many of the Bombay Fire Brigade lost their lives in the second explosion.

1947 The last British troops leave India, through the archway known as the Gateway of India

1947 Bombay becomes the capital of the state of Bombay. Independence changes the political life of the city. The new decision makers affect trade restrictions, licensing laws and import quotas. These changes lead to a growth in the Black Economy

1960 Bombay state is split into Maharashtra and Gujarat states again, on the basis of language. Maharashtra retains Bombay as its capital

1971 Population 6 million. Sahar International Airport opened

1974 Oil discovered off-shore

Early 1970s Following the success of the back-bay reclamation scheme, Nariman Point becomes the hub of business activity. Several offices shift from the Ballard Estate to Nariman Point, which becomes one of the most expensive real estates in the world, as high demand pushes land prices yet higher

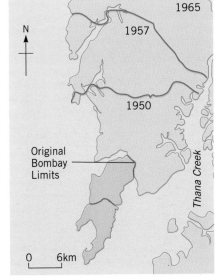

Figure 18.11 The growth of Bombay, northwards.

1 How far did political events of this period affect Bombay? Were they local, regional, national and/or international?

Figure 18.12 The port of Mumbai.

Mumbai 1980 – 2002

1981 CIDCO – the City and Industrial Development Corporation of Maharashtra – plans New Bombay in twenty satellite towns on the mainland.

1990s Bombay is the largest textile and pharmaceutical centre in India.

1991 Population 12 570 000. 67% live on Mumbai Island – home to the Stock Exchange, and one of Asia's busiest ports.

1991 Prime Minister introduces reforms to reduce the government debts, cut inflation and attract foreign investment in to a global free market.

December 1992 to March 1993 Riots between Muslims and Hindus end with bombing campaigns that wreck the city centre. 150 000 people fled the city and over 800 died. Rapid population growth, (4 million extra people between 1981 and 1991), had worsened housing shortages, increased competition for resources like drinking water, and led to deterioration of the environment.

1995 Bombay Stock Exchange computerised; now Bombay is the financial and business capital of India.

1 May 1995 Bombay is known as Mumbai, within India, after the Hindu goddess Mumba.

2000 Each day, several thousands of migrants arrive from rural areas of India in hope of a better economic future.

Year	Population
1661	10 000
1675	60 000
1780	114 000
1806	200 000
1814	240 000
1864	817 000
1872	644 000
1881	773 000
1891	822 000
1901	813 000
1911	1 018 000
1921	1 245 000
1931	1 268 000
1941	1 686 000
1951	2 967 000
1961	4 152 000
1971	5 971 000
1981	8 227 000
1991	12 750 000
1994	14 500 000
1999	16 000 000

Figure 18.13 Population growth in Mumbai.

Migrants to Mumbai

The population of Mumbai consists of various religious and racial groups, e.g. Hindus, Muslims, Christians, Buddhists, Jains, Parsis, Sikhs and Jews. Some of these people have migrated from parts of India (intra-migration) including Maharashtra, Gujarat, Goa, Kerala, Punjab, Sind, Rajasthan and Uttar Padesh. Often these migrants are single males. Other people have been involved in international migration from places such as Iran, China and Africa.

There is evidence of social segregation in where these people choose to live. Many migrants need the support of their local community when they first arrive at a new place. The community can provide the services that are needed, such as special food, clothing, religious worship, etc. People from Maharashtra move to Parel and Dadar. Gujarati migrants live in Kalbadevi and Bhuleshwar. Muslims are concentrated in Mohammed Ali Road and Mahim. Christians are found in Byculla and Mazagaon. The more desirable residential areas tend to be multi-cultural, e.g. Malabar Hill.

1 Draw a graph to illustrate the population growth in Bombay from 1661. Make sure the horizontal axis is accurately scaled.
2 How does its growth pattern compare to that of Manchester (Chapter 17)?
3 Identify the states and regions of India from which people have come to Mumbai. Draw a sketch map to show this migration. Entitle it Migration to Mumbai.

The daily commute

Most of the businesses in Mumbai are concentrated in the south of the island. This leads to a daily migration north to south in the morning (07.30 to 09.00) and south to north in the evening (1630 to 1830). During peak rush hour times the average speed of traffic can be as low as 7km an hour; commuting can take up to five hours a day. The congestion from millions of commuters has a negative effect on economic activity and health. In an effort to alleviate the congestion, an off-shore highway has been constructed and dozens of flyovers are in the process of being built. There are 800 000 cars on the road in Mumbai. Traffic accidents claim over 300 lives a year; Maharashtra as a whole has over 8 000 road fatalities a year.

Can Mumbai's growth be analysed using models?

In LEDCs, many rural migrants move to the cities in search of an improved quality of life. Ravenstein's laws of migration and Rostow's stages of economic growth may be applied to Mumbai. Refer back to these two models in Chapter 17. These can be supplemented by the Lee and Todaro models, described in the theory box.

Mumbai approaches the end of the millennium destined to become the second biggest city in the world, though half the city's inhabitants still live without water or electricity. Nothing demonstrates its deteriorating environment better than a recent report which claims that just breathing the air in Mumbai is equivalent to smoking 20 cigarettes a day. It is hoped that the satellite city of New Bombay, which is taking place on the mainland, will relieve some of the pressures on the urban environment.

Figure 18.14 From the Lonely Planet web site, 1999.

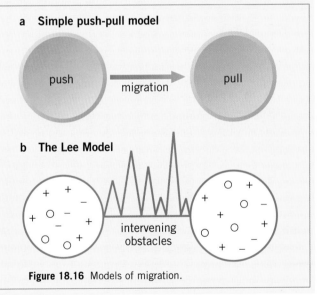

Figure 18.15 Slum housing in Mumbai.

Theory

The Lee model (1966)

The Lee model was developed from the simple push-pull model of migration. Push factors are reasons for leaving a place, e.g. war, famine, unemployment, lack of suitable partners. Pull factors are the attractions of the place to which you are moving, e.g. nightlife, better education, potential employment opportunities. In his model, Lee believed each individual would have his own views and perceptions about his place of origin and his possible destination. These perceptions would be different depending on personal characteristics, e.g. age, gender, marital status, socio-economic class, education. Each individual would then evaluate the pros and cons of each place. Before making a decision whether or not to migrate, the individual would also consider 'intervening obstacles', e.g. cost of moving, family ties, ease of selling existing home, etc.

Figure 18.16 Models of migration.

1 How can personal characteristics, such as age, gender, marital status, socio-economic class, education, affect a person's decision to migrate?
2 a) Consider each characteristic in turn (e.g. age) and decide who, within that category, is most likely to migrate, e.g. 18-40 year old. Justify your views.
  b) What are the characteristics that would create a stereotypical migrant to a city?

Theory

The Todaro Model

Todaro's model is concerned with rural to urban migration. He believed people's expectations about a city are economically realistic and rational. People moving to the city may not find work immediately. However, Todaro believed the migrants are prepared to stay in the city rather than return home. In this way, they take a risk and stay in the city because if they are eventually successful, wages are much higher than in the rural areas.

1 Todaro's model could be linked to a typical 18 year old student moving to go to university. How does Todaro's model fit this situation?
2 Working in groups, prepare a presentation on one of the four models studied so far: Ravenstein, Rostow, Todaro and Lee. Include a brief overview of the model and evaluate how useful the model is in explaining
   • Mumbai's development
   • Manchester's development.
3 Produce a summary comparison table using the outline below.

Characteristic	Mumbai	Manchester
Natural advantages of the site		
Social reasons for population growth		
Economic reasons for growth		
Political reasons for growth		
Links with other countries		
Famous landmarks		
Human disasters		
Supports urban models		
Present day population size		
Present day population density		
Period of fastest population growth		

Summary

You have learned that:
• Mumbai is of major importance within India in terms of its size, economic importance, cultural significance, and as a magnet to people.
• The city is ethnically and culturally mixed, as a result of migration and different phases of growth over time.
• Its population is rising steeply at the moment, both as a result of natural increase and of rural-urban migration.
• Like Manchester, its economic significance is the result of industrial growth, together with its increasing significance as a centre for service industries such as finance.
• Its growth, like that of many cities, can be seen to fit patterns of models of development.

Ideas for further study

1 Investigate the growth and development of your nearest urban area. To what extent does it:
• show similarities or differences from Mumbai and Manchester?
• fit any of the models shown in this or Chapter 17?
2 Compare the growth and development of Mumbai with another city in a LEDC, such as Bangkok. To what extent do cities in LEDCs show similarities in terms of when and how they have grown?

References and further reading

Lonely Planet Mumbai, 1999, Lonely Planet Publications
http://www.lonelyplanet.com
http://theory.tifr.res.in/bombay/persons/jamsetji-tata.html
http://www.lonelyplanet.com/dest/ind/bom.htm"
http://www.mumbainet.com/

Comparing functions in Manchester and Mumbai

This chapter will focus on the city centre of Manchester, and its functions. It will then explore these functions in detail, comparing the city centre with outer suburbs and parts of Greater Manchester.

It then compares a range of functions that occur in both Manchester and Mumbai, and how these are arranged in spatial patterns. As each pattern is unfolded, you will learn how geographers have studied these patterns of functions to create land-use models.

Investigating Manchester's CBD

A Central Business District (CBD) is an area of the city where the shops, financial centres and offices are located. It is the centre of the local economy. Figure 19.1 shows the main functions found in different parts of the CBD, e.g. retail, legal system and offices. This important area of Manchester is bounded by communication routes (motorway, inner ring road, train stations) as shown in Figure 19.2. The shape of Manchester's CBD is roughly triangular. The high density of streets reflects two features: this is the historical core of the city, and the space where, traditionally, everyone has wanted to be. Offices, shops, administration centres, leisure and recreation, education all have space here.

Advantages of locating in the CBD

Selling products or services requires customers. If customers can easily reach a service, there is a greater chance of selling. The CBD is traditionally the most accessible area of a town or city in the centre. It is also the part of the city where property is most expensive, as companies seek land in a small space. This is described in the theory box, *Bid-rent curves*.

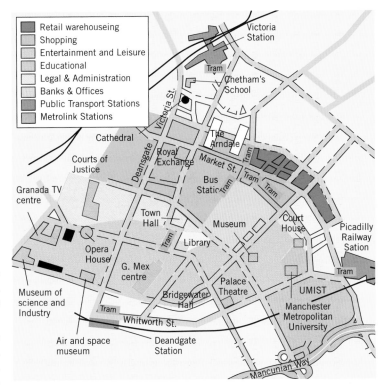

Figure 19.1 Land use in Manchester's city centre.

1 Study Figure 19.2 of Manchester and Salford. On a plain sheet of A4, use different colours to draw a sketch map of the following:
   - the extent of the CBD. Mark and label 5 pieces of evidence
   - Manchester and Salford's universities, colleges and schools
   - road and rail links to the city and other towns/cities to which they lead
   - inner 19th century suburbs of the city, shown by high-density housing
   - lower density suburbs built in the 20th century
   - key leisure and entertainment facilities in the city.
2 Explain the different distribution of:
   a) universities, colleges and schools
   b) 19th and 20th century suburbs

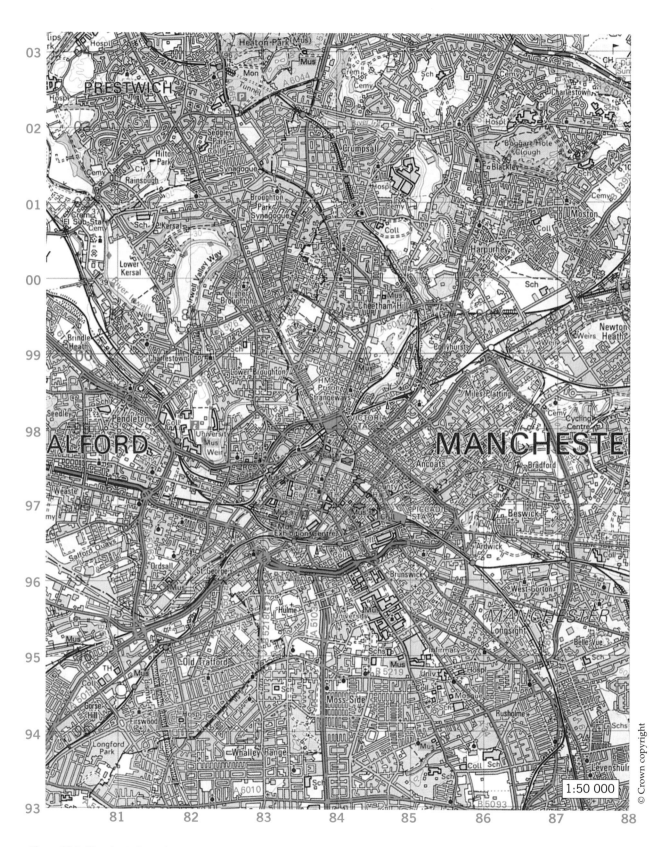

Figure 19.2 Manchester's road network and inner suburbs.

1:50 000

Bid-rent curves

Some areas of a city are more desirable than others. This affects people's willingness to pay high or low land prices. It costs much the same to build a house or office in one place as it does in another, and so cost of land affects whether a building is valuable or not. The factors that affect land values are desirability and accessibility. Accessibility is determined by transport routes. Routes usually converge on city centres, making them the most accessible parts of a city. Hence they become most desirable.

Those most able to pay usually pay the highest prices in competition with each other. Compare residents, industrialists, and office or retail developers. Individual residents are least able to afford high land prices. Industrialists usually want large land areas, so, although they are wealthier than individual residents, they are restricted to places where land is plentiful. However, office and retail developers are usually able to pay much higher prices for smaller amounts of land. Faced with small areas, they are most likely to 'build upwards' if they cannot 'build outwards'. In most cities, buildings appear as high-rise blocks on the skyline, though some residential blocks may also be high-rise where space is short. The ability of different groups to pay high costs is shown in Figure 19.3 and is known as a **bid-rent curve**, i.e. what users are willing to pay for their rents.

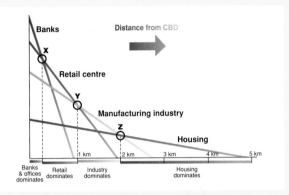

Figure 19.3 Bid-rent lines for four different users.

Figure 19.3 shows the following features:

- Office and retail developers are willing to pay highest prices but only for land they most want. A bank requires less space than a retail centre. It is willing to pay to be in the heart of a CBD. Its theoretical bid-rent is high, but falls rapidly away from the CBD. At point X on Figure 19.3, its price is outbid by a retail centre, which in turn is able to outbid other competing land users. Point X is close to the city centre. Hence demand is highest at the city centre, but falls off rapidly with distance.
- Industrial users may be more flexible. Their demand for accessibility and land enables them to outbid residential users, but only to outbid banks and retail centres at certain points. On Figure 19.3, this point is at Y, 1km from the city centre.
- Between Y and Z on Figure 19.3, industrialists are able to outbid other users. At Z, the desirability of land for industry has fallen, so that housing is now able to outbid all other users.

This is the theory behind all concentric circle models – outbidding of one land use by another. The pattern of land use shows a progression from the city centre – banks, retail, industry and housing.

Bid-rent theory also explains variation in prices between different places. Your own home area will show differences in house prices between places that may be very close physically. This is simply a reflection of what people are willing to pay, and is shown in Figure 19.4, based upon the city of Topeka, Kansas, USA.

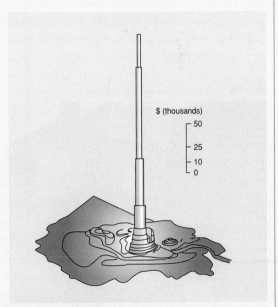

Figure 19.4 Land values in Topeka, Kansas, USA. Notice how, although the CBD stands out in rents, other minor centres stand out too. These might be local retail areas, industrial or science parks, or desirable suburbs.

In most cities, road networks lead to, and radiate from, the CBD. Although Manchester's largest motorway circles the city, two major motorways (M602 and A57 (M)) lead towards the CBD. Figure 19.2 shows the network of major roads that lead to the CBD.

Manchester's CBD is served by two main-line railway stations, Piccadilly and Victoria, which link Manchester to the rest of the UK. On the edges of the CBD are local railway stations such as Deansgate and Oxford Road, providing access to the city centre for surrounding areas. The Metrolink, a tram or light-rail system, was added to the transport network in 1992, bringing commuters and shoppers into the city centre from Bury (in the north) and Altrincham (in the south). All parts of Greater Manchester are linked to the city centre by road.

Geographical patterns exist in the CBD

Within the CBD, similar functions often locate together, or cluster, as they are competing for the same customers. Examples within Manchester's CBD include:

- retail, where similar kinds of shops, such as shoe shops, cluster together
- Chinese restaurants, in the Chinatown area of Princess Street and Faulkner Street
- office functions, such as estate agents, banks and building societies, solicitors and insurance companies, all based upon the sale of property
- nightclubs, where several clubs may be located in the same area of a CBD.

Sometimes functions locate together as they need each other's services, such as bureau de change with travel agents. The benefit to customers of functions clustering together is the ease in finding what you want.

The future of Manchester's CBD

Manchester's land use zones are not haphazardly arranged but occur in defined areas. Each has developed separately for specific reasons, and each faces different issues. This section considers various functions within Manchester's CBD and the challenges facing them.

Shopper's heaven?

In recent years, shopping has become a leisure activity. Within Greater Manchester residents have a variety of shopping locations, shown in Figure 19.5. Each serves a different purpose, and each faces different issues.

---

1 Consider a CBD in a town or city known to you. How far does the clustering described for Manchester apply?
2 Suggest reasons why the following services prefer to locate in the CBD, rather than the outskirts, and why they cluster in the same areas of a CBD; a) retail, b) restaurants, c) offices, d) nightclubs.
3 Why do banks and offices traditionally prefer to locate in the CBD? How might the use of telephone or internet banking change this?

Figure 19.5 Different retail services in different parts of Manchester.

a Local corner shops are the backbone of local services.

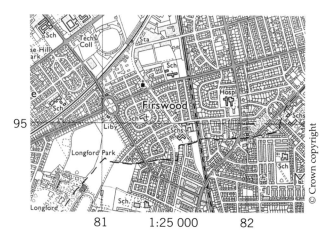

© Crown copyright

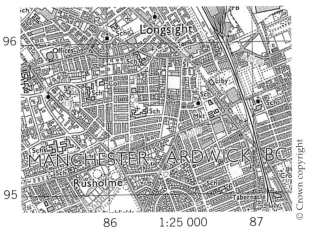

b City neighbourhood centres – Longsight, an inner suburb of Manchester.

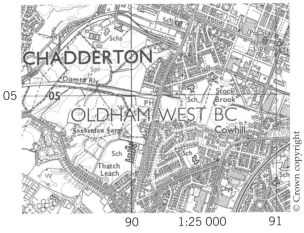

c Small suburban town centres – Chadderton, closer to Oldham than to Manchester, is a typical suburban shopping centre.

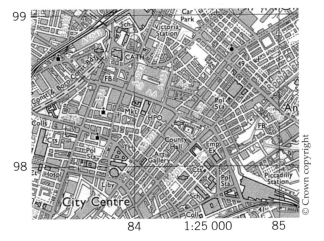

d CBD – Marks & Spencer in Manchester City Centre.

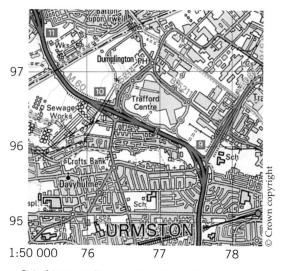

e Out-of-town retail parks – The Trafford Centre, a large out-of-town shopping centre, which also has a cinema and entertainment complex.

Since 1985, out-of-town shopping centres and superstores have been growing in the UK, while the number of smaller shops, such as Figure 19.5a, has halved. When the Trafford Centre opened in 1998, many people were concerned about the impact it would have on other shopping centres within Greater Manchester, especially the CBD. Concerns were also raised about accessibility for non-car owners. Many out-of-town shopping centres have public transport facilities and coach parking but are primarily designed for car owners. Like others in the UK the Trafford Centre has been designed with leisure in mind, with a 1600-seater food court and 20-screen, 4130-seater cinema.

1 Using all the evidence, describe the location of each shopping centre and its surrounding environment.
2 Consider the shops shown in Figure 19.5. Make a large copy of the table below and complete it for each of the shops shown. Consider the people most likely to shop there in terms of age, income group and other characteristics such as car or house owners.

Shop location	Opening hours	Advantages	Disadvantages	How often people shop there	Types of purchases	Distances people travel & methods of transport
Local 'corner shops'						
City neighbourhood centres						
Small suburban town centres						
CBD shops						
Out-of-town retail parks						

3 Find out about out-of-town shopping centres, e.g. Lakeside (Thurrock), Meadowhall (Sheffield), Merry Hill (Dudley), Metro Centre (Gateshead), White Rose Centre (Leeds).
   • What are the advantages of the site to shoppers? and to retailers?
   • What impact has the centre had on shopping habits elsewhere in the region?

Advantages of out-of-town shopping	Disadvantages
• Car parking • Free 10 000 spaces • Disabled facilities • Shop Mobility unit offering scooters and wheelchairs, Meeters and Greeters to help, lifts and toilets designed for wheelchair users • Weatherproof – regulated environment, protected from weather • 'Safe' – Meeting point for lost people and children; security officers • Full range of services, including post office, banks, travel agents • Fully pedestrianised, plenty of safe space.	• Motorway hold-ups along M60 due to Trafford Centre • Environment – circulated air, artificial lighting • Lack of local, home-grown stores; all big chains and names • No supermarket for food shopping • Access mainly restricted to those who can afford to shop there or to get there; access is often denied to the elderly and poor (non-car owners) or the homeless (often kept out by security)

Advantages of city centre shopping	Disadvantages
• History and atmosphere connected with buildings, streets • Small, specialist shops, e.g. The Model Shop, antiques, bookstores • Largest Marks and Spencer store in the world • Easy access via public transport • Weather permitting, great to sit outside • Compact city centre, approximately 30-minutes walk north to south • Choices of routes to walk around and between places • Choice of shops, eating places and entertainment • Theatres, concert halls, opera houses, Laser Quest, etc.	• Expensive parking • Perception of more crime • Walk greater distances with shopping to the car • Exposure to the weather

Figure 19.6 Comparing city centre and out-of-town shopping

Whose space is it?

Not all residents have equal access to the full range of shops in Manchester, which may be affected through lack of money, poor public transport, or access to a car. Some shops provide low-order goods (day-to-day necessities) but larger shopping centres provide high-order, comparison goods like hi-fi equipment, furniture and clothing. Those without access to larger shopping centres are denied access to services found there. In their book *A tale of two cities*, the authors, Taylor, Evans and Fraser, describe how this can be seen.

'People of different social backgrounds (young people in groups, subsistence shoppers looking for a bargain, young professionals en route to a wine bar, unemployed men of different ages looking for company to pass the time of day) routinely make more or less heavy use of quite different parts of the city centre'.

Figure 19.7 A-Z extract of Manchester's CBD.

They define:
- a 'poor people's' Manchester (Oldham Road and the markets)
- space dominated by young professionals (Market Street and St Ann's Square)
- family and middle-class shoppers (the streets between St Ann's Square, Deansgate and Market Street.

> **1** How easily can similar patterns in socio-economic groups in the CBD be identified in a city or town known to you?

Office functions in Manchester

The traditional CBD included office-based functions such as estate agents, solicitors, banks and building societies, insurance, advertising and marketing (Figure 19.8). These businesses were related to the workings of property and of the economy. Manchester is also a centre for large companies and regional headquarters, e.g. CIS, BT, Big Issue, Bank of England, Bank of Ireland, Bank of India. Other accounting and consultancy companies are also attracted to CBD locations, such as Deloitte and Touche, and KPMG.

Recent employment opportunities have been created, such as call centres and direct banking, which are footloose and can locate anywhere. Manchester has new office locations southwards towards the airport, where flights into London or Europe might be more important in future than having access to the CBD. Some office-based companies have been attracted to this location from the CBD as a result, such as Anderson Consulting.

Figure 19.8 Office zone of Manchester's CBD.

High-technology industries

Manchester is home to a Science Park, the first of which were built in the 1971 in Cambridge and Edinburgh. Science parks aim to knit university research closely with high-technology companies. Increasingly, whole departments in universities are being funded by research grants from companies, which locate close by in landscaped surroundings, often on brownfield sites, or former derelict land. The Manchester Science Park (MSP) was opened in 1986 and is located in the Education Precinct of Manchester (Figure 19.9). Manchester and Salford have many higher education institutions including the University of Manchester, University of Manchester Institute of Science and Technology, Manchester Metropolitan University and Salford University.

The Manchester Science Park, shown in Figure 19.10, is a new function of the city and is located close to Manchester University. It advertises its advantages as *'situated to the south of Manchester city centre and is well served by the local, national and international infrastructure. The M56 and M60, Manchester's motorway ring road are in easy reach and the Manchester International airport is only 20 minutes away'*.

The Science Park is a short walk from the shops, offices and nightlife of the CBD. It offers new employment opportunities in medical and computer-based technology. This type of employment is classified by geographers as Quaternary employment, where specialist advice and consultancy draw on and fund research. In this location, MSP has attractive accommodation, space to expand, parking, optical fibre links to university computers and use of the university library, conference facilities and sport centre.

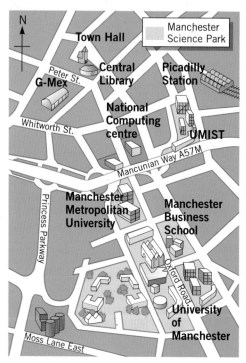

Figure 19.9 Location of Manchester Science Park.

Figure 19.10 Manchester Science Park.

Manchester as a leisure centre

Within Manchester, restaurants and fun pubs attract young Mancunians to the city centre at night, particularly at weekends. The city also offers a variety of cultural experiences such as opera, ballet, classical music and theatre. In sport, the city already has a global reputation through Manchester United, and now hosts events such as the European Cup (1996) and the Commonwealth Games (in 2002). Its Granada Studios tours and Coronation Street attract visitors from around the world. There are alternative attractions to mainstream culture, e.g. the Cornerhouse cinema complex, the Gay Village, comedy clubs, and Mardi Gras Festival.

Leisure facilities in the CBD

Like retail, space in the CBD varies. In the evenings, it is predominantly 'young space' with the youth culture of themed pubs, and clubs. However, the Bridgewater Hall, Opera House and Cornerhouse Cinema may attract a different clientele. During the day, different populations may tour the city centre, or visit Museums, Art Galleries, or the Roman site of Castlefield. Available both day and evening are theatres, cinemas, restaurants, gyms, and guest authors and speakers at bookshops like Waterstone's.

What are the threats to leisure in the CBD?

An increase in the number of people living in or close to the CBD has given life to the city centre as an entertainment centre. Out-of-town leisure complexes exist, e.g. multi-screen cinemas along motorways (M66 Bury) and within outlying towns (Stockport). Multi-screens attract families with cars. The CBD has facilities for a drink with friends before the film, a meal in a restaurant afterwards and public transport. The compactness of Greater Manchester reduces travel times into the city centre in the evening, so more people use it for leisure and entertainment, e.g. Hyde to Piccadilly by train takes 15 minutes, Bury to Manchester by tram takes 25 minutes. Some people avoid the CBD due to a perceived fear of crime in the CBD at night, but as the city centre becomes more popular, people feel safer in numbers. Manchester City Council are developing partnership schemes such as CCTV (Figure 19.11) with the police and health organisations, in order to prevent the development of 'fear of the CBD'.

Home from home – reclaiming the city as living space

As a result of the Industrial revolution, Manchester now has vast numbers of Victorian nineteenth century warehouses, whose companies have long since closed or moved out. Their conversion into loft apartments has revitalised inner city areas such as Castlefields (Figure 19.12), though prices are expensive compared to elsewhere in the region.

> How far are there different people's 'spaces' in a town or city known to you?
> a) Do age and gender of people there vary between the day and evening?
> b) Do social variations exist in leisure space, e.g. pubs, clubs?
> c) Do economic variations exist in leisure space, e.g. shopping, bars?

Figure 19.11 Closed-circuit TV, designed to enhance feelings of safety in the CBD.

Offers the true loft – the open plan space and the original structure of the warehouse such as high ceilings, large windows, exposed brickwork, highlighted by the well-considered modern inserts of the designers that make home life comfortable. With its own landscaped courtyard and canal inlet coming into the site, Britannia Mills has secluded outdoor areas unlike most city centre loft developments. An on-site gymnasium means leisure time can be filled without leaving the gates of the site.

Figure 19.12 Britannia Mills at Castlefields – advertising lofts for sale.

In January 2000:
- a one-bedroomed flat in Flint Warehouse cost £77 000
- a two-bedroomed flat in Sand Warehouse cost £106 500
- the most expensive apartment cost £195 000 in a city where it is still possible to buy a terraced house for under £30 000.

Central Manchester has residential developments costing up to £1 million. These do not suit everyone, particularly families. However, professional people without children see the advantages of living in the city centre, within walking distance of work, and the advantage of having nightlife on the doorstep. Demand is high with most developments sold before conversion is completed. Social inequality between homeless people and million-pound apartments is increasing. Housing Associations and developers have been building low-cost housing for rent or to buy in other parts of the city such as Hulme.

For a city that you know, compare its present-day functions with Manchester's. You may wish to use the following headings as set out in the table.

Function	Manchester	Your choice of city
Housing	Britannia Mills (conversion); Housing association estates	
Shopping	CBD, Trafford Centre	
Education	Cheethams School of Music (specialist), Manchester Metropolitan University, UMIST, University of Manchester (Higher)	
Open space	Heaton Park	
Established industry	Chloride Batteries	
New industry	Science Park	
City Centre public buildings	Royal Exchange	
Communications	Airport, Metrolink, Piccadilly and Victoria stations, M60 and inner relief road	
Environmental schemes	Tree planting, pedestrianisation, paving for outdoor cafes	
Cultural areas	Chinatown (ethnic), Gay Village (social), Roman Fort (historical), Castlefield (redevelopment)	
Health provision	St Mary's, Manchester Royal Infirmary	

Ancoats – changing the inner-city

Ancoats (Figure 19.13), is an inner city area of Manchester. It is located to the north of Piccadilly Railway Station, 500m north-east of the city centre. Ancoats and Miles Platting have a combination of 3 000 council properties (public sector housing), including houses, sheltered accommodation, maisonettes and multi-storey flats. The two areas also have 100 owner-occupied terraced houses. Council properties were improved in the early 1990s. The majority of public sector houses are in good condition and nearly all have central heating. While local housing officers find it difficult to let properties, as most areas in North Manchester are considered undesirable, those who live there feel it is a stable, close-knit community. The Crime and Safety Audit (1997) found that 79 per cent had lived there for more than ten years and 88 per cent said they had no intention of moving in the next year. It is not well-off; in 1991, 83.7 per cent of households in the area did not have a car, compared to an average of 32.4 per cent for England and Wales.

Since 1950, Ancoats has declined in population and few now visit the area. Most local people are concerned about rising crime, unemployment and poor environmental quality. The Rochdale Canal runs through the estates, but derelict land on either side attracts youths and fly-tipping. The area was considered particularly unsafe in the Crime and Safety Audit (1997). Miles Platting and Ancoats form part of the North Manchester Division within Greater Manchester Police, which accounted for over 10 per cent of all Greater Manchester's recorded crimes 1997-1998. Various projects now attempt to tackle residents' worries such as:

- Safe as Houses – aims to improve home security
- Oldham Road Business Watch – a pilot security patrol and watch group set up with local businesses and police
- Car Crime Initiative – works with young people to examine consequences of driving illegally
- Drug Education Programme.

Ancoats was the busiest area of Manchester during the Industrial Revolution, with many huge mills. Its heart became known as Little Italy, after Italian families who migrated there and found work in the nearby mills. Extra income was provided by small businesses run from home, such as ice-cream manufacture. Roman Catholic churches became focal points. The community was severely affected by the Second World War and by slum clearance programmes in the 1960s. Many families were split up as poor, bomb-damaged housing was cleared.

Now, the Ancoats Building Preservation Trust, the Urban Village project and the Italian community hope to recreate Little Italy. A new square is proposed in front of St Peter's church and possibly a second piazza (square) to the north. The buildings around the square will have flats above shops, bars and restaurants.

Ancoats has many listed Victorian buildings. Anita Street – originally Sanitary Street but a few letters were removed after sanitary became associated with bad odours! – has survived as an example of 1890s housing. Victoria Square is a recently restored structure of five-storey flats around a central courtyard. Ancoats Building Preservation Trust is using English Heritage money to maintain and restore buildings. The Urban Village Company is working towards a mixed-use neighbourhood by infill development and conversion projects. The Rochdale canal (opened in 1804) links Ancoats to Castlefield, and some of its sections pass beneath the CBD. Once regeneration work is complete, the Castlefield, Rochdale Canal and Ancoats area will be a World Heritage Site.

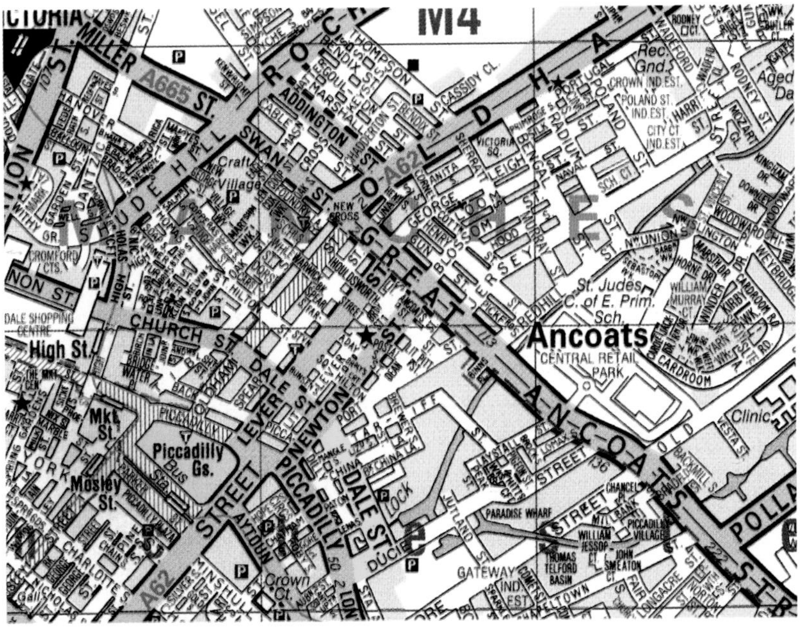

Summarise life in present day Ancoats, under the following headings: Social, Economic, Environmental, Political. You will use these headings later to compare it with conditions in other locations.

Figure 19.13 The Ancoats area of inner Manchester.

Shaw

Shaw is an outer suburb 11km north-east of Manchester. During the nineteenth century it was reputed to be the richest town in the country with the greatest proportion of millionaires. Its money at that time came from the cotton industry. Its population in 1981 was 9368, growing by 8 per cent to 10 119 in 1991 (Figure 19.15). During the same period, Oldham as a whole showed a population decline. Shaw now has a higher percentage of people between mid-20s and mid-40s compared to the rest of Oldham.

Combined, ethnic minority groups make up 2.5 per cent of Shaw's population, half of whom are Bangladeshi. Levels of economic activity are relatively high in Shaw, shown by the high percentage of owner-occupied housing (72 per cent in 1991). Within Shaw, 5304 (82 per cent of the population) are economically active, i.e. aged 18-65. 3550 people are in full-time employment (66.9 per cent). In 1991, 7.8 per cent of Shaw's population were unemployed. Quality of life improved between 1981 and 1991, shown in census data in Figure 19.16.

Today, Shaw is the 'Home of Home Shopping'. Following the decline of the cotton industry, large mills have had a new lease of life as storage warehouses and distribution plants for catalogue companies, such as JD Williams and Littlewoods. Up to 2 000 people are employed in home shopping in Shaw. Shaw also has manufacturing industry, although the Osram light-bulb factory closed in 2000.

Shaw town centre is where Market Street crosses Beal Lane. It includes shops and services such as Woolworths, Boots the Chemist, Halifax Estate Agents, a mini supermarket, Iceland, shoe repairs, and dress shops. It has a local train station, linking Shaw to Manchester Victoria and Rochdale. At this point are the catalogue warehouses for JD Williams and Littlewoods. Littlewoods also has a large new, modern warehouse that opened in 1999. Away from the town centre, an area of terraced housing rises up the hill. This gives way to semi-detached houses, and on the highest slopes is a detached housing estate known as Harden Hills.

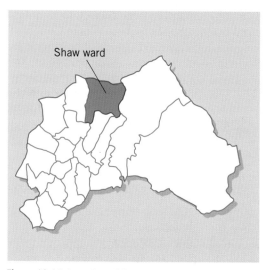

Figure 19.14 Location of Shaw within Oldham.

Males	Age	Females
377	0-4	333
327	5-9	329
318	10-14	328
345	15-19	328
369	20-24	381
481	25-29	479
382	30-34	394
338	35-39	388
443	40-44	424
317	45-49	332
272	50-54	240
230	55-59	229
173	60-64	210
200	65-69	232
136	70-74	230
122	75-79	213
71	80-84	164
20	85-89	98
12	90+	45
4933	Total	5377

Figure 19.15 Population data for Shaw, from 1991 census.

1 Using Figure 19.15, draw a population pyramid for the ward of Shaw.
2 Compare the population structure with:
- the ward you live in
- the national average for the UK.

	Number of households 1991	% 1991	% 1981
More than one person per room	72	1.7	7.3
Lacking or sharing bath / shower and / or inside toilet	15	0.4	7.9
In houses without central heating	444	10.7	Not available
Without a car	1479	35.8	69.8

Figure 19.16 Quality of life indicators for Shaw in 1991.

1 Explain how each of the conditions in the table may indicate the quality of life for residents in Shaw?
2 Evaluate how reliable these indicators are. What other indicators might you choose?

Figure 19.17 Shaw town centre.

Figure 19.18 Housing in Shaw.

The wealthy suburbs of Manchester

Manchester's inner city areas remain poorer than outer areas, in spite of redevelopment in areas such as Ancoats. Generally, outer suburbs tend to be wealthier, particularly those to the south and south-west (Figure 19.19), including Bramhall and Cheadle Hulme. Though part of Stockport, these are considered to be part of Manchester and many people commute into central Manchester each day.

Further away, Alderley Edge and Wilmslow are wealthier still. Each contains relatively large proportions of high income-earners. In 2000, houses are found which sell for over £300 000. Although lower in price than similar properties in London, its value far exceeds houses sold in inner-city areas of Manchester and Salford, where it is possible to buy houses at £30 000 or less.

Houses such as these are built at lower densities than inner Manchester. The impression of wealth is

3 Draw up a summary table of life in present day Shaw, under the following headings: Social (e.g. housing conditions), Economic, Environmental, Political (e.g. Government initiatives).
4 Compare and contrast the summary for Shaw with the one for Ancoats. Suggest reasons for differences between the two.

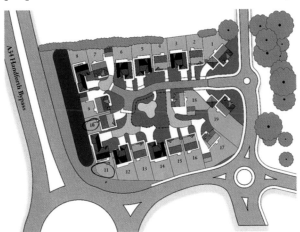

Figure 19.19 Examples of housing in Cheadle Hume, Cheshire, located on the borders with Greater Manchester.

Area	Population	Male unemployed	% White	% Owner occupied	% No Bathroom	% No Car
Outer Greater Manchester						
Cheadle Hulme North (Stockport)	14 862	4.8	97.1	79.3	0.1	24.1
Cheadle Hulme South (Stockport)	14 362	3.7	97.7	88.6	0.1	14.5
Didsbury (Manchester)	13 386	7.8	97.3	72.1	2.1	19.8
Inner city Manchester						
Ardwick (Manchester)	9 654	25.8	75	7.6	1.6	71.2
Central Manchester	9 037	29.2	92	9.5	0.2	79.5
Longsight (Manchester)	14 931	23	58.5	37.5	3.7	55.2

Figure 19.20 Indicators of wealth in outer Manchester and the inner city

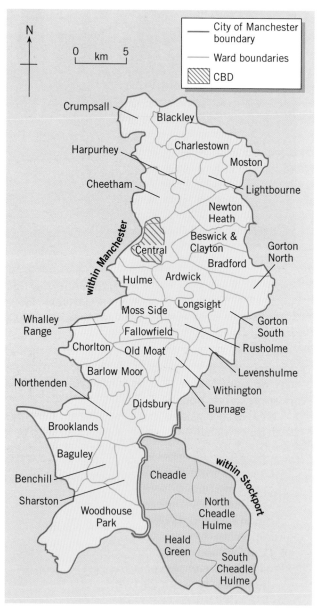

Figure 19.21 Wealthier areas of Greater Manchester, located on the borders of northern Cheshire.

borne out by census data. Figure 19.22 shows some of the census recordings taken from Cheadle Hulme in 1991. Most urban areas have such affluent suburbs, normally located towards the periphery of the city. Their location, and that of other areas of Manchester described in this chapter are shown in the theory box, *Urban models and land use in cities.*

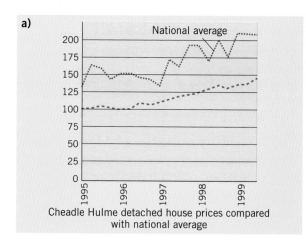

a) Cheadle Hulme detached house prices compared with national average

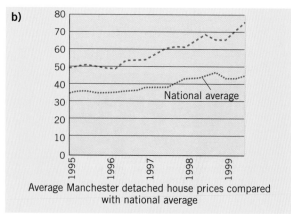

b) Average Manchester detached house prices compared with national average

Figure 19.22 Indicators of wealth in Cheadle Hulme compared to elsewhere.

Theory

Urban models and land use in cities

Urban models are ways of simplifying reality. Cities are complex places; models help to focus on the character of places, and their dominant qualities. Different models have tried to simplify reality by showing how cities grow and how their different functions are arranged spatially.

The Burgess Concentric Zone Model

The earliest urban land use model was created by Burgess, in a book called The City (1925). He based his model on Chicago in the USA and argued that certain functions competed for a particular space in the city. His model contains five concentric functional zones which are based on differences in land use and culture in Chicago. These are outlined below.

Figure 19.23 The concentric model.

- In the city centre is the Central Business District (CBD) of offices, financial services, shops and entertainment.
- The zone of transition is where migrant communities first lived and worked on arriving in Chicago. Increased prosperity allowed them to move out - hence the transitional nature of the population.
- Lower quality housing was to be found in zone 3.
- Higher quality housing was found in zone 4.
- The ability of businesses to pay high costs automatically allowed them to survive in the most competitive place - the CBD.

Review the information in this chapter so far. How far do Burgess' assumptions about five different zones fit Manchester?

Mumbai

Mumbai is a major port and India's largest and wealthiest consumer market. International and national trade are therefore very important. This section focuses on the CBD of Mumbai, comparing its function with outer suburbs.

Figure 19.24 Mumbai's skyline.

Investigating Mumbai's CBD

The shortage of space in Mumbai means that land prices are some of the most expensive in the world. High prices are caused by a shortage of land because the main city is built on an island. The skyline is full of skyscrapers, many of which are occupied by overseas companies that can afford the prices.

The Town Hall and State Central Library, the Assembly Hall, the Bombay Stock Exchange, the University, the High Court, the commercial district, Victoria railway station and the City government are located around the Fort area of the city. This area is located in the south-east of Mumbai and includes the historic quarter of the city. To the east of the CBD are the docks, while to the south lies the University, to the west Back Bay, and to the north the main railway station.

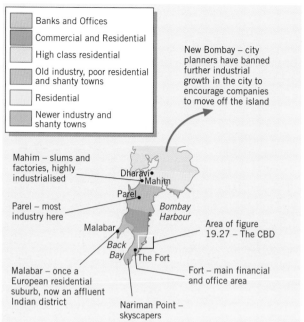

Banks and Offices

Commercial and Residential

High class residential

Old industry, poor residential and shanty towns

Residential

Newer industry and shanty towns

New Bombay – city planners have banned further industrial growth in the city to encourage companies to move off the island

Mahim – slums and factories, highly industrialised

Dharavi

Mahim

Parel

Parel – most industry here

Bombay Harbour

Malabar

Area of figure 19.27 – The CBD

Back Bay

The Fort

Malabar – once a European residential suburb, now an affluent Indian district

Fort – main financial and office area

Nariman Point – skyscapers

Figure 19.25 Land use in Mumbai.

Figure 19.26 Images of Mumbai's CBD.

Functions of Mumbai's CBD

Mumbai's CBD has similarities to that of Manchester; each has banks, courts, and a Town Hall. Both have main railway termini and a University on the periphery. Mumbai however, differs from Manchester because it also contains India's main Stock Exchange, the largest in Asia after Tokyo. Mumbai is the largest commercial centre between Europe and Singapore. In the UK, the Stock Exchange is in London. Mumbai is also the home for the Mint, where money is printed.

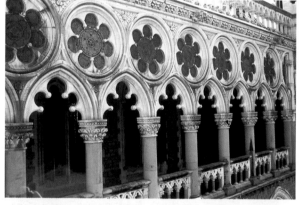

b) Traders at the Stock Exchange

a) Mumbai – State Central Library and Town Hall

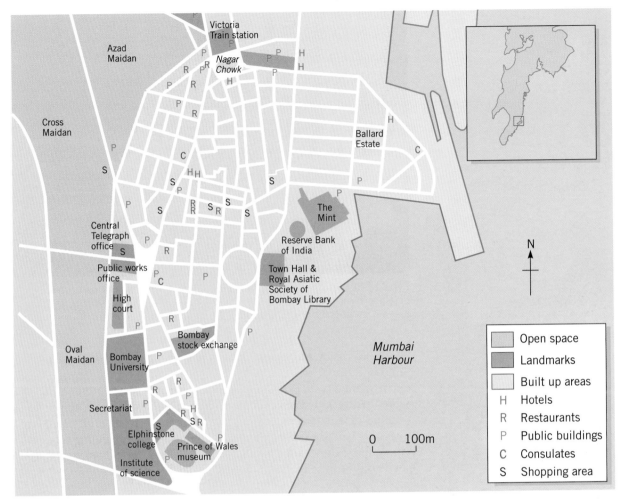

Figure 19.27 Mumbai's CBD.

As in Manchester, Mumbai's CBD is a mixture of old and new development. Part of Mumbai's CBD is sited on reclaimed land, constructed from waste material when the Alexandra Docks were built. Manchester was also an important port with its ship canal used for transporting coal and cotton. Manchester's role as a shipping port has finished, whereas Mumbai's port is increasing in importance.

Shopping

Mumbai is India's largest and wealthiest city for shopping. The main shopping areas are Colaba causeway (approximately 500m to the south of the University), Kemp's Corner and Phulabhai Desai Road (both to the north of Malabar Hill) and Linking Road (north-west of Mahim, approximately 20km north-west of the CBD). However, within this two kinds of retail exist. Like any major city around the world, Mumbai sells everything from designer clothes to artwork; shopping hours are from

Figure 19.28 Twilight shopping in Mumbai.

Figure 19.29 Customers in a Mumbai shopping centre. Such centres are common in LEDCs and resemble American malls. The difference is that they tend to be located in the CBD rather than suburban locations.

1 In pairs, suggest the advantages and disadvantages of shopping:
- in government fixed-price department stores
- with street traders.

10 a.m. to 7 p.m. Monday to Saturday. Some large stores also open on Sunday but markets are usually closed. Within formal shops, such as department stores, prices are usually fixed and shopping is much the same there as in London or New York. On the streets, stalls sell as wide a range of goods, varying between genuine articles and fake items. These are usually known as 'informal trading', and haggling is normal, where the shopper tries to talk the price down. Street traders are open seven days a week from morning to night.

Employment prospects in Mumbai

India is now the sixth largest economy in the world and Maharashtra is its leading industrial region, responsible for almost a quarter of the country's industrial production. Mumbai is the major industrial centre within the region and a major port, generating approximately 40 per cent of the country's revenue from imports and exports. The city's population also provides a huge potential market for goods. It acts as a regional headquarters for much of India's wealth, and is therefore on a par with most major global economic centres, such as Singapore, Tokyo, London, or New York. The textile industry grew up in the region originally but this expanded to include other industries such as engineering, petro-chemicals and food processing. Between 1991 and 1996 the state of Maharashtra received over US$64 million of foreign investment.

Almost 200 of India's top 500 companies have their corporate headquarters in Mumbai. Many banks and financial services have also been established in the city; the Fort area of Mumbai contains many banks and financial services such as Citibank, American Express and Chase Manhattan.

Nariman Point is also a centre for finance, including multinational companies such as Global Trust Bank Limited, Bank of America National Trust and Savings Association, Bank of Nova Scotia, and Bank of Tokyo. The Rajan Raheja group, based in Mumbai, has interests in real estate, pharmaceuticals and media.

Figure 19.30 The offices of Mumbai's CBD.

Industrial employment in Mumbai

Industry has developed here for many reasons. Mumbai is the financial centre of Maharashtra and India, and the infrastructure (oil, nuclear and HEP power supplies, communications, etc.) is amongst the best in the country. In recent decades Mumbai has changed its employment structure from manufacturing to become an important financial and commercial centre. Over 66 per cent of its workers are in the service sector. Almost a third are still employed in manufacturing such as chemicals, textiles and engineering.

More recently, the fortunes of the state and the city have been changing. Success has meant that wages, land prices and office rents have risen, so much so that some companies have decided to move out of Mumbai, e.g. Coca-Cola who moved to Delhi to reduce production costs. The restriction of land availability on the islands has forced companies to relocate and expand on the outskirts of the city.

In contrast, some high-technology industries have moved into Bombay from Pune, Maharashtra's second largest city. In 1996, 57 per cent of India's software exports were to the USA; companies such as Onward Computer Technologies in India were working in partnership with Novell. Nucleus Software Offshore Limited operates from its own premises in Mumbai. Nucleus has two development centres. One development centre is a dedicated unit for Citibank support and is located next to Citibank's premises. The other centre is located in the suburb of Mumbai-Andheri. Hi-tech companies are attracted by the low labour costs, and quality and skills of the Indian workforce.

Formal and informal exployment in Mumbai

Employment in Mumbai can be classified into two types: informal and formal; the difference lies in how they are recognised by the government. Those who collect a pay slip and are registered for taxes are in the formal sector. Those outside the formal economy and hidden from the paperwork of pay slips and taxes are in the informal sector, e.g. shoe shiners. Up to 70 per cent of jobs in Mumbai could be considered informal; some residents of Dharavi (Figure 18.2) have up to five informal jobs. Informal workers have no employment rights. Migrants play an important role in supporting the city, working in construction, service industries and recycling of the city's waste.

Housing

Mumbai is a city of contrasts, from the very wealthy to the 13 million who live on the streets or in slums. Mumbai has one of the highest population densities in the world. It struggles to meet the demands of its growing population. The overall population density was 580 people per square kilometre in 1996. Within the city there is a huge difference between the lifestyles of the wealthy and those of the poor:

- approximately 50% live in slums or on the streets
- more than two million do not have access to a toilet
- six million go without clean drinking water.

At the same time, houses on Malabar Hill (Figure 19.26) sold at over three million Rupees ($US 74,800) per square metre in the mid-1990s.

Dharavi

Dharavi, shown in Figure 19.33, is a slum district to the east of Mahim Bay, 11km to the north of the CBD. It is probably Asia's largest slum. It is an example of a shanty town that has become a permanent area of the city. In 1999, its population was 750 000 people, with a population density of 230 000 people per square kilometre. Within Dharavi, there are many industries including leather, metalwork, ceramics and furniture. Some of these industries are said to produce some of the highest quality products in

Mumbai. The district has multi-storey buildings and a range of community self-help groups. Dharavi is one of the better slum areas to live in.

Living in the slum districts of Mumbai is not easy, for example 200 families may have to share one standpipe for clean water and two toilets. The city has plans to develop permanent tenement buildings, like those in Mahim, to the south of Dharavi, as shown in Figure 19.33. These are permanent buildings and some argue that they would be better than the temporary slums in which some are now forced to live. However, residents of Dharavi have resisted government plans to allow developers to build tenements, and have instead pushed for a railway station to be built at the edge of the settlement, providing access to the suburban rail network.

Recycling is an important industry in Dharavi. A migrant truck driver from Gujarat now living in Dharavi estimates that he earns twice as much recycling than he did driving. Recycled goods include aluminium cans, cardboard boxes, plastic bags and steel barrels. These goods are collected or found around Mumbai, and sold on.

One space, two worlds: on Bombay

Rahul Mehrotra

In Bombay, the rich and the poor live in distinct physical environments and locations, one static, monumental, and on the high areas of the city, the other sprawling along the transport lines and into any available interstices or crevices. Poor rural migrants, in huge numbers, are shaping the culture and form of the city – today it is like a bazaar, a kaleidoscope of snapshots and symbols overlapping to create an often incomprehensible mosaic. In fact, the bazaar, a chaotic marketplace of shops, stalls and hawkers, can be seen as the metaphor for the physical state of the contemporary Indian City – an informal enterprise zone expressing energy, optimism, and a will to survive outside any formal system.

Figure 19.31 From Mehrotra 1997 Harvard Design Magazine, Winter/Spring.

Figure 19.33 Tenements in Mahim.

1 Study Figure 19.32. Write a list of five questions you would ask people in the photograph to find out more about what you see.
2 Using Figure 19.33, comment on the range of living conditions in the picture, for example building materials, size, suitability for weather conditions like monsoons, facilities such as television and drying clothes.
3 If you lived in Dharavi, would you choose the tenement developments or the tempory slums? Justify your answer.

Figure 19.32 Wash day in Dharavi.

The Brown Agenda

Cities in LEDCs are aware of the environmental problems they face and the Brown Agenda is an attempt to resolve some of these issues. The Brown Agenda is a set of proposals designed to enhance environmental quality in LEDC cities. In 1997, the World Health Organisation (WHO) reported that environmental quality had both a direct and indirect influence on human health. The WHO stated that deteriorating environmental conditions were a major contributory factor to poor health and poor quality of life. Poor environmental quality was directly responsible for around 25 per cent of all preventable illness in the world, particularly diarrhoea and respiratory infections. The urban poor suffered from illness due to a lack of sanitation, poor water supply, poor food safety, air pollution and poor housing. For example, 2.9 billion people were without adequate sanitation in 1994 and 1.1 billion did not have access to safe water supplies.

Within Mumbai, the main problems are those of air pollution, slum developments, poor access to clean water and the disposal of waste, including sewage. All were raised in Agenda 21, a blueprint for saving Planet Earth in 1992. The rapid eco-nomic growth of Mumbai has led to many of its environmental problems; even the local newspapers have referred to it as 'the filthiest city in the world'. In 1997, the WHO ranked Mumbai the fifteenth most polluted city in the world.

Half of Mumbai's population lives in slums; it is not surprising that there are problems with access to clean water and poor sewage disposal. High population densities and a lack of open space, such as in Dharavi, have contributed to environmental quality. Poorly maintained vehicles and leniency in controlling industrial pollution means Mumbai's air quality is dangerously low. Industrial waste and domestic sewage flow into the sea, threatening the marine environment. These are summarised in Figure 19.34.

Mumbai is making progress in the efforts to improve the environment, particularly with regard to air pollution. Unleaded petrol was introduced in 1996 and all new cars are fitted with catalytic converters. Motorised rickshaws have been banned from the city centre due to their emissions. Water quality is also being addressed as two sewage treatment works have recently been completed. Ocean outflows now consist of treated waste; the pipe stretches 3km out into the sea to reduce the impact along the shoreline. Local volunteers have taken a role in improving their environment, with groups such as I Love Mumbai and activities such as tree-planting. However, as in most cities around the world, environmental concerns sometimes take second place to addressing the social and economic problems facing its poorest residents.

Since 1980, the City and Industrial Development Corporation have built 20 satellite towns on the mainland around Mumbai, to help reduce congestion and pollution. Vashi is one such town, situated between Mumbai and New Bombay, close to the road bridge across Thane Creek. The town is built on high ground surrounded by marshland. Vashi has

Third World Lament

The First World has had its fun
The Third World's just begun
But fluorocarbons from the fridge
Make ozone holes we cannot bridge
So poverty must be our lot
And development, it seems, must stop.

The First World now has made its gold
With grand new plans it makes so bold
Fudging figures to give skewed statistics
(Developed folk are adept at these antics!)
They tell us:
'Rice fields pollute more than our cars!'
So pressure is put on poor countries like ours
For decades of indulgence, they pay no price,
But in such matters, we have no rights.
Loans are withheld;
Hard bargains are driven.
With backs against the wall,
We toe the line.
The First World goes all out for their kill
And the poorer nations become poorer still.
A change of heart must take place
If the poorer world is to see better days.

Rekha Menon, 14, India

Figure 19.34 Third World Lament.

Research the Earth Summit in Rio de Janeiro in 1992 which led to Agenda 21.
  a) What were the main issues covered?
  b) Which countries attended?
  c) What time limits were set on achieving the goals?

Chapter 6 Human health depends on a healthy environment, clean water supply, sanitary waste disposal, adequate shelter and a good supply of healthy food.

Chapter 7 A growing number of cities are showing symptoms of the global environment and development crisis, ranging from air pollution to homeless street dwellers.

Figure 19.35 From Agenda 21, the Rio Earth Summit in 1992.

provided homes for 10 000 people. The homes have one room of 14 square metres, an inside toilet and a tap providing fresh water. Within the first year of completion, 12 500 people had been attracted to the area.

Malabar Hill

This area, shown in Figure 19.36, lies to the south-west of Mumbai CBD and is an expensive residential area (Figure 18.4 on page 246). The coastal location provides cooling breezes and the elevation gives stunning views. Originally the hill was forested but some was cleared for residential development. Malabar Hill was once inhabited by colonial Europeans; today it is the home of Mumbai's rich. Compared to the rest of Mumbai, it is less densely populated. It has open space in its parks, the heritage area of Banganga Tank and around the temples. There is also a covered reservoir on the summit, which provides water to the city and the nearby Hanging Gardens (Pherozeshah Mehta Gardens).

North of the city centre, the broad thoroughfares splinter into a maze of chaotic streets. The central bazaar districts afford the glimpses of the sprawling Muslim neighbourhoods, as well as exotic shopping possibilities, while Bombay is at its most exuberant along Chowpatty Beach, which laps against exclusive Malabar Hill.

Figure 19.37 From Rough Guide to Bombay (Mumbai) – the city's main districts.

Figure 19.36 Malabar Hill in Mumbai – one of the wealthiest districts.

Leisure and entertainment

Bollywood, India's film capital, has existed for over a hundred years. By 1925 there were ten film production companies. By the mid-1930s, film was competing with textiles in its importance to the economy. In the 1970s, the film industry expanded to a state of the art studio called Film City, a 140-hectare site of scrubland, 30km north of the CBD. At its peak, in the 1980s, half a million people were employed in the film industry of Mumbai, Calcutta and Madras. The government took 60 per cent of box office receipts, and sales averaged one hundred million tickets a week. Since then, cable television and video has reduced its popularity, but it remains a popular pastime for the homeless and very poor. Film City is now surrounded by slums and its technology is out-dated. The international appeal of Bollywood has grown, in 1998 *Kuch Kuch Hota Hai* was one of the top ten selling movies in the UK.

Film City is possibly going to be turned into a film studio theme park. Would you recommend such a proposal, based upon the success of other similar theme parks, e.g. Granada Studio Tours in Manchester, Universal Studios in Los Angeles?

Conflict over land use in Mumbai

Mumbai's large and increasing population means that space is at a premium. According to the BEAG (Bombay Environmental Action Group) over 86 000 migrants arriving in the city, and in need of a home, have started to build shanty towns on the edges of the Sanjay Gandhi Park in the north of Greater Mumbai. BEAG aims to protect the National Park and have challenged the Maharashtra State government to have the settlers removed.

However, the State government has a duty to provide space for any settlers who arrived before 1995, but finds it difficult to find space for people. It sees an easier solution to the problem by disbanding some of the National Park land. Mumbai's open spaces are shrinking each year; at present there is approximately 1 hectare of open space for 1400 residents, 30 times less than a typical European city. The conflict has now spread to include wealthier groups in the city who wish to preserve the National Park area.

Development in Mumbai is affected by environmental groups. Father D'Britto, a Roman Catholic priest, has led marches to protest against further destruction of arable land on the mainland to the north. Apartment blocks built on the land have provided much needed housing. However, housing has placed more pressure on nearby water supplies. As more people now use the wells, salinity levels have increased and in some cases the wells have dried up completely.

One example of conflict over land use involved the Koli fishing community, one of the oldest communities on the island of Bombay. As the city grew, the group became sandwiched between office blocks and apartments in the CBD and along the shoreline. The Koli people succeeded in halting a land reclamation project at Back Bay (Figure 19.25). The scheme would have limited their access to the water, using land for offices and apartments. The group was given 8500 square metres of land, which allowed them to continue fishing in the same way, although the government has prevented new migrants joining the community.

Figure 19.38 Leisure facilities in Mumbai.

1 Summarise life in present day poorer areas of Mumbai such as Back Bay (the Koli community) or in Dharavi, under the following headings: Social, Economic, Environmental, Political.
2 How does this compare with conditions in Ancoats and Shaw?

Theory

Urban models in LEDCs

The Griffin and Ford urban model (1970s)

The Griffin and Ford urban model (Figure 19.39), was designed 40 to 50 years after the models of American cities developed by Burgess. It was based on studies of Latin American cities, which are similar to other LEDC cities. The key difference between this model and the American models is in the location of housing types. In the LEDC model, the higher quality housing is found near the CBD, whereas in the MEDC models the lower quality housing is found around the CBD. In Mumbai, richer suburbs include Malabar Hill, close to the city centre. Newer squatter settlements on Sanjay National Park are on the rural–urban fringe, 30km north.

Within LEDCs, there are some areas of higher quality housing built on land of a high value towards the edge of the city. This move from the CBD has been made possible by increasing car ownership, which relieves some of the pressure in already crowded, central residential areas. The model allows for this development.

The Sjoberg model (1960s)

The Sjoberg model from the 1960s is American and is based on land use patterns and values before industry is

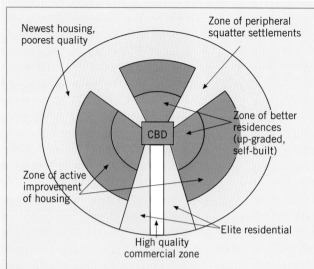

Figure 19.39 Model of a Latin American City, Griffin and Ford (1980).

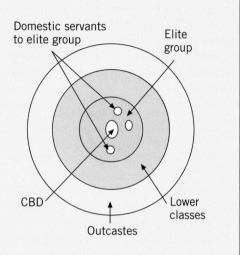

Figure 19.40 Model of a Pre-Industrial City as developed by Sjoberg.

established, similar to the pre-take-off stage of Rostow's model in Chapter 18. Sjoberg felt that the main mode of transport in the pre-industrial city would be on foot. Thus good access to the city centre would create high land values around the CBD.

The Hoyt sector model (1939)

The Hoyt model (Figure 19.41) considers direction and desirability of land, as well as distance. Hoyt found that cities rarely contain identical land use within the same radial zones, but vary, with industry and housing found together. 'Wedges' of residential land use grow out from the centre, with higher value housing in the most attractive location. Hoyt assumed that cities vary in landscape and relief, and desirable areas would develop, such as Malabar Hill in Mumbai. High-income earners are unlikely to choose a location close to industry. Once the industrial area is defined - usually along transport links - it influences the location of other zones which fit around it.

Cheapest housing is found in less popular, less marketable areas, adversely affected by industrial noise or fumes, and occupied by low-income earners. However, they benefit from low transport costs to and from work. In Manchester, cheaper housing and heavy industry were located to the north and west, while higher income areas developed to the south. Industry's ability to pay higher land costs attracted it to accessible points such as along transport links, producing a 'linear' pattern of development (Figure 19.41).

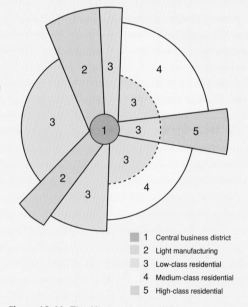

1 Central business district
2 Light manufacturing
3 Low-class residential
4 Medium-class residential
5 High-class residential

Figure 19.41 The Hoyt sector model.

Harris and Ullman's multiple nuclei model (1945)

This model (Figure 19.42) developed in response to criticisms of the others. It is more complex. Harris and Ullman saw the CBD as just one nucleus for development because only some land uses can afford the high rents. Elsewhere, suburban centres, or nearby villages may be absorbed into the growing settlement, and these form nuclei for further development. Cities develop from a number of points, therefore, rather than a single point as suggested by other models. Each nucleus grows until it merges with others, producing a larger urban area.

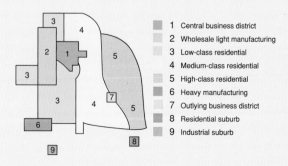

1 Central business district
2 Wholesale light manufacturing
3 Low-class residential
4 Medium-class residential
5 High-class residential
6 Heavy manufacturing
7 Outlying business district
8 Residential suburb
9 Industrial suburb

Figure 19.42 The Harris and Ullman multiple nuclei model.

1 Make your own copy of the Griffin and Ford model. Annotate it using your knowledge and understanding of Mumbai.
2 How far is the model similar to and different from that of Burgess?
3 What similarities and differences can you see in the urban structure between Manchester and Mumbai?
4 How relevant is Sjoberg's model now? Consider how people get around in the CBD, within LEDCs and MEDCs.
5 Select which model fits a) Manchester and b) Mumbai. Justify your selection using named examples from this chapter.

Summary

You have learned that:
- Cities such as Manchester and Mumbai contain a range of functions, both within the CBD and elsewhere within the city.
- There are both similarities and differences between the CBDs of Manchester and Mumbai.
- CBDs have distinct advantages and disadvantages for different functions. Retail and office functions have traditionally located in CBDs, but now are changing as lifestyles change and land values rise.
- Housing varies within cities and reflects the variety of socio-economic groups living there.
- Traditionally, poorer housing in MEDC cities has tended to be located close to city centres; now, urban regeneration and a preference for urban lifestyles is favouring the growth of some inner city areas. Wealthier housing has tended to be located on the urban periphery.
- Conflicts exist over urban land use both in Mumbai and in Manchester.

Ideas for further study

1 Rapid population growth is a problem for many LEDC countries. Some have adopted self-help schemes to ease the burden for the local council in providing housing and basic amenities. Investigate some examples of these schemes.
2 Your nearest large urban area provides many opportunities for fieldwork about how cities are changing. Examples include:
- Environmental improvements, e.g. street lighting, street furniture; architecture; public sector development, e.g. leisure centres, trams; building materials, e.g. tinted glass, metals, UPVC
- Safety, e.g. CCTV, security shutters
- Entertainment
- Retail, e.g. PC World and varied urban retailers
- New functions and businesses, e.g. cyber cafes; changes in open spaces.

Work as a groups or as individuals, each focusing on one aspect of change. Present your results using photos, sketches, data from primary and secondary sources. Emphasise how the urban area is changing, who is responsible for the changes and your views on the changes.

References for further study

Witherick Michael (1999) *The Urban World*, Stanley Thornes

Nagle, Garrett (1998) *Development and Underdevelopment*, Nelson

Nagle, Garrett and Spencer, Chris (1997) *Geographical Enquiries*, Stanley Thornes

Bowen, Ann and Pallister, John (1997) *Tackling Geography Coursework*, Hodder & Staughton

Warn, Sue and Bottomley, Christine (1986) *Fieldwork Investigations 4: Towns and Cities*, Arnold Wheaton

Taylor, Ian, Evans, Karen, and Fraser, Penny, (1996) *A tale of two cities*, Routledge

Web site

http://www.mumbai-central.com/

20 Movement in cities

Consider the headlines in Figure 20.1. Most are from Manchester newspapers, but could apply to many cities in the UK, in other MEDCs, or increasingly, in the world. The twentieth century saw a revolution in transport, allowing large numbers of people in MEDCs to own cars. This chapter shows how continued growth in car usage in cities is affecting urban populations, economies, and environments. Road traffic in many cities has reached saturation point, and this chapter looks at attempts to manage traffic in Bangkok and Melbourne. It questions whether travel by car any time and anywhere, can be allowed to continue.

How and why is urban traffic increasing?

Car ownership is increasing throughout the world. The number of motor vehicles globally could grow from 580 million in 1990 to 816 million by 2010. Most of these are concentrated in MEDCs and in urban areas. In 1993, MEDCs had 70 per cent of the world's cars, and car ownership is still rising. In the USA alone, where car ownership rates are highest, 58.8 per cent of households own two or more cars and 20 per cent own three or more. The average for other MEDCs, excluding the USA, is 36.6 per cent with one or more cars.

In LEDCs, car ownership rates are far lower, ranging from 68 cars per 1000 people in Latin America and the Caribbean in 1993, to 29 per 1000 in East Asia and the Pacific, to 14 cars per 1000 in Africa. LEDCs will see the greatest increases in motor vehicles in the first three decades of the twenty-first century. Growth rates in the Newly Industrialising Countries (NICs) of south-east Asia (Singapore, Thailand, Malaysia, Indonesia, Taiwan, and Korea) will be particularly high. Again, most growth will take place in urban areas.

The result of this is increased road traffic congestion, most of it in cities. There are several reasons for traffic problems in many cities; many of these are interrelated.

a) Growing urban populations

Urban populations are growing rapidly in LEDCs. In MEDCs, urban populations are growing less rapidly, and in some cases, are declining as more people

---

Rail victims' safety plea
(The Observer, 7 November 1999).

No fault on death crash bus brakes
(The Guardian, 16 October 1999)

Traffic chaos due to M66 work is to get worse
(Manchester Evening News)

Road tolls announced for the city centre
(pre-budget speech, November 1999)

£3000 per person for work-based car parking space
(pre-budget speech, November 1999)

City bosses in gridlock talks as traffic chaos grips centre
(Manchester Metro News, 26 November 1999)

Bypass plan back on track
(The Advertiser 28-10-99)

Figure 20.1 Traffic in the news.

move out to rural areas. However high proportions of those who move out continue to work in cities and commute each day. In both MEDCs and LEDCs, therefore, cities become increasingly congested.

Primate cities draw the largest concentrations of vehicles; in Thailand, nearly half of the country's cars are in Bangkok. In Asia, most vehicle growth is due to increases in motorised two- and three-wheel vehicles, which are more affordable than cars. In Thailand, two and three wheelers – such as tuk-tuks in Figure 20.2 – make up half of all motor vehicles.

Figure 20.2 Tuk-tuks in Bangkok. These vehicles are much cheaper than cars, and are powered by 2-stroke engines. Acting as taxis, journeys in them can be hair-raising!

b) Growth in urban incomes

Incomes in urban areas are usually higher than in rural areas. In both LEDCs and MEDCs, increasing household incomes among urban populations have led to a rise in car ownership. Although cars are more expensive than they were in 1970, average incomes have risen faster, especially among the better-paid. Cars are therefore relatively cheaper. Increases in the number of smaller households also mean that it takes more cars to serve the same number of people.

c) The growth in number of journeys

As car ownership increases, so does the number of journeys that people make. In London, a household without a car makes three trips per day on average, whereas a household with a car makes more than five. The two additional journeys are new or replace those formerly made by foot, bicycle or public transport. Public transport usage decreases. More people use the car on longer trips for social and leisure purposes. In the USA in the 1980s, the number of cars increased by 14 per cent, the number of trips grew by 25 per cent and the number of car miles travelled grew by 40 per cent.

d) Economic growth

Since 1980, the largest economic growth has taken place in retailing and service industries, such as finance. The vast majority of this has been concentrated in urban areas. Increasing economic activity has sent more service vehicles on to city streets and has produced more freight traffic, such as delivery lorries.

e) Moving the workplace out of town

Cities have become more dispersed in the past century. In LEDCs, unplanned and uncoordinated land development has led to rapid urban expansion on the periphery of cities (e.g. urban growth in Bangkok, Figure 20.3). In MEDCs, more and more commercial activities are taking place away from city centres. This has two effects.

- Many commuter journeys now take place between one suburb and another, rather than suburb-to-CBD. Most public transport systems were developed to travel from suburb-to-city, rather than suburb-to-suburb, so that most suburb-to-suburb journeys have to be made by private rather than public transport. As a result, suburban roads are often as congested as the CBD.
- As cities become dispersed, the cost of building and operating public transportation systems is prohibitive. Dispersed residential patterns make public transportation systems less convenient for the average commuter. In New York, despite ten years of investment, the number of passengers using public transport decreased from 4.8 million per day in 1980 to 4.3 million per day in 1992.

The sprawling nature of many cities has resulted in a greater demand for roads, longer journeys to work or school, and in greater congestion, fuel consumption and pollution. The rest of this chapter explores how different cities – Bangkok, Manchester and Melbourne – are attempting to manage traffic problems.

Choking to death in Bangkok

Traffic in Bangkok is a major issue. Central Bangkok is shown in Figure 20.4, part of the larger Bangkok Metropolitan Region. Congestion is made worse by poor planning and inadequate funding. Bangkok has one of the lowest proportions of road surface to total area of any major city (around 11 per cent, compared to London's 22 per cent and New York's 24 per cent) making traffic density a major problem. Most of the city lies at no more than two metres above sea level, and flooding during the monsoon is a big problem.

About 1285 new vehicles reach the streets each day. In the first three months of 1995 alone, motorcycle sales increased 11.3 per cent with sales of 356 700, while

Figure 20.3 Urban growth on the periphery of Bangkok.

car sales jumped 23 per cent to 128 500; 45 per cent of new cars in Thailand that year were registered in Bangkok alone. Figure 20.5 shows the composition of vehicles registered in Bangkok in 1993.

Figure 20.4 City of Bangkok and its road networks. The population of this central area is about 4 million; the larger Metropolitan Region has about 11 million. Central Bangkok is shaded in orange.

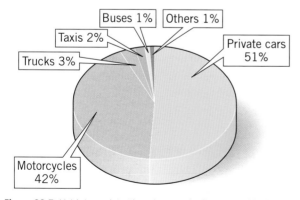

Figure 20.5 Vehicle registrations by type in Bangkok, 1993.

One author describes Bangkok's traffic problems thus:

On one holiday weekend in 1995, the tailback of vehicles stretched 200km and lasted 18 hours. The city's most popular radio station features non-stop traffic news and has developed a sideline selling standby in-car urinals. ... The majority of traffic policemen on Bangkok streets suffer from respiratory diseases. Mahidol University reported that the air contained bacteria, fungi, and three to five times the acceptable level of dust particles.

Even the city's governor announced: 'Bangkok is not a place to live ... there are fumes, disasters, accidents. Truck drivers drive like hell through the city. This is a jungle, an unorganised place.'

Journey times become slower every year; the average speed of journeys in Bangkok was estimated in 1998 to be 5km per hour. This varies. At 6 a.m. journeys are fast on the new freeways built above the city rooftops. During evening rush hours and the monsoon, roads become congested and flooded, especially at ground level. At such times, traffic chaos develops. Traffic conditions in Bangkok cause vehicles to wear faster than normal. Engines spend a long time idling in traffic and accelerating away from traffic junctions, with wear on the engine and on gearboxes. The consumption of unleaded and super benzene petrol increased from 3.73 million litres a day in 1992 to 6 million litres in the first quarter of 1995. The traffic crisis causes Thailand about 60 billion Baht (£1 billion) a year in terms of petrol consumption, the impacts of traffic pollution upon health, and costs of traffic accidents.

Impacts of traffic on Bangkok

Traffic costs both Thailand and Bangkok itself in terms of wasted time and health-care expenditure for traffic-related illnesses. Traffic conditions affect most people's daily lives in Bangkok. People have to get up early to avoid being stuck in traffic jams, children and parents have their breakfast in their cars, and some families even move to temporary city residences to avoid commuting during the weekdays.

Pollution and health

Figure 20.6 is a view taken over Bangkok in 1999. Visibility is so poor so that only buildings in the near distance can be seen. This is the result of pollution, of which there are many sources in Bangkok. Air quality is one of Bangkok's major problems. Road traffic accounts for over half of the air pollution in the Metropolitan Region. The problem has compelled the government to take several urgent measures. There is now a daily monitoring of air pollution on a city web-page; its reference can be found at the end of this chapter. Blood lead levels of children and adults are among the highest in the world. A study of 82 infants during 1989-90 found

Figure 20.6 Air pollution over Bangkok in December 1999.

average lead levels of 18.5 micrograms per decilitre, nearly twice the level considered dangerous.

A study by the World Bank identified air pollution from particulates, lead and traffic congestion as Bangkok's most serious urban environmental problem. It showed that even small reductions in air pollution and congestion would provide significant benefits. Reducing concentrations of pollutants by 20 per cent would provide health benefits estimated at between £250 million and £1 billion for particulates and between £200 million and £1 billion for lead. For congestion, the study estimates that a 10 per cent reduction in peak-hour trips would provide benefits of about £250 million annually. Leaded petrol has now been phased out. Pollutant gases from traffic contribute to urban smog, shown in Figure 20.7.

Following success in phasing out leaded petrol, the next important air quality measure is to reduce the amount of particulate matter in the atmosphere, which cause the greatest pollution threat to Bangkok residents. Small particulate matter penetrate deep into the lungs to cause major health disorders such as asthma and bronchitis; there are possible links with cancer. Public exposure to air pollution in Bangkok is estimated to cause thousands of premature deaths, and several million cases of work- and school-days lost through sickness every year. The major sources are motor vehicles (23 per cent) and dust from roads (33 per cent).

Figure 20.7 The causes of photochemical smog. Ultra-violet light causes natural nitrogen dioxide (NO_2) to break up into nitric oxide (NO) and free oxygen. Free oxygen (O) combines with natural oxygen (O_2) to form ozone. Ozone reacts with nitric oxide to produce more nitrogen dioxide, and oxygen. Photochemical smog results from this reaction and other chemical reactions of hydro-carbons from fossil fuels with free oxygen. This 'cocktail' is linked with increases in respiratory illnesses. The way in which the gases react with one another and with other pollutants is not fully known.

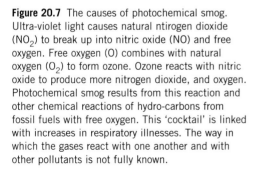

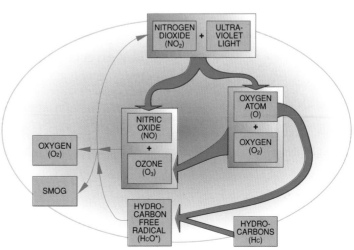

Vehicles which contribute most to this are two-stroke motorcycles (especially those that are not maintained and which use adulterated lubricating oil) and ageing, poorly maintained diesel buses and trucks. Other sources of dust particles include industrial boilers (29 per cent), power plants (12 per cent) and construction (3 per cent).

Accidents

Road accidents have grown at a rate far greater than that of the population. Figure 20.8 shows the increase in road casualties between 1983 and 1996. For people, these mean tragedy of lost or injured family or friends; for society, the cost of hospital treatment and health care, and the loss of young, economically active people is a considerable burden.

	1983	1992	1996
Accidents	13 674	46 743	56 208
Fatalities	708	983	1509
Injured	4551	11 025	23 814

Figure 20.8 Road casualties in Bangkok Metropolitan Area, 1983-1996.

A report in *The Economist* in 1990 showed how, on 24 September at 10.30 p.m. a truck carrying liquefied petroleum gas collided with a tuk-tuk. The resulting fire killed 50 people and severely burned another hundred or so. Most of the victims were stuck bumper-to-bumper in six lanes of traffic – even at 10.30 p.m. – and were unable to escape. Emergency vehicles were unable to get through to help.

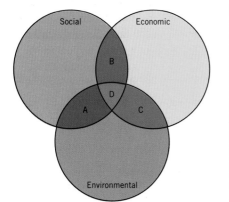

A = Social and Environmental
B = Social and Economic
C = Economic and Environmental
D = all three

1 Plot the data shown in Figure 20.8 on a graph. Is the rate of increase in accidents, fatalities and injuries steady, rising, or falling?

2 Suggest reasons for the increase in accidents, fatalities and injuries, based on this chapter so far.

3 Consider all the effects of Bangkok's traffic congestion.
  • List all the effects.
  • Make a copy of the Venn diagram, left, and classify each effect that you have listed, according to whether it is social, economic or environmental.

The effects of congestion on retailing

Traffic problems have led to changes in retail business. Suburban department stores and shopping malls already attract people living in the outskirts to avoid having to drive through the traffic jams of Bangkok. Over 550 new convenience stores provide fast-food services as changes in people's journey to work affects eating habits. With less time in an everyday life, people use fast- or ready-to-eat food, which has an annual value of 16 billion Baht (£250 million). Frozen meals now have a market sale of 1 billion Baht (£160 million) and are growing at 15 to 20 per cent a year.

Car parking

Parking has become a headache for everyone. An inner-city parking area can earn as much as 100 000 Baht (over £1600) per day for its owners. In a city where £10 is a good daily wage (likely to be earned by an university graduate engineer first beginning work), this is good money. Residents claim that parking lots outside the city would be in more demand if there were a car-restricted zone together with a quality mass transit system between outskirts and inner Bangkok. Now that the new Skytrain has opened this may be possible.

Solving Bangkok's traffic problems

As GDP in Thailand increases, so more people buy cars. Car ownership was made cheaper by a government decision in 1991 to cut all import duties on cars. However, cars are not an easy option for getting to work because of the time spent in jams.

The Bangkok Mass Transit Authority (BMTA) is responsible for managing traffic in the city, and is exploring different options for traffic management and control. In the medium term, solutions to Bangkok's traffic problems lie in creating mass transport systems, capable of transporting large numbers of people to and from work every day. Two strategies are being attempted:

- to create mass rapid transport systems, capable of moving traffic more quickly, by both road and rail
- to develop a range of means by which people can travel into and across the city, including the development of private as well as public transport.

Developing the city road system

Roads at ground level in Bangkok are poor. State investment on road improvements is low and often based on expensive foreign loans. Given the state of congestion at ground level, the Expressway and Rapid Transit Authority have developed a number of elevated roads to cross the city, using private finance. One is shown in its construction stage in 1999 in Figure 20.9, with a plan view of how these work in Figure 20.10. These are mostly three to four-lane dual carriageways, and carry traffic at up to 50m above ground level. They are accessed by ramps at which there are toll gates. Road charges have become a common means of financing road development in Bangkok; a typical toll for the road shown in Figure 20.9 was 10 Baht for a car in 1999, while the expressway was half-completed, and 20 Baht when the road is finally completed in mid-2000.

Figure 20.9 Construction in 1999 of a new elevated expressway in Bangkok.

Expressways have been developed where traffic is able to bear the cost of tolls, such as taxi services between Bangkok's International Airport and the CBD or hotels. Although tolls first began in the late 1980s for the two expressways to vehicles entering the CBD, they are now standard on all elevated sections. In some cases, early expressways are being over-passed by newer ones above them.

Expressways are not problem-free. At rush-hours, traffic queues build up on the ramps and at toll booths. Many freeway junctions have converging cross-overs, where traffic negotiates scissors-like changes of lane at very high speeds of 100 km per hour or more. Accidents are common.

---

1 Consider a city or urban area that you know.
  a) How has the use of elevated roads helped to manage traffic flow? With what effects?
  b) Would it be possible to create expressways like those in Bangkok? Why?
2 To what extent do you think that expressways are effective solutions for Bangkok?

Developing mass transit rail systems

A plan for mass rail transit has been needed in Bangkok for decades; a combination of financial controls and poor government planning have resulted in delay. There is a rail system serving part of the city, shown in Figure 20.4. Although cheap, it is slow; the journey from the International Airport, 25km from the CBD, costs 15 Baht (25p) but takes 45 minutes. At least five government agencies, who rarely co-ordinate each other's work, are

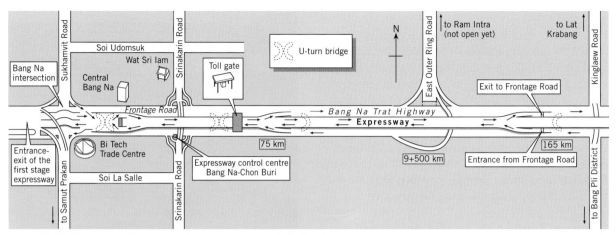

Figure 20.10 Plan showing how elevated expressways operate in Bangkok.

responsible for Bangkok's public transport. However, the government is now imposing co-ordination so that Mass Transit Rail Systems can be developed. These include an elevated rail system, known as the Skytrain, a proposed tram system, and a new subway in order to reduce traffic congestion and improve journey times.

The Skytrain

Bangkok Mass Transit System Corporation has built and developed an elevated rail system (Figure 20.11). It opened in December 1999. The line is 28.5km long, between the south-western edge of the city at the south-eastern end of Sukhumvit Road and the northern bus terminal (Figure 20.4), linking the northern and eastern suburbs to the CBD. It has 23 stations. In its first month, it encountered opposition; on average, 160 000 people were using it each weekday and 200 000 at weekends against 600 000 predicted users each day. It raised complaints from the elderly and from parents, who claimed that access to the elevated stations was difficult, without escalators. However, its comfort and speed have won many compliments.

A new tramway system

At the same time, Bangkok Elevated Transit Systems (BETS) plan to develop a 60km elevated tramway system. Construction began in 1993 and plans to accommodate over 50km of six-lane road expressways alongside. 42 stations are planned for the system, with a daily planned passenger load of three million. However, the Asian economic downturn in 1997-98 created problems for the project, and it is not yet open. Battles between different government departments create delays, and the Thai currency collapse in 1997 made the project much more expensive.

A subway system

Bangkok has no subway system at present, but its first line – a 20-km system between Bangkok's main central railway station at Hualamphong and Bang Sue in the northern part of the city – is currently being built and is planned to open some time between 2002-2003. Again, the work is contracted out to a company – Bangkok Metro Co – and financed by overseas Asian sources.

Figure 20.11 The Skytrain, shown here above Sukhumvit Road, one of the main streets in Bangkok CBD. Users claim that stairs to the stations make access difficult.

1 Identify the routes of the Skytrain and of the proposed subway system on Figure 20.4. How far do you think that these will reduce traffic congestion in the city centre?
2 Draw a sketch map of central Bangkok, using Figure 20.4. Draw on it:
   * the main expressways
   * the rail links that currently exist as shown on the map
   * the route of the new Skytrain
   * the proposed route of the subway.

Private sector or public transport?

Cheap public transport is an effective travel option in Bangkok in terms of cost. A bus journey anywhere in the city usually costs little more than 5 Baht (8p) in a public-service, non-air-conditioned bus, and 16 Baht (26p) in one which is air-conditioned. However, many people earn very low wages; over half of Bangkok's factory workers in 1998 received 162 Baht (£2.65) as their daily wage. For them, the open windows and polluted air on one of Bangkok's public buses is the only option.

Public transport has not kept pace with the city's population and economic growth. As a result, people are opting increasingly for private-sector transport, and 82 per cent of Bangkok's population now travel to work by private means.

Motorbike taxis

Motorbike taxis have become popular. They are faster because they can move between traffic lanes, but are also very dangerous. Nonetheless, many commuters prefer a dangerous journey to sitting in carbon monoxide fumes for long periods. It is estimated that motorcycle hire services generated more than 5 billion Baht (£80 million) of income in 1994. There are now 40 000 motorcycles with 1600 different firms in Bangkok, providing short to medium-distant services.

Microbus

The Bangkok Microbus Company operates air-conditioned bus services in Bangkok city and Metropolitan district. The cost is a flat fare anywhere on the route for 35 Baht (55p). Even though Microbus fares are up to seven times more expensive than those of BMTA buses, frequent services and convenience have led to increasing popularity.

Commuting van services

Private van services provide a convenient alternative to crowded buses. The vans often use expressways, and take short cuts in Bangkok to beat traffic jams. There are about 2000 vans at present with a total income of about 1 billion Baht per year (£17 million).

Boat and ferry services

In recent years, travelling by boat has been regaining popularity as the traffic problem has been getting worse on the roads of Bangkok. In 1994, there were 339 186 passengers of boats and ferry services from over 100 piers in the Bangkok Metropolitan Region every day. These include the

Figure 20.12 The Chao Phraya express ferry service along the Chao Phraya river.

Figure 20.13 Local ferries cross the Chao Phraya river each day.

fast Chao Phraya Express, shown in Figure 20.12, which runs between Nonthaburi (Figure 20.4) to the north of the CBD and the main commercial areas close to the river in the south. It is affordable and efficient; in 1999, it cost 8 Baht (12p) for a journey. Other smaller ferries cross the river, such as those shown in Figure 20.13, and link any points where roads meet the water. In addition, a private-hire service operates at great speed – and noise! – along the river for anyone wanting a water taxi.

Which way forward?

Part of the traffic problem is its volume. Any solution must deal with volumes of people first; hence the mass transit systems proposed. However, government action could also reduce some of the effects of traffic, such as air pollution. A priority in Bangkok in the late 1990s has been to set new emissions standards for two-stroke motorcycles, to try and reduce emissions (by 90 per cent for particulates) to the level of four-stroke motorcycles. Higher taxes on transport fuels might also restrain the growth of private transportation, but penalise low-income earners. Congestion problems also could be reduced by expanding flexible work hours, upgrading bus services, and improving traffic management.

1 Copy the following table. On it, compare the advantages and disadvantages of each of the types of transport shown.

Type of transport	Advantages	Disadvantages
Tuk-tuk		
Private car at ground level		
Private car on expressways		
Existing rail		
Skytrain		
Proposed subway		
Motorbike		
Micro-bus		
Private van services		
River ferries		

2 Based on this:
- which forms of transport should be developed further in order to manage Bangkok's traffic in the next 20 years? Why?
- which forms should be discouraged? Why?

3 Form groups of three or four. Consider the likely impacts of the following on Bangkok's traffic problems:
- setting new emissions standards
- expanding flexible work hours
- upgrading bus services
- improving traffic management.

Which, if any, do you regard as a 'best' solution? Are there others you think should be considered?

4 Should there be an overall co-ordinator for traffic in Bangkok? Why?

Transport issues in Melbourne

Bangkok has shown how unplanned growth requires overall planning. This study of Melbourne, the largest city in Victoria, Australia, shows how changes from public to private ownership can affect transport provision.

Melbourne has approximately 3.5 million people (1996), who live in a large area in southern Victoria. The city has grown more to the east and south-east than to the west, and now extends some 30–35 km to the outskirts towards Dandenong. It is well provided for in terms of transport networks.

- Its road system is extensive and includes several freeways which give access to the south-east. Freeways head south-east to Dandenong and then to Sydney; another follows the coast to Adelaide; and the airport is served by a freeway to the north-west (Figure 20.14).
- Its tram system was among the best early examples of public light-rail development in the southern hemisphere, giving access to the city as it was up to about 1920. However, no additional lines have been built since, with the result that only the inner- and middle-city areas of Melbourne are now well served.

- The rail system, shown in Figure 20.15, is extensive, cheap and serves most suburbs, including the outer parts of the city.

The problem is that increasing numbers of journeys are made by private car. People use public transport less and less. In recent years, the government of Victoria has tried to cut public expenditure, and has raised fares on public transport to try and increase the proportion of money earned by buses, trains and trams. In the end, a vicious circle has developed; people use public transport less because it is more expensive; as a result, greater losses are created and the government raised fares to try and cut these.

As a result, some parts of the city trains and buses have been reduced to frequencies of one hour between services during the day, so that people feel less and less able to rely on them. One outer suburb – Keysborough (Figure 20.16) – has had its fares rise and frequency of service reduced substantially. Staff cutbacks at rail stations increased a sense of fear, especially at night. In Melbourne, public transport raises only 33 per cent of its costs through fares; all the rest comes from taxes. Yet in Toronto, 68 per cent of operating costs are recovered from

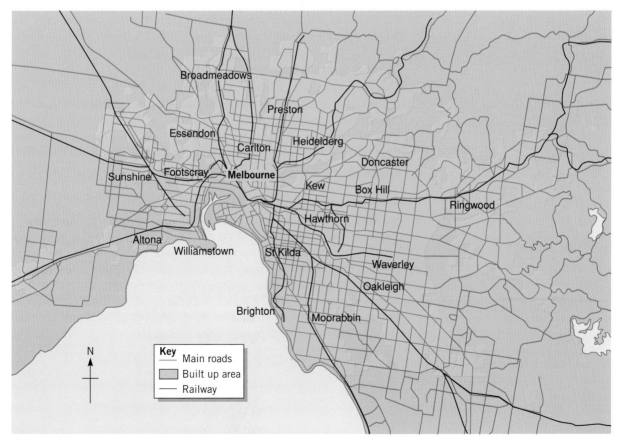

Figure 20.14 Melbourne's road networks.

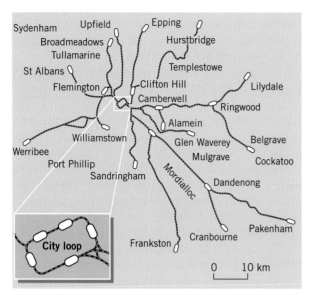

Figure 20.15 The rail system of Melbourne.

City	Melbourne
Suburb	Keysborough
Distance from city centre in km	25
Population density per hecture	32
Bus route no.	815
Bus service frequency in minutes	
Peak period	60
Off-peak daytime	60
Evening	no service
Last bus at	6.30 p.m.
Saturday a.m.	60
Saturday afternoon	no service
Sunday	no service
Fare to city (Bus and train) in 1999	£3.70

Figure 20.16 Public transport frequency in Keysborough, an outer-city suburb of Melbourne.

fares, unlike Melbourne's 33 per cent. It is a matter of encouraging people to use them more.

Now, the rail system, buses and trams have been sold off into private companies, who have been given the job of persuading people back. In addition, a new road system is creating a great debate in Melbourne; for the first time, people are being asked to pay for road journeys into the city. This poses questions about how transport and access to city centres might be managed in future, not only for Melbourne, but for other cities.

The 'Linking Melbourne' plan

Figures 20.17 and 20.18 show two opinions of a plan to extend road networks in Melbourne, with 'the world's biggest urban freeway programme'. One – Figure 20.17 – is from the Public Transport Users' Association in Melbourne, while the second is from Transurban City Link Ltd, the people responsible for its construction and management.

The CityLink plan for Melbourne

Figure 20.19 shows a map of the new Freeway system for Melbourne, which opened in January 2000. Not all the freeway is new; the Tullamarine Freeway existed before, together with the Westgate Freeway and the bridge over the Yarra. However, these ran into the CBD and stopped. On the south-eastern side of the city, the south-eastern Freeway is now connected to the western link by means of two large tunnels (Figure 20.19a). The CityLink plan therefore links existing freeways and provides routes through the city.

Wrong Way Go Back

A freeway revolt is beginning as the world's most liveable city begins to wake up to 'Linking Melbourne', the world's biggest urban freeway programme. The Victorian State government's road plan will build 211km of new freeway and widen many existing freeways, more than doubling the size what is already Australia's biggest freeway network. The plan includes City Link (a tunnel under the Botanical Gardens, extension of the Tullamarine Freeway through North Melbourne and freeway widenings), a $2.5 billion ring-freeway through scenic areas on Melbourne's outskirts, extension of the Eastern Freeway and many more projects.

The financial cost of Linking Melbourne is $6.5 billion; the cost to the environment and quality of life still greater. Melbourne is set to become the Los Angeles of the southern hemisphere.

If implemented, Linking Melbourne will increase traffic levels, worsening pollution which damages health and contributes to the greenhouse effect. Communities will be divided, homes will be demolished and scenic areas destroyed to lay hundreds of kilometres of bitumen and concrete. Public transport will be undermined, setting off a downward spiral of patronage decline, service cuts and fare rises.

There are alternatives. Across the world, cities are turning to public transport for lasting solutions to traffic problems. First class public transport – with fast, frequent, integrated, safe, clean and economical services – is winning customers away from cars across Europe and North America. Even Perth and Brisbane are getting in on the act! Melbourne can join this trend to environmentally sustainable urban transport. The time to start is now.

Figure 20.17 The view of the Public Transport Users' Association in Melbourne on the new road proposals for Melbourne.

Eyes of the world on CityLink

Congestion costs the state economy around $5 million a day and compromises our ability to compete in international markets – meaning less activity and fewer jobs. The infamous 'car parks' that build up on our freeways have resulted in Victorians spending too much time caught in traffic, cutting into their family and leisure time. In a unique partnership with the Victorian government, Transurban is delivering a project of national significance that has drawn the eyes of the world to Melbourne.

CityLink is a major engineering feat, applying technology that will deliver real benefits to Victorians and their businesses. It will significantly reduce travel times, improve the sfaety of our roads and allow Melbourne to retain its boast of being Australia's most 'liveable city'. CityLink will also have a positive impact on the Victorian economy, delivering saving to business of around $265 million each year as a result of reduced travel times and vehicle operating costs, fewer accidents and more efficient movement of goods. It has been estimated these savings will deliver a permanent increase in Gross State Product of about $382 million a year.

Thanks to CityLink, Melbourne will be a great place to do business. CityLink will increase connectivity and access to major road networks and improve Melbourne's competitive edge as the place in Australia to do business.

Figure 20.18 The view of Kim Edwards, Managing Director of CityLink.

Yet the link has created huge uproar in Melbourne. Although its supporters claim that it offers quicker journey times, its opponents believe that it represents poor value for money, not least because people now have to pay to travel on roads which were free before. As Figure 20.19b shows, tolls are charged for using different sections of the road. The further you travel, the more you pay, either in sections or by using the stretch of Freeway in its entirety. The company believes that it will offer faster journey times; opponents believe that it will only do so by displacing traffic on to side-roads, thus creating rat-runs for people who try to avoid paying charges.

How it operates

Payment is made using a system known as E-Tag, mounted to the car windscreen above the rear-view mirror, which transmits a signal each

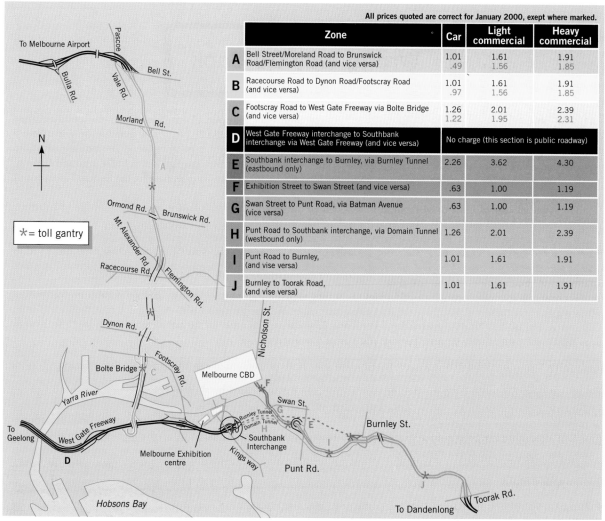

All prices quoted are correct for January 2000, exept where marked.

Zone		Car	Light commercial	Heavy commercial
A	Bell Street/Moreland Road to Brunswick Road/Flemington Road (and vice versa)	1.01 .49	1.61 1.56	1.91 1.85
B	Racecourse Road to Dynon Road/Footscray Road (and vice versa)	1.01 .97	1.61 1.56	1.91 1.85
C	Footscray Road to West Gate Freeway via Bolte Bridge (and vice versa)	1.26 1.22	2.01 1.95	2.39 2.31
D	West Gate Freeway interchange to Southbank interchange via West Gate Freeway (and vice versa)	No charge (this section is public roadway)		
E	Southbank interchange to Burnley, via Burnley Tunnel (eastbound only)	2.26	3.62	4.30
F	Exhibition Street to Swan Street (and vice versa)	.63	1.00	1.19
G	Swan Street to Punt Road, via Batman Avenue (vice versa)	.63	1.00	1.19
H	Punt Road to Southbank interchange, via Domain Tunnel (westbound only)	1.26	2.01	2.39
I	Punt Road to Burnley, (and vise versa)	1.01	1.61	1.91
J	Burnley to Toorak Road, (and vise versa)	1.01	1.61	1.91

Figure 20.19a The CityLink in Melbourne.

Figure 20.19b CityLink toll charges in Melbourne.

time the user's car passes beneath one of the gantries on the freeway. Charges are then debited against the user. Charges have to be pre-paid, or alternatively settled at a post office by noon the day after usage, otherwise fines are payable. In January 2000, fines were running at 1000 per day as people struggled to come to terms with this system.

In theory, the argument goes, people become conscious of how often they use their cars, and become aware of the 'real' cost of journeys made by car. In this way, they might use public transport for some of their journeys. However, privatising the road network means that the company, Transurban CityLink, makes a profit by sustained or increased use of their freeways. Opponents claim that this will encourage people to use their cars more.

1 In small groups, discuss the advantages and disadvantages of asking motorists to pay for road usage:
   • in Melbourne
   • in an urban area known to you.
2 The UK government has stated that charging for access to city centres is likely in Britain in coming years. How similar or different do you think that people in the UK would feel about this, compared to people in Melbourne?

Privatisation of public transport

In 1999 rail and tram services were privatised. Half of each system, shown in Figure 20.20 and 20.21, was sold to National Express, the UK coach and rail company.

Rail services

Figure 20.20 shows the rail network in Melbourne; compare this with the map of Melbourne in Figure 20.14. Notice that the rail system covers most of the urban area as far as the outer suburbs. Two companies have been created – Bayside trains, purchased by National Express, and Hillside Trains, purchased by Vivendi, a French consortium. Hillside claim that they can offer the following improvements in services to Melbourne's eastern and north-eastern suburbs in the 15-year franchise by:

- investing $314 million in new trains to enter service from 2002
- improved service frequencies with 130 extra services per week
- improvements in punctuality and reliability in each year
- revitalisation of Melbourne's main Flinders Street Station as a tram-train interchange, with upgraded passenger security
- saving taxpayers AU$389 million in reduced subsidies over 15 years
- introducing new car-parking spaces

National Express make similar claims for their Bayside franchise.

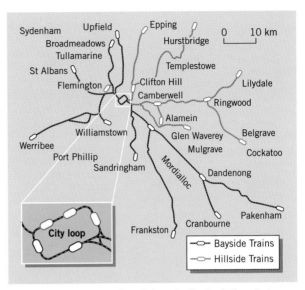

Figure 20.20 The allocation of the privatised rail lines to two overseas companies, National Express and Vivendi.

Each company claims to be able to increase rail traffic substantially. National Express has already achieved growth of 10 per cent on some of its operations in the UK, and claims that to do so in Melbourne will not be difficult. It also makes lines more profitable by staff cuts and by increased revenue.

Trams

Trams are part of Melbourne's character; it is the only Australian city to have them (Figure 20.21). Like the rail network, lack of investment by the Victoria state government has led to reduced services over the network shown in Figure 20.22. Compare the map in Figure 20.22 with the map of Melbourne (Figure 20.14). Trams cover a relatively small area of the city and are mostly confined to the older inner city. This network has been privatised; it too has been sold to National Express as Yarra Trams.

Recent population trends may bring about a change in fortunes. The population of Melbourne's inner city areas, such as South Melbourne and Carlton (Figure 20.14), has risen dramatically since 1991. Urban regeneration and a new generation of 20-29 year-olds who prefer living near the CBD has led to rapid population growth and demand for public rather than private transport to get to work. Increasing tram patronage should not prove difficult. Most sports and entertainment complexes, such as Melbourne's MCG Cricket Ground and Tennis Centre, lie next to tram routes.

Figure 20.21 Tram in Melbourne's CBD.

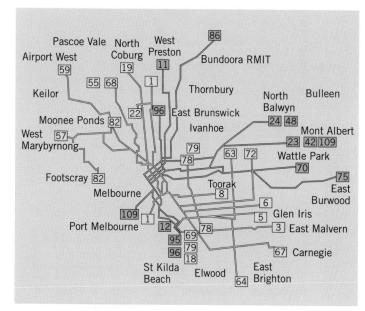

Figure 20.22 The extent of Melbourne's tram network.

IT'LL ENCOURAGE COMPETITION

BETWEEN NATIONAL EXPRESS, NATIONAL EXPRESS, AND NATIONAL EXPRESS

Figure 20.23 From *The Age*, 26 June 1999; reactions to privatisation and greater competition on Melbourne's transport network!

Does the future lie in public transport?

Privatisation seems, on the surface, to result in investment which governments rarely make. However, investment is often run down over a number of years prior to privatisation, so that anything offered by a private company seems better. Similarly, service frequency improves when companies run trains rather than letting them stand idle. Most rail costs are fixed; that is, they must be paid for whether the train is idle or working, so that any revenue that is taken is an extra which improves profitability. The question in Melbourne is why the state government did not make the investment now being introduced by National Express and Vivendi.

The Public Transport Users' Association in Melbourne considers that three new rail lines are needed to provide people with better transport links. Their proposals are as follows:

In the north of the city:

- extend the Upfield service (Figure 20.20) north to Craigieburn, to serve housing estates in the area
- extend the St Albans line (Figure 20.20) to Sunbury
- extend the Broadmeadows line (Figure 20.20) by 6km to Tullamarine Airport, and serve the city centre with trains running every 10 minutes across to the eastern suburbs.

To the east of the city:

- develop a rail link to Doncaster by extending either the Belgrave or the Lilydale lines (Figure 20.20). This, it is argued, would cost AU$300 million, compared with the cost of extending the Eastern Freeway of AU$700 million.

The total cost of these developments would be AU$500 million, compared to the cost of AU$6.5 billion for new Freeway development in the 'Linking Melbourne' plan. They argue that improved access would result in increased usage, thereby cutting the deficit between passenger costs and revenue.

---

1 Form pairs. Take one of the questions below, and prepare an answer to it for discussion.
- Should investment take place in public transport, or cars take over?
- Should public transport be owned by private companies, who provide a service for profit?
- Should public transport be owned by the public sector, who provide services for people?
- Should investment in roads be made by private companies who seek increased usage and income?
- Should people be charged for road usage, just as they are for other forms of transport?
- Should governments pay for roads for the public good from taxation?

2 Write a 1000-word essay entitled *Transporting people in cities of the future*. In it, explore how the movement of people in urban areas might take place in 2025.

Summary

You have learned that:

- Road traffic is increasing across the world, the result of increasing incomes and expectations.
- Urban concentrations of traffic show the greatest growth and have the greatest effects; this growth is likely to continue as standards of living rise in LEDCs.
- Road traffic especially has considerable effects; these may be social, economic and environmental. Environmental consequences are great, resulting in air pollution which in turn has serious effects on people's health.
- Cities such as Bangkok lack co-ordinated attempts to manage traffic; each strategy is tried piecemeal. However, there are several attempts being made, involving roads (elevated expressways), rail and subway links.
- City authorities can co-ordinate transport provision if the will is there to develop service provision for people.
- Recent privatisation in Melbourne has resulted in likely growth in rail and tram traffic, but the privatisation of the freeways is more problematic.

Ideas for further study

Many urban areas offer great potential for study and are debating which way to progress in managing traffic.

1 What are the key issues in an urban area known to you? How are these being resolved? With what effects?

2 Keep track of current issues in urban transport, e.g. the privatisation proposals for London's underground. Which options are being considered? Who supports them, and why? Who does not, and why?

3 Identify key traffic issues in a LEDC city, such as Delhi, Mumbai, São Paulo. To what extent are the issues similar to or different from those in Bangkok?

References for further study

Greater Manchester PTE web page
http://www.GMPTE/gov.uk/aboutus/aboutus.htm

21 New York – environmental quality or inequality?

New York City is often described as a melting pot of humanity, but as one commentator has remarked, 'if New York is a melting pot then someone forgot to light the fire'. New York is a city of very distinct neighbourhoods that over the years have become segregated on the basis of wealth and ethnicity. It is a city of extremes, home to the wealthiest postal code in the USA – 10021 on the Upper East Side of Manhattan – as well as one of the poorest, Melrose Commons in the South Bronx. This chapter explores some of the issues of environmental quality in the city.

New York City

New York City (1990 population: 7 322 091) is distinct from New York state (Figure 21.1). The city itself is divided into five boroughs. The most famous is probably Manhattan, which gives the popular image of New York City. New York City is made up of five boroughs – The Bronx (1.2 million), Brooklyn (2.3 million), Manhattan (1.5 million), Queens (2 million), and Staten Island (375 000). Each is large enough to be, and is like, a city in its own right. Each is sub-divided into Community Districts, which are small

Figure 21.2 Looking along Manhattan, from the southern tip towards 'mid-town' around the Empire State building and central park beyond.

geographical areas with responsibility for administering the needs of that community. There are 59 Community Districts for the whole of New York City, acting as liaison between residents and city agencies. These have an important role in urban environmental quality.

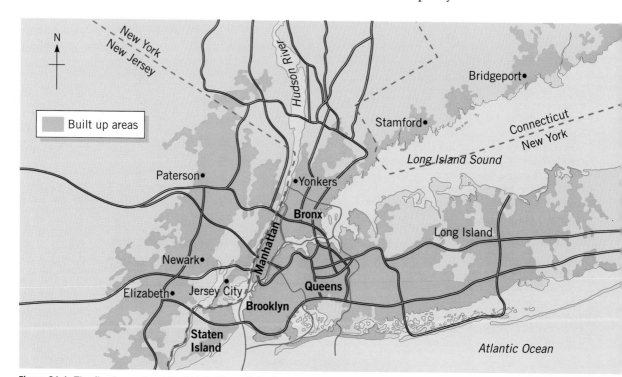

Figure 21.1 The five boroughs of New York City and the State of New York

The Bronx

The Bronx is the only borough of New York City connected to the mainland. It is divided into 12 Community Districts, all having their own characteristics. During the 1970s and 80s, the Bronx attracted world-wide attention for all the wrong reasons; it was a scene of dereliction and burnt-out buildings, many the result of arson, as landlords tried to 'get out' of the Bronx by claiming insurance. The photo in Figure 21.3 was taken in the early 1980s at the height of the worst burnings. Although the Bronx still faces many problems, the South Bronx has improved, and once-shattered neighbourhoods are being rebuilt. This has been achieved through the co-operation of citizen groups, neighbourhood organisations and local police. Figure 21.4 shows the progress that has been made in the South Bronx over the 1990s, as well as highlighting the obstacles the area still faces.

Figure 21.3 The South Bronx from the early 1980s, showing some of the worst burnt-out buildings.

Study Figure 21.4. Form groups of two or three, and summarise the following:
1 What has prevented commerce and other services following the development of 57 000 new homes in the South Bronx?
2 How could the siting of the European American Bank help to bridge the inequality gap that exists between the South Bronx and other parts of New York City? Construct a diagram such as a flow chart to help you think about the possible impacts.

Bank Dips Its Toe in South Bronx

Hilda Hernandez stared, stepped back, stepped up. She stared some more. The look on her face suggested that a flying saucer had just landed in her neighborhood. In fact, only a bank had, but for Ms. Hernandez, who has lived through her neighborhood's seesawing fortunes for 30 years, that was news enough. Last week, the European American Bank, or E.A.B., officially opened its first branch in the South Bronx, making it the first bank in the area served by the borough's Community Board 3 in more than a decade.

In the early 1980's, the building of the ranch homes on Charlotte Street, a block away from the bank, helped start the borough's huge housing renaissance, in which 57 000 apartments and houses have been built or rehabilitated. But even as the Charlotte Street homes were followed by dozens and dozens of new multifamily homes on nearby streets, stores and services were slow to follow. The perception that poverty and crime were more abundant than customers kept businesses away. Today, though, commercial life is slowly creeping back to areas that became retail wastelands as the Bronx burned. "When I moved here, there was nothing," said Abdullah Sabuur, a neighborhood resident since 1987, and a new account holder at E.A.B. Now, he said, "I don't think I can get a better neighborhood, with what they're bringing here."

There is, of course, still a long way to go. Most of this stretch of Southern Boulevard is still ghostly, with most stores boarded up or vacant. And there are those who worry that the airy new South Bronx – where single-, two- or three-family homes have often replaced giant apartment buildings – may not have a high enough concentration of free-spending customers to sustain much retail business, particularly when welfare revision has also reduced the cash flowing to the area.

Bodegas and other small businesses run by immigrant entrepreneurs were among the first to return. But gradually, the businesses coming in are diversifying, both in their nature and in their size. There is a growing sector of insurers and realtors serving the healthy real estate market in the Bronx. E.A.B.'s statistics show an average annual household income of $18 000 for the 19 000 households around the bank, and a home ownership rate of 8 per cent.

Armin Krauss, a butcher who kept his business going on Jennings Street for 41 years, said he was skeptical that the bank or the new businesses would do well. But the bigger problem is that while the Charlotte Street houses are beautiful, far fewer people live in them than lived in the apartment buildings that once shadowed the street. There are now about 6 families per acre in Charlotte Gardens where there were once close to 200.

Fewer homes mean fewer customers. And that, said Mr. Krauss, "is why the neighborhood has come back, but not businesswise."

Figure 21.4 From the *New York Times*, 22 July 1999; adapted from an article by Amy Waldman.

The Bronx is also home to the Bronx Bombers, better known now as the Yankees Baseball team. Their stadium is located in the South Bronx, though the owner of the team repeatedly threatens to move the club across the Hudson River to New Jersey if it does not get a new stadium.

Manhattan

Manhattan (New York County) has perhaps the most famous skyline in the world. It is the third largest borough of New York City in terms of population, and an island of extremes in terms of wealth, poverty, crime and open space. It is ethnically diverse, and divided into neighbourhoods, ethnic areas, quarters, or 'barrios'; these include East Harlem (mainly Puerto Rican and Mexican), Washington Heights (Dominican), West Harlem (dominated by African Americans), to the Lower East Side where the Chinese community is concentrated. Housing stock varies, from penthouse apartments overlooking Central Park (Figure 21.5a), the Brownstones of Greenwich Village (Figure 21.5b), to high-rise housing projects on the Lower East Side (Figure 21.5c). Manhattan gets a good deal of the limelight and financial investment and often this results in a 'Manhattan-centric' attitude which leaves poorer neighbourhoods feeling marginalised. Each neighbourhood has its own environmental concerns, but most commonly includes the availability of open space.

Figure 21.5 Varied neighbourhoods of Manhattan.

B 'Brownstones' of Greenwich Village.

A Penthouse apartments overlooking Central Park.

C High-rise housing projects on the Lower East Side, built in the 1950s.

Figure 21.6 Brooklyn Bridge, crossing the Hudson River. Brooklyn is in the distance.

Brooklyn

Brooklyn is by far the most populous borough in New York City and is located on the western part of Long Island. It is equivalent to the fourth largest city in the USA after New York, Los Angeles and Chicago. It has always received, and still receives, thousands of immigrants each year. Figure 21.6 shows one of the most famous bridges in the world, Brooklyn Bridge; opened in 1883, it was the first direct connection between Brooklyn and Manhattan. Today numerous bridges and tunnels connect Manhattan with the boroughs of Brooklyn and Queens, further east on Long Island. One of the major environmental issues facing Brooklyn and other boroughs today is that of waste disposal. Brooklyn is already the site for a number of waste transfer stations and is being targeted for new stations in the near future.

Queens

Queens is the largest borough in area and probably the most ethnically diverse. Distinct neighbourhoods exist, from the Greek community in Astoria, the Asians of Flushing to the Jewish of Forest Hills (Figure 21.7).

New York City attracts over 33 million tourists a year; most arrive by air into either JFK or La Guardia airport; both are in Queens. Queens is mostly residential, now that many of its waterfront industries along the East River are gone. Most have moved to New Jersey, where taxes are lower. Two urban environmental issues are of concern to the residents of Queens; housing quality and health on one hand, and transport on the other. Traffic congestion on freeways which lead to the airports and elsewhere on Long Island is the result of lack of investment in public transport.

Figure 21.7 Forest Hills, an ethnically diverse residential area.

Staten Island

Staten Island is the smallest and most isolated borough, connected to lower Manhattan by the Staten Island Ferry (Figure 21.8), and to Brooklyn and New Jersey via the Verrazano Narrows Bridge. It is the least diverse of the five boroughs, mainly white middle-class with a significant Italian influence. There are poorer Black areas, particularly around the ferry port. Staten Island witnessed rapid suburban development by white middle classes; escaping high prices and cramped conditions in Manhattan, after the opening of the bridge in 1965. One of the main environmental issues on Staten Island is the issue of the Fresh Kills Landfill Site, described later.

Figure 21.8 The Staten Island ferry, linking Manhattan with Staten Island.

People and inequality in New York City

It is expected that New York City will have grown in population by the census in April 2000, perhaps the only city to do so in the north-eastern United States. Other older, industrial cities have seen sharp declines in population since 1960. Suburbanisation and decentralisation of industry, together with a rapid expansion of the highway system, has fed the car-culture mentality of American people, and driven more and more people away from large urban areas. Like most cities, the city witnessed rapid suburbanisation and urban sprawl after 1945 in upstate New York, and in the neighbouring states of New Jersey and Connecticut.

Why is New York growing?

The reason for New York City's increase in population in the 1990s is the continuous arrival of immigrants, without whom the City would decline. Before 1960, most overseas-born migrants originated from Europe. In recent years, immigrants from the former USSR have joined others from Bosnia and eastern European countries. However, the majority have come from Latin America and Asia. By the 2000 census, New York City will consist of a majority of ethnic minorities. However, it has to face major issues of inequality.

Although the population of New York City is growing, de-industrialisation has resulted in the loss of thousands of manufacturing jobs. The impact has been greatest on the urban poor, who are unable to

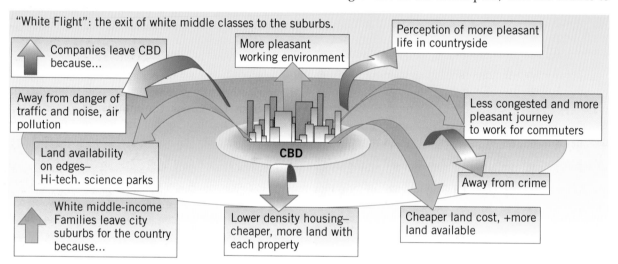

Figure 21.9 White Flight from New York City.

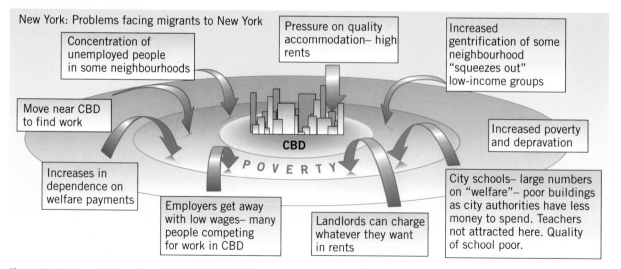

Figure 21.10 Issues facing migrants into New York City.

compete in the new high-technology job market. Growth of high-technology complexes has taken place along highways on the outer edge of the city. 'White flight' from the city continues, as wealthier income groups migrate further out to Long Island, to up-state New York, or to New England (Figure 21.9). This has left behind a minority population who face increasing unemployment, poverty, crime, worsening conditions of health, and limited access to open space (Figure 21.10).

Urban environmental issues in NYC

This section investigates environmental inequality within New York City, attempts to explain the patterns, and finally poses solutions about managing some of the problems created by such inequalities. Three areas of environmental inequality in this section are:
- open space
- urban environmental health
- disposal of urban waste.

Inequalities of open space in New York

New York City has a long tradition of providing parks for its residents, two of the most loved being Central Park in central Manhattan (Figure 21.12) and Prospect Park in Brooklyn. The City of New York Parks and Recreation Department (Figure 21.11) manage and provide a service to the City, by ensuring over 28 000 acres of public space is maintained and accessible.

However, the distribution of open space in the whole of New York City shows glaring inequalities. Many residents live in built-up areas with little or no access to open space, especially in less fashionable areas, such as the Lower East Side and the South Bronx. Throughout the city many vacant and derelict lots exist, creating eyesores and sources of environmental health problems for inhabitants.

Open-space provision in the City

New York planners claim that there is a two-tier park system of privatised and public in the city, creating inequality in provision. 'Flagship' parks such as Central Park benefit from privatisation, while hundreds of small parks, such as Lower East Side Park, suffer from neglect due to scarce City Council funds and inefficient operations. Privatisation has created unequal quality. Privatised parks are run as businesses, where investment is freely available if it produces profit;

> The Municipal Park System totals 28 000 acres
> - 854 playgrounds
> - 700 playing fields
> - 500 tennis courts
> - 33 outdoor swimming pools
> - 10 indoor swimming pools
> - 33 recreation and senior centres
> - 15 miles of beaches
> - 13 golf courses
> - 6 ice rinks
> - 4 major stadiums
> - 4 zoos
>
> Also responsible for 500 000 street trees and 2 million park trees

Figure 21.11 Fact file on open space in New York City.

public parks are bound by the same management and labour practices established many years ago, and are often under strict financial controls. Compared against the private sector, the public park system is under threat.

Figure 21.12 Central Park in Manhattan.
First donated to the people of New York City in 1876, it is described by the New York Rough Guide as the thing that 'makes New York City a just-about bearable place to live – their only contact with nature'.

The Green Guerillas

Managing open space is not only to be found in the public sector or privatised companies. Partnerships between the public and private sectors can also achieve desired results. One example of a scheme that started out as a private venture is the Green Guerillas, a non-profit making organisation who work with city officials to reclaim the numerous vacant and derelict lots in an attempt to 'Green the City'.

Figure 21.13 A community garden in Manhattan.

In 1973, Lizzy Christy began a volunteer group called the Green Guerrillas to help residents of New York City turn abandoned lots into an asset for the community. It began in the Lower East Side and today there are over 700 'Green Thumb' gardens located throughout New York City (Figure 21.13). There is now a permanent staff of six, with 700 members and 200 volunteers who acquire plants and equipment to create and maintain gardens that help to establish a sense of community. In the early days, their first attempts to 'green the city' were to launch peat 'grenades' into vacant lots packed with wildflower seeds. They now take a more relaxed approach in waiting for communities to approach them with possible sites (Figure 21.14), believing that it is more likely to succeed if the community takes ownership of the idea from the start.

Unfortunately, increasing pressure from the city to sell off such sites for development has tainted their success. Three reasons explain this:
- Once-derelict lots are now in demand for property development.
- There is a desperate need for low-income housing in the city. The Green Guerrillas are not against low-income housing, as most of their work is focussed in low-income neighbourhoods. However, they feel housing without open space is a recipe for disaster.
- Increasingly, brown-field sites for development are sought after in preference to green-field sites on the city edge.

Community organisations have organised demonstrations to protest about selling off gardens which they see as focal points of the community. In May 1999, when gardens were to be destroyed, actress Bette Midler bought 112 gardens to ensure their survival! In spite of this, other gardens are being bulldozed every week.

How does the programme work?
- Individuals or groups call the Green Guerillas express interest in establishing a community garden.
- The Guerillas conduct a site visit, make recommendations and draw up a design.
- Volunteers provide expertise and labour, and free plants, shrubs, trees and soil.
- Volunteers conduct follow-up visits help gardeners work against vandalism, insects and disease.
- The Guerillas provide assistance to any gardener, as long as the individual or organisation provides public access to the garden or provides a social service to the community.

Projects
- The Green Guerrillas collect and distribute donated plants and garden supplies to community gardens. Donations come from a variety of sources: shrubbery from a museum, perennials and large planters from the Rockefeller Centre, trees from a Park Avenue penthouse, lumber from a closing disco and bulbs from the Parks department.
- The Green Guerrillas offer an 8-week job training programme in horticulture, providing the homeless with marketable skills.
- The group publishes handbooks, newsletters and fact sheets on gardening.
- The group runs workshops on tree planting, the garden in winter, signmaking, wildflower planting and fence making.
- The Guerillas recycle Christmas trees, delivering mulch to gardens all over the city.
- The group assists an AIDS resource centre with its garden.
- The group is helping residents of a co-operative apartment for formerly homeless families plant trees and put in tree guards around their new home.

Figure 21.14 Greening the city – how the Green Guerrilla scheme works.

The benefits of community gardens

The Green Guerillas see two benefits:

- **social benefits.** Community gardens are a part of a strong neighbourhood, giving neighbourhoods an improved quality of life. In low-income neighbourhoods, access to open space may be restricted, so that space closer to home has added importance. Streets become safer and more appealing places, resulting in lower crime rates. The gardens have also been a source of food for other city agencies that run soup kitchens and food distribution centres.
- **environmental benefits.** More than one hundred different types of birds and animals use the gardens as their habitat. The gardens create their own micro-climate; the air temperature in community gardens is on average 7°C cooler than on a city street during summer. Finally, trees and plants act as natural filters of the air, absorbing pounds of unwanted pollutants created by urban life.

---

1 Summarise the importance of 'Greening the City' and of community gardens. What other benefits of community gardens might there be?
2 In groups, debate the motion that 'The Community Gardens should be sold off for development'. Consider the attitudes of the following people:
  - the Mayor of New York City
  - a property developer for low-income housing
  - an upper-market real estate developer
  - a Green Guerilla representative
  - a local resident.

Inequality in urban environmental health

One of the most disturbing factors of urban inequality is in the health of the population. Reports show that asthma hospitalisation rates in New York City are far greater in poor, ethnic minority neighbourhoods. The Centres for Disease Control and Prevention report asthma occurs disproportionately in African-American and Hispanic populations, and that it is most severe in urban inner cities.

The disease affects people according to their economic and social status. Although more research is needed to prove a link between asthma and environment, poverty, quality of housing and pollution seem to aggravate it. Poorer patients rely on care in hospital emergency rooms where little time is available to educate sufferers. In wealthier areas, hospitalisation rates are lower; people have resources to deal with it more effectively and seem able to control its severity.

Some research shows that asthma is now the most common chronic disease of childhood, particularly in the Western world. Why should it especially occur in inner cities? Changing lifestyles of inner city youth may have an effect. Issues of crime and safety in inner cities have led to an increase in home-based entertainment for young people, who exercise less and are exposed to sources of allergies in the home, e.g. dust mite, pets and cockroach allergens.

One way of managing the alarming incidence of asthma cases is to reduce exposure to the cockroach allergen. Patient education, roach traps and child-safe insecticides, need to form part of the city's attempts to reduce asthma cases. Despite the existence of asthma therapies, deaths among individuals aged under 25 in the USA increased 118 per cent between 1980 and 1993. These were disproportionately greater in poorer ethnic populations in inner city areas, such as The Bronx and East Harlem.

Urban waste: Is it just New York City's problem?

The inhabitants of New York City create 26 000 tons of waste every day; half is commercial and the remainder residential. Private companies collect commercial waste, leaving residential waste to New York City's Department of Sanitation. The sole destination for 13 000 tons of residential waste is the Fresh Kills Landfill on Staten Island. It is reported that the highest waste mound to be found at Fresh Kills is 180 feet high, twenty inches higher than the Statue of Liberty!

Fresh Kills Landfill

Fresh Kills Landfill, on the western side of Staten Island, (Figure 21.15) is the only remaining in-City landfill site that remains open. Having served New York City for over half a century, it is due to close on 31 December 2001. It receives 12–14 000 tons of solid household waste per day, by barge from the other four boroughs, Staten Island's waste being transported directly by lorry. Some argue that it is unfair that one borough should suffer the burden of the city's waste alone, when it has only 5 per cent of the city's population and the smallest space.

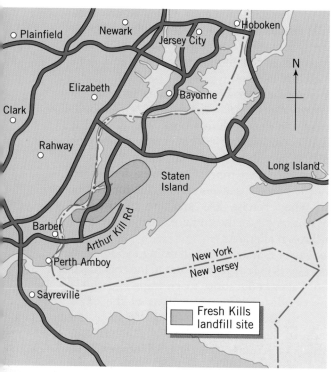

Figure 21.15 Fresh Kills Landfill and Marine Transfer Stations.

The sweeter side of being New York City's dumping ground

There are benefits in the garbage industry; economic development, jobs and increased spending power follow. To operate the landfill requires 500 employees, whose jobs range from crane and tractor operators to chemists and geologists. Although the daily volume of waste handled there is high, there has been a fall from 21 200 tons per day in 1986, to 13 000 tons per day in 1995.

Can the garbage industry be environmentally friendly?

The Fresh Kills Landfill attempts every precaution to ensure that it is non-threatening to both human health and the environment. Seven management techniques to implement the precautions are shown in Figure 21.16.

Management technique	Description
Monitoring system	Allows checks to be made of the build-up of landfill gas, as well as the effects of the operation on ground and surface water supplies. Landfill slope stability is examined regularly.
Litter prevention	The barges are covered with nets during transport, while booms contain litter around the waters of the landfill site. Fences around the perimeter of the operation have the same effect on land.
Landscaping	Ecologists are working to introduce native plant varieties tolerant of the conditions in an attempt to establish the area as a wildlife habitat.
Leachate Treatment Plant	To prevent percolating rainwater from becoming contaminated the plant removes the pollutants prior to the water being returned to the waterways.
Construction Debris Recycling Area	Processes 750 tons per day of debris, e.g. concrete, asphalt and soil. Steel is extracted from the process and sold for recycling. The product is used to construct the roads throughout the landfill site.
Composting facility	Garden waste is collected and turned into compost. It is used in the landscaping process around the site. It is also made available to the public free of charge.
Landfill Gas Recovery System	Consisting of primarily methane and carbon dioxide, landfill gas is collected and customized at a facility onsite. The methane is purified into pipeline quality gas, which is then sold to a local company.

Figure 21.16 Management techniques at Fresh Kills Landfill.

In Chapter 4, the following principles were established about sustainability – Features which act against a sustainable future include anything which:

- is environmentally destructive - e.g. something which pollutes; which removes a resource permanently, or damages it for a period of time.
- uses a resource at a rate which cannot be renewed before current stocks run out
- is short-term, and uses high energy-consuming materials, whose futures are finite. Not only does the raw material itself decline in stock, but may use other exhaustible resources in the process.

On this basis, how well would you evaluate New York City's attempts to manage household waste?

In pairs, suggest ways of making New York City's household waste problem more sustainable.

Managing the city's waste in future

In spite of predictions in 1996 that Fresh Kills Landfill could operate for another two decades, Mayor Giuliani announced its closure for 2001. There was to be a new approach in dealing with the City's waste, and he announced the creation of a Task Force on Fresh Kills closure. Five recommendations covered issues for households across the whole city such as:

- increase recycling
- promote waste reduction
- encourage waste prevention
- refuse exportation of waste elsewhere
- support education about waste and recycling.

Figure 21.17 Barges carrying waste, including paper, to be recycled on Staten Island.

Economic development and waste

A recycling drive throughout the City now means that every household is part of the Curbside Program, where waste for recycling is collected from the roadside. A contract was negotiated with Visy Paper (NY) Inc., who constructed and now operate a $150 million recycled paper mill on Staten Island (Figure 21.17). It is the largest manufacturing project in New York City in 50 years. It employed 1 000 people during construction and now has a labour force of 115.

New York State Governor and New York City's Mayor may seem like environmental heroes in tackling the waste problem. The last landfill is to close, there is a pledge not to build nor renovate any incinerators in the city, and many recommendations have been put forward by the Task Force they commissioned to manage the city's solid waste differently in the future.

Half of New York's waste is household; what is to happen to commercial waste that accounts for the other half? After 2001, will recycling, waste prevention and waste reduction schemes, as good as they may be, really absorb all 13 000 tons of waste a day currently handled by Fresh Kills? One person who is not convinced of this is the State Governor of Pennsylvania, Tom Ridge (Figure 21.18).

Dear Governor Pataki and Mayor Giuliani

Having received a copy of the Report of the Fresh Kills Task Force, I wanted to commend both of you for a thoughtful report on a difficult subject. I am concerned that the solution to New York City's garbage disposal still relies too heavily on the use of disposal facilities in neighbouring states.

New Yorkers have been dumping increasing amounts of their garbage in Pennsylvania. Imports of waste from New York have risen from 680 000 tons in 1989 to over 2.7 million tons in 1995. Since the announcement of the closure of Fresh Kills, a growing number of Pennsylvanians have become alarmed over the prospect of additional garbage from New York.

I am encouraged by the Task Force's recognition that some communities do not want New York City's garbage. I find it troubling that the Task Force did not examine efforts that could be taken by New York to find suitable in-state locations to accept waste now going to Fresh Kills. It would appear that little or no effort was made to see if changes could be made to increase in-state capacity. Years ago, Pennsylvania took a decision to become self-sufficient for disposing of municipal waste, and New York should do the same. Your neighbours would be more willing to work with you if there was more of a commitment on New York's part to site more disposal facilities. New York should consider all the options, including incentives to local government and private industry to locate environmentally sound disposal sites within New York.

Pennsylvania will not allow itself to become the waste disposal option for people in New York City. Your Task Force acknowledged that many New Yorkers have an 'out-of-sight, out-of-mind' mentality, that allowed Fresh Kills to grow far beyond its original planned use at the expense of residents of Staten Island. The best way to combat that is not to redirect garbage from Fresh Kills to other states - where it will be even further 'out of sight' - but for New Yorkers to take responsibility for their own garbage.

Sincerely,

Tom Ridge, Governor

Figure 21.18 Extract of letter from Governor Tom Ridge, 27 January 1997.

1 What is Tom Ridge's letter, (Figure 21.18) really saying? What issues does it present for New York City? What issues does it present for other states?
2 List policies or issues that you have studied in urban areas throughout the world, or in the area where you live, that could be accused of being in the 'out-of-sight, out-of-mind' category.

Garbage importers and exporters

The state of Virginia is number two importer of trash after Pennsylvania, with New York accounting for 60 per cent of Virginia's imports. A survey by Virginia Commonwealth University found that 87 per cent of people were keen to limit garbage imports, with only 9 per cent against. Those against argue that loss of revenue would hurt communities where garbage is big business. With imports totalling 4 million tons per year, there are fears that the state at some time in the future will have exhausted its landfill capacity and be unable to cope with its own trash, never mind imports. However, the waste business is big in Virginia; the state is home to 10 mega-landfills generating valuable dollars and employment.

The largest exporter of waste in the USA is New York State, followed by New Jersey, Illinois and Maryland. Mayor Giuliani argues that New York cannot deal with its own garbage plus that created by 3 million visitors each day. 'Virginians were part of that crowd, Virginia (should) take some of the trash'. His views incensed some of Virginia's officials. The rivalry between New York City and most places in the United States does not help this situation. The issue of waste is therefore not just a problem for the urban area that generates it.

Figure 21.19 Landfill site in Virginia, USA. Much of this waste is from New York City.

Environmental racism

In a study from 1987 entitled *Toxic Wastes and Race*, Weintraub believed that cities were often guilty of environmental racism. Environmental racism is 'the intentional siting of hazardous waste sites, landfills, incinerators, and polluting industries in communities inhabited mainly by African-American, Hispanics, Native Americans, Asians, migrant farm workers, and the working poor.' A summary of his ideas can be found in Figure 21.20. Minorities are particularly vulnerable because they are perceived as weak and passive citizens, who will not fight back against the poisoning of their neighbourhoods, in fear that it may jeopardise jobs and economic survival.

- Race is the most significant variable associated with the location of hazardous waste sites.
- Most commercial hazardous facilities are located in communities with highest concentrations of racial and ethnic minorities.
- The average minority population in communities where there is one commercial hazardous waste facility is twice the average minority percentage in communities without such facilities.
- Although socio-economic status was an important variable in the location of these sites, race was the most significant.

Figure 21.20 Weintraub's ideas on environmental racism.

The case of Williamsburg/Greenpoint, in Brooklyn

Brooklyn's Community District 1 is home to the highest concentration of garbage transfer stations in the city. It is densely populated with limited open space and little or no access to the waterfront along the East River, which offers spectacular views of the Manhattan skyline. Large sections of the waterfront have been zoned for industrial use and residents resort to sneaking through holes in fences to gain access to the waterfront. The issue of waste is so complex and controversial that even within this one Community District in Brooklyn, tensions exist between communities who feel they are being treated unfairly. Williamsburg is the larger of the two neighbourhoods with characteristics that overshadow Greenpoint.

Between 1990 and 1994, Greenpoint was the destination of some 7000 new immigrants, the majority Polish. In spite of a 19 per cent growth in population, Greenpoint is still the smaller of the two neighbourhoods. Its population characteristics are shown in Figure 21.22. The population of Greenpoint is new and transient. For many years, the area has been a starting point for new immigrants, who have aspirations to move to Long Island or Queens. Lack of community stability makes it hard to organise people to take action on environmental issues such as waste; people have more immediate concerns when they first arrive, such as employment and housing. In addition, many recent immigrants are not eligible to vote, a major hurdle when trying to convince local politicians.

Williamsburg on the other hand is a more stable, well-established Hasidic community. Its population characteristics are also shown in Figure 21.22. People from Greenpoint accuse the local Community Board of showering resources on to Williamsburg while neglecting Greenpoint. $10 million has been targeted to transform an old industrial site in Williamsburg into a waterfront park. However, there were also plans to give permission for yet another garbage transfer station on a parcel of land on the same property. This highlights the conflicts of land use that can occur; recreation and garbage disposal are not compatible.

Study the data for Brooklyn's Community District 1 in Figure 21.22. Do the data support the idea that environmental racism is at work in this area?

Backlash to the Environmental Racism lobby

Not everyone supports the idea that environmental racism exists. In an editorial published in a New York tabloid there was an attack on the environmental racism lobby, charging it with threatening job opportunities. It is particularly scathing about the fact that it will, in particular, hurt the minority groups for which it is supposed to be advocating.

Total population	153 951
Polish ancestry	16 759 (10.9%)
Italian ancestry	11 396 (7.8%)
Dominican ancestry	11 207 (7.3%)
Population born overseas	
Of these, % who had migrated to the USA in the past 10 years	41 612 (27%) 21 105 (50.7%)
Language other than English spoken at home	97 637 (69.7%)
Households with income less than US$10 000	16 642 (32%)
People who were determined to be living below poverty level	54 780 (35.7%)
Families with children under 18 living below the poverty level	10 916 (43.5%)
% of households without a car	68.9%

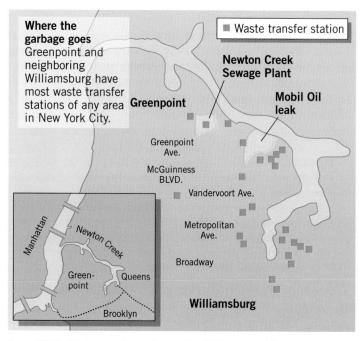

Figure 21.21 Waste transfer stations in the Williamsburg/ Greenpoint area of Brooklyn.

Figure 21.22 Showing Selected demographic data for Community District 1, Brooklyn.

1 Using information supplied in this chapter, investigate the validity of the findings of the study, *Toxic Wastes and Race in the United States* in Figure 21.20, when applied to areas of New York City.
2 List and then explain why certain interest groups would support the editorial that appeared in the *New York Post*, 18 May 1998.
3 Divide your class into two, and prepare statements that argue a) for and b) against the existence of Environmental Racism in New York City.

Fighting Environmental Racism

The Clinton administration, through the Environmental Protection Agency, has come up with a new and interesting way to destroy job opportunities for racial minorities. It has taken up arms in the fight against something called 'environmental racism.'

'Environmental racism refers to any policy, practice or directive that disproportionately affects or disadvantages individuals, groups or communities based on race or color,' according to Dr. Robert D. Ballard, the leading exponent of the idea.

President Clinton is a believer. One of his first orders, in February 1993, directed all agencies to take environmental racism into consideration when making a decision. In response to that directive, the EPA recently issued a guideline telling states and localities that they could face loss of federal funds or even a federal civil-rights lawsuit if they are found to be issuing permits that 'disproportionately' burden minority communities with incinerators, dumps or industries thought to be polluting.

What's 'disproportionate?' Anything greater than the statewide average, according to EPA. Needless to say, any corporation in its right mind would avoid areas with large concentrations of minorities - like New York City - like the proverbial plague. Job opportunities? Forget 'em.

There is no evidence that such a thing as 'environmental racism' even exists. A study by the Dallas-based National Center for Policy Research found that there is simply no link between high concentrations of minorities and concentrations of hazardous waste facilities. In New York City, there is a major example of what appears to be the exact opposite of environmental racism. Staten Islanders, 85 per cent of whom are white, have long demanded that the city close the Fresh Kills landfill. Fresh Kills is the destination for virtually all the trash from heavily minority neighborhoods in Brooklyn, Queens and the Bronx.

The best environmental policy is to create jobs for people who don't have them. Sadly, that is just what the EPA is likely to kill.

Figure 21.23 From the *New York Post*, 18 May 1998.

Summary

You have learned that:

- New York City is, like most urban areas, a city of inequalities.
- Inequalities may be social, economic, or environmental.
- Many environmental indicators can be used to measure inequality, such as health, open space and waste.
- Land pressures in cities often lead to land pressure within city boundaries to find space on which to dump waste, and 'exports' of waste to other areas.
- Areas in which waste is dumped are often those in which lower income groups or ethnic minorities are likely to live; this is referred to as 'environmental racism'.

Ideas for further study

New York City offers a wealth of urban issues to study. Good sources include the NYC LINK, the *New York Times*, and Regional Plan Association (RPA) Web Sites. The RPA offers links to other sites in studying urban issues in New York, New Jersey and Connecticut. Possible topics include:

- Transport issues, e.g. improving access to JFK and La Guardia Airports; Metrolink.
- Health issues such as HIV/AIDS, a growing problem amongst minority communities in the city, particularly among women and within prison populations.
- The geography of crime in New York City, the home of Mayor Giulliani's 'zero tolerance' campaign. It is praised by some as the reason for falling crime rates, while criticised by others for its infringement of civil liberties.

References for further reading

Web sites

Census data for Counties in New York State. (The five counties of New York City are Bronx, Kings (Brooklyn), Queens, New York (Manhattan), Richmond (Staten Island).)
http://govinfo.library.orst.edu/cgi-bin/usaco-state?New+York

Greater New York Chamber of Commerce
http://www.chamber.com/

Metropolitan Transportation Authority
http://www.mta.nyc.ny.us/

New York Times on the Web
http://www.nytimes.com

NYC LINK - New York City Official Web Site
http://www.ci.nyc.ny.us/

Regional Plan Association (RPA)
http://www.rpa.org

USA Census Bureau http://www.census.gov/

Films

West Side Story
Manhattan
Wall Street

22 Green cities of the future – The Sydney Olympics

Figure 22.1 shows celebrations on the Harbour bridge in Sydney on the night of 23 September 1993. Sydney had won the bid to host the Olympics in 2000. For many, there were several reasons to celebrate – the honour of hosting a world event, a celebration of Australia as a country, a recognition that sport is a part of Australian culture, and a boost to economic activity.

This was to be the Olympics of great hope in a new millennium. Sydney promised the world's first 'green' Olympics; the Games were designed to be environmentally friendly and Greenpeace helped to write much of the Olympics bid. After the studies in this book, you might wonder whether cities could ever be 'environmentally friendly'. This chapter explores some of the economic, social and environmental urban issues about the Olympics, and explores how well the environmental promises made in 1993 have been upheld. The chapter will try to assess the economic, environmental and social sustainability of the Olympics and their implication for cities.

The Sydney Olympics as an economic activity

The Olympics are basically an economic activity, because sport is an industry. The Olympics and sport are as much about costs and income as any other economic activity. Costs have to be borne and income sought to balance it. If an activity can be maintained without huge economic cost then it can be said to be economically sustainable. Chapter 4 described a sustainable development as 'one which does not compromise quality of life for future generations by current practice'. Sustainability may be judged on social, economic or environmental criteria. Economic sustainability means whether it brings an improvement in the economy, locally as well as nationally, and that all people will be better off.

The costs of hosting the Olympics

The Olympics are expensive to host. The New South Wales government is responsible for bearing all costs; they estimated in 1995 that the Olympics would cost AU$1.25 billion (1999 conversion = £500 million). The Olympic Park is situated in Homebush Bay, about 14km west from central Sydney (Figure 22.2). The site marks a point close to the geographical centre of the city; the CBD lies to the east, around the Harbour Bridge. Homebush Bay is well served by roads and the home of the former Sydney showground. 15 of the 27 Olympic sports were held in Sydney Olympic Park. Consider the cost of developing:

- a new Olympic Stadium - seating 80 000
  - an Olympic pool at the Sydney Aquatic Centre
  - a range of water facilities for rowing
  - specialist arenas for sports such as tennis, and hockey
  - an Olympic Village, to house 13 000 athletes and team officials.

The land is owned by the New South Wales State government, and all costs had to be borne by them. The Olympics had a huge impact upon space, how it is used, and the movement of people. Figure 22.2 shows a map of the different Olympic facilities that had to be built; these included a new Stadium, Aquatic Centre and space for sports such as archery.

Figure 22.1 Firework celebrations from the Harbour Bridge, Sydney.

Uniquely, all sports were located on one site, together with the Olympic Village, which houses all competitors and their coaches. The Barcelona Olympics in 1992 had brought the city to a traffic standstill because of the number of journeys generated by coaches and athletes travelling to venues. In Sydney, all amenities and accommodation were on site. The only exceptions included Volleyball, held on Bondi Beach in a temporary stadium, and rowing, held at Penrith Lakes (Figure 22.3), some 25km west of Sydney.

Developing the infrastructure of Sydney

The infrastructure of the city has had to be developed to handle both competitors and visitors to the Olympics site. These include construction of a new airport runway, needed to absorb the huge increase in international flights to Australia since 1993. The city has therefore gained an improved infrastructure; its transport links were enhanced considerably.

- In addition to the new airport runway, new terminals were built for both international and domestic passengers.
- A new light rail and road link was built to the airport; unlike New York, in Chapter 21, Sydney now has a full rail link between the airport and the CBD.
- A new rail link was built to the Olympic Park at Homebush Bay. It works on a one-way system of continuous access to trains which draw alongside platforms on both sides of the train, so that passengers were able to get on and off trains quickly.

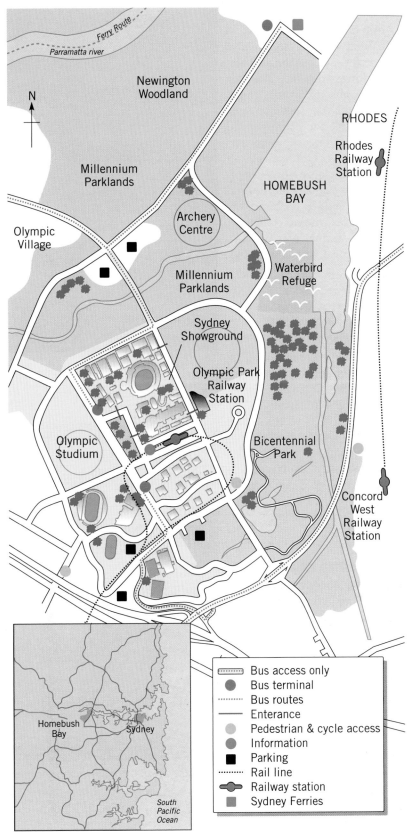

Figure 22.2 The location of Homebush Bay and the Olympic park.

- Road widening and improvements have disrupted the city for nearly seven years, but the result is an infrastructure that is far more able to handle mass travel. Dual carriageways now link the airport with central Sydney 12km away.
- The network of ferries that transport large numbers of tourists and commuters across Sydney harbour has also been enhanced. A new ferry terminal was built at Homebush Bay close to the Olympics site, shown in Figure 22.5.

Economic benefits of hosting the Olympics

The potential economic benefits to Australia and to Sydney of hosting the Olympics were huge. Some were tangible, that is, measured in terms of income and cost. Others were less tangible but nonetheless beneficial, such as improved infrastructure. It is not easy to assess the economic benefit of good airport links in actual dollars.

The boost to employment

The boost to employment is huge. The construction of hotels and of Olympics facilities took place over a 7-year period; 1000 jobs were created in the construction of the multi-use Arena alone. During its operation, this centre alone will require 300 casual and permanent staff. Many other jobs were directly and indirectly created.

Figure 22.3 The rowing lakes at Penrith, 25km west of Sydney.

Figure 22.4 The Olympics Park in Homebush Bay.

The Olympics were also a huge boost to tourism. The number of quality hotels in Sydney increased three times between 1993 and 2000 in expectation of the increase in international visitors. 1.3 million tourists were expected to come to Australia directly as a result of the Olympic Games. Experience shows that many others were likely to join those attending, together with those attracted because of Australia's increased publicity during 2000. In fact, the number of expected overseas tourists for 2000 is 9 million, way above the 1.2 million who visited in 1993, and 2 million greater than the estimates made in 1993.

Figure 22.5 New ferry terminal at Homebush Bay close to the Olympics park.

There were other benefits derived from advertising, the sale of Sydney and Olympics souvenirs, revenue from sales of programmes, logo design (Figure 22.6) and franchising, as well as tourist goods; popular goods include Australian wine, soft toys, photos and artwork, and trinkets.

These provide a boost to spending – hotels, restaurants, souvenirs, shops and stores, and visits to, for instance, Sydney Opera House. In addition, domestic travel receives a boost by plane, car hire and rail. From all businesses involved, the Australian government receives increased revenue from taxes on company profits and taxes from increased employment. A study in 1993 estimated that between 1994 and 2000, the Olympics will add AU$7.3 billion to Australia's GDP, and will create 150 000 full- and part-time jobs.

Revenue from broadcasting

Broadcasting Olympics events to the rest of the world via satellite links with other countries is a major income earner for the host nation. Channel 7 in Australia has many of the rights to broadcast pictures from the Olympics, together with the BBC (UK), CBS (USA) and Sky (News International). Each company purchases the right to transmit pictures, which it is then able to sell on to other channels. Because television profits are high through the transmission of sport, the potential revenue from television sponsorship was enormous. Up to half the direct costs of the Olympics could be recouped through television rights.

Other revenue

Direct income is earned from ticket sales. However, the largest source of revenue after television comes from sponsorship. The advertising industry gains a surge of projects, advertising the Games themselves, and various products associated with sponsorship deals. Many companies stand to gain hugely from the Olympics; sales of Coca-Cola, McDonalds and Pizza Hut rocketed during the Atlanta Olympics in 1996, especially on 'big event' days when people stayed at home and ate take-away food. Together with sports clothing, the revenue from programme sales, logo design and souvenir franchising is huge.

Figure 22.6 The Olympics logo on an Ansett plane – the 'official' national Olympics carrier, for which the airline had to bid and pay!

1 Assess the economic benefits to Australia. How far would you predict that income from the Olympics would cover the costs of the Olympics facilities and improved infrastructure?
2 How economically sustainable do you judge the Olympics to be so far?
3 How would you try to convince someone that the economic benefits to Australia would last longer than the year 2000?

The Green Olympics

Environmental sustainability means that there are no environmental impacts that create short- or long-term environmental damage. For the Olympics Co-ordination Authority (OCA) this was the main challenge. The Sydney bid promised that the 2000 Games would be the 'Green' Games. The process which led to the Olympic proposal began in early 1992 with a design competition for the Olympic Village. The competition was organised by Sydney Olympics 2000 Bid Ltd, with architects, planners, housing associations and the local Auburn Municipal Council. One of five winning entries was a Greenpeace Australia submission. This led to the adoption of Greenpeace Australia Guidelines for the Olympic Village.

Building the Games facilities

One of the early principles in all new Olympic projects was that, wherever possible, previously used industrial and commercial sites – or brown-field land – should be used for development so that undeveloped land remains untouched. Homebush

Bay is currently an industrial area, though many past industries have closed. However, as Figure 22.2 shows, the head of Homebush Bay is mangrove and salt-marsh. Water quality of mangrove, estuarine and salt-marsh environments near Sydney Olympic Park was protected during construction by erosion and run-off controls.

Figure 22.7 The athletic centre, where earth excavated during construction was reused for grassed embankments.

In addition, firm principles were established in building and design. Base materials for building foundations were recycled concrete and masonry which resulted from the demolition of an old abattoir on the site. During the construction of Sydney showground, 95 per cent of waste was recycled. Another example is shown in Figure 22.7. Wherever practicable, non-toxic materials and processes were to be used, such as natural fibre insulation, and non-toxic paints, glues, varnishes, polishes, solvents and cleaning products. Building techniques and interior design were such as to minimise the need for chemical pest control (mosquitoes, spiders, cockroaches) and maximise pest management. The use of CFCs, HFC and HCFC-free refrigerants and processes was banned, and the use of chlorine-based products (organo-chlorines) such as PVC and chlorine-bleached paper was minimised.

Travel to and from the Games

The Olympics park station and ferry terminal (Figure 22.5) are the chief ways in which OCA tried to deter people from travelling by car to the Olympic events. However, there were limits; the Sydney ferries are limited in capacity and frequency. The organisers were therefore dependent upon rail to do the job of mass transit. Road travel was publicly discouraged, though there were 10 000 parking spaces, and satellite car-parking venues were established so people could transfer to trains, buses and ferries for access to Olympic sites. That said, public transport was the only means by which spectators could directly access events at major Olympic sites, and the sale of admission tickets and public transport tickets occurred at the same outlets.

Recycling and waste

Several characteristics of 'green' lifestyles were employed in the design of waste, water and energy usage at the Olympic park. These were focused upon saving water and energy, and reducing waste. Wherever appropriate, solar building design was used, selecting building materials for their thermal performance, insulating and ventilating naturally rather than, for instance, using air-conditioning. A wide range of renewable sources of energy was employed, including high-efficiency lighting systems with maximised use of natural light. The Aquatic Centre, for instance, needed only ten artificial lights to light the centre by day. At night, control systems were used to minimise energy requirements; ventilation flow was switched off when spaces are unoccupied. Energy-efficient appliances and the use of recycled and recyclable building materials complete the menu of energy efficiency. Pool water is ozone-filtered to reduce chlorine, and half of the water on some parts of the site will be storm- or recycled water, used on flushing toilets or in irrigating landscaped areas.

Figure 22.8 The Aquatic Centre pool is ozone-filtered to reduce chlorine usage, and pool backwash is recycled.

How far do the Olympics seem to match the aim of a 'green' Olympics?

Homebush Bay – the big clean-up

So far, the Olympics seem to provide an urban model for the future, in which green, or environmental principles, can be made to work. However, there have been great concerns about escalating costs of development.

Part of the explanation lies in Homebush Bay itself. The planned Olympic Village is on the site of the former Newington Naval Armaments depot. For years it was subjected to widespread illegal dumping of industrial waste. The site of the Stadium was formerly a dump for household garbage. Until June 1995, it was owned by the Federal Department of Defence; it was then sold for A$70 million to the New South Wales government. Although the site for the village itself is relatively clean, part of the land is contaminated with a toxic mixture of chemical compounds, whose source is an old gas plant up-river at Wilson's Park in Silverwater. Homebush Bay is enclosed and flushed only by tidal movement of the Parramatta River, a major inlet from Sydney Harbour. In the 1970s, over 200 000 tonnes of toxic waste were covered up, and in 1994 Wilson's Park had to be fenced off to protect the public from chemical contamination. Some of the waste now seeps into the Parramatta River and collects in Homebush Bay.

Toxic waste has accumulated from this and a former industrial site owned by Union Carbide Australia Ltd. Its American parent company was allegedly responsible for one of the world's worst contamination disasters at Bhopal in India in 1984. Their former site is on Homebush Bay. The site was expanded several times between 1937 and 1974, using landfill from industrial waste. Herbicides and pesticides were manufactured there. Among its by-products are a group of chemicals known as dioxins. Dioxins are known carcinogens – or cancer-causing agents – and are also known to affect sexual development, damage the immune system and to arrest intellectual development. Seepage of groundwater from this site has led to the accumulation of toxic dioxin sediments in the Bay. Consultants from the USA were invited to sample sediments and water, and found that the site was among the worst-polluted 'ever seen'. Dioxins and other compounds are still seeping into the Bay, so that the level of dioxin on the site proposed for housing is 6500 times the safe limit for people.

Two significant points stand out here.

- The site was formerly owned by Union Carbide, allegedly responsible for the world's worst pollution incident to date in 1984 that killed and injured thousands of people at Bhopal in India.
- The proposal to redevelop the Olympics village afterwards as a residential development is significant. Figure 22.9 shows that tonnes of dioxin contamination would be left in the middle of any development. Only in early 1999 did the New South Wales government take control of the site, allowing the Union Carbide dioxin legacy to be cleaned up.

Now that the site is no longer owned by Union Carbide (Australia) Ltd, it is impossible to charge them with pollution. The principle that 'the polluter pays' cannot work under Australian or international law. The site is owned by the New South Wales government, who are likely to pay for any clean-up using tax-payers money.

Clearly, Homebush Bay and the surrounding land could not be cleaned up in time for the Olympic Games in September 2000. There were other landfills on the Olympic site, though far less hazardous and safe for Games visitors. Homebush Bay itself is a toxic mess and a much more pressing issue. Greenpeace considered the hazard posed to the community from the Bay to be totally unacceptable and were pushing the New South Wales government to clean it up.

Toxic problems won't just go away

HOMEBUSH BAY, situated next to the Sydney 2000 Olympic Games site, is one of the most polluted waterways in the world. It and the lands adjacent to it contain over 500 000 tonnes of dioxin-contaminated soils and sediment which pose both a human and environmental threat.

After decades of industrial dumping it is heavily contaminated with dioxins, organo-chlorine pesticides, heavy metals and other toxic chemicals which are finding their way into the wider environment. This is clearly demonstrated by the high levels of dioxins, DDT and other chemicals found in fish in the Bay. The toxic chemicals in Homebush Bay are unconfined, dangerous and cast a dark shadow on Sydney's reputation as host of the first ever 'Green' Olympic Games.

The multinational chemical companies Union Carbide and ICI (now called Orica) were responsible for this toxic mess. From the 1950s to the 1970s Union Carbide manufactured organo-chlorine pesticides such as DDT and the ingredients of Agent Orange at Homebush and disposed of the wastes in the Bay and surrounding lands.

The extensive contamination in the area was ignored by the NSW government and the Olympic authorities for years. It was only after Greenpeace safely secured an abandoned stockpile of 69 drums of dioxin waste adjacent to the former Union Carbide factory site in 1997, that the New South Wales government committed $21 million to clean-up the Bay in time for the 2000 Olympic Games. The commitment was to remove all the contamination and use innovative Australian non-incineration processes to treat the dioxins - a commitment that was widely praised by community groups, including Greenpeace.

However, the clean-up plans ran into a problem. The New South Wales government was responsible for the sediments, Bankers Trust owned the land adjacent to the former Union Carbide factory site, and a holding company - Lednez Australia Ltd - owned the factory site. Union Carbide left Australia in the early 1990s leaving the site in the hands of Lednez and they weren't willing participants in the clean up.

Figure 22.9 Homebush Bay's industrial past, from the *Canberra Times*, 14 October 1999.

In 1997, the Olympic Co-ordination Authority (OCA) uncovered a stockpile of 400 tonnes of dioxin-contaminated waste on the Olympic site. The usual method of dealing with this type of waste is either to incinerate, bury or place it in secure storage. Incineration itself generates dioxin waste and simply redistributes the pollution. Land-fill was originally used but has only resulted in the present problem. Long-term secure storage only leaves the problem for the future. In Australia incineration of hazardous waste is not an option, as it was banned in 1992 by the Australian government. To the OCA's credit, it did not take the cheaper option of storage.

Australia has now developed the technology to destroy hazardous wastes. A non-incineration treatment process called ADOX, developed by Australian Defence Industries (ADI) reduces the 400 tonnes of waste to common table salt, carbon and oil. The treatment of the waste will mean that the toxic threat is eliminated forever.

The Olympics – social impacts

The term 'socially sustainable' is used to determine whether a development brings social benefits for all people, rather than just a few. Already, it is easy to see how employment was generated, and how the 'clean-up' of Homebush Bay has taken effect, but it still needs to be determined how far people were affected.

Strathfield and Homebush, the two local suburbs that adjoin the Olympic park, are low-middle income suburbs of Sydney. What impact the Olympics would have upon them remains in question. Greenpeace commissioned a report to investigate the social sustainability of the Olympics. They found that in previous Olympics, especially those in Atlanta in 1996, the Olympics resulted in a rapid increase in property prices. Investors bought property, expecting to rent it out. The result was an upsurge in both house prices and rents. This had greatest effects on the poorest and homeless. Those who already owned their homes benefited from increased property values. Figure 22.10 shows how fast prices rose in Sydney during the three months after the bid was won; since then, prices have risen by about 200 per cent!

For those who rented property – the majority in this suburb - the situation is made worse by house price increases that result from the sale of the Olympic Village. It looks as though the original plan to turn the Olympic Village into affordable rented property will fail, as financial pressure grows to maximise income from the property by selling it. Rent increases follow on from the increased value of property. Conversion of boarding houses into tourist accommodation has taken place in many areas, especially close to central Sydney, which has increased the cost of staying as hotel owners seek to recoup costs. It has accelerated the process of 'gentrification' in areas where social displacement of older and lower-income tenants has occurred. There were cases in Atlanta of public and private harassment towards homeless persons, as they became crowded out of afford-able housing, due to increased pressure on prices of hotels and apartments.

Sydney's housing market 1993. Graph A shows the average property price of properties sold in Sydney, 1989–93. Graph B shows the percentage of properties available for sale which sold during each year 1989–93. 1993 was a significant year – before the result of the Olympic bid was known, both the sale price and the proportion of properties selling was levelling off or even falling.

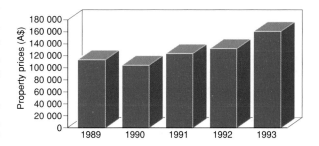

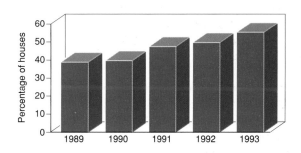

Figure 22.10 Property price rises in Sydney within three months of winning the bid for the Olympics.

1 How far do the price rises shown in Figure 22.10 affect your judgement of whether or not the Olympics development is 'sustainable'?
2 In November 1999, Green Games Watch awarded the OCA an 'A' grade for cleaning up the Homebush Bay site. Form groups of three to four.
   • How far would you agree with this grade?
   • How would you rate the other aspects of the development – its economic sustainability, social sustainability, the design and attempts to incorporate 'green' principles?
3 What do the Sydney Olympics tell us about:
   • the possibilities of generating environmentally-friendly urban development?
   • the costs of generating environmentally-friendly urban development?
   • whether these costs are worth bearing?
   • whether the world's global urban areas have to live with toxic pollution?

How sustainable are current trends in urbanisation?

Having studied the previous chapters, you should now understand some of the challenges that face planners and decision makers in urban areas. You will have some ideas about whether – or how – these issues can be resolved. Both private and public sectors have a role to play as decision-makers and problem-solvers within urban areas. Increasingly, global decisions are being made that affect the environments in which we live. Agenda 21 is one example of such an international agreement that aims to improve the quality of life for people around the world, without harming the environment for future generations. This final section will enable you to consider issues about sustainable urban areas locally, and then globally. Three different views are given- in Figures 22.11 and 22.12 – to identify different people's views of sustainability.

Thinking local, thinking global

Interdependence and globalisation have become buzzwords of a generation. Interdependence is concerned with the links between people around the world, e.g. through trading patterns. Globalisation explains the way industry, communications, travel, etc. are organised on a world scale. It has been said that the world is shrinking because we are now inter-connected by email, the internet and satellites. We are increasingly aware of the impact our actions can have on the rest of the world, e.g. coral reefs being damaged by tourists, oil spills when transporting energy, nuclear accidents.

The consequences of human actions are not limited to their immediate environments; air pollution in the UK has been shown to lead to acid rain in Scandinavia. Clothes bought in Manchester may originate in sweat-shops in Bangkok. The Rio Summit, the Kyoto Summit and others, demonstrate a need for world leaders to regularly meet to help manage the planet.

Following consultation with local residents, Oldham Metropolitan Borough Council in Greater Manchester decided that the following issues should be tackled to improve the quality of life. These were:

Biodiversity, Pollution, Wildlife habitats, Recreation, Access to open space, Quality of the human environment, Transport, Energy, Waste Disposal, Recycling, Access to Information, Consumer power, Equality, Food, Employment, Planning for the future, Education and awareness, Democracy, Poverty, Health, Natural Resources.

These complied with Local Agenda 21, an initiative which came from the Rio Summit in 1992, a global conference that aimed to establish principles for sustainable living in local areas. It became a world-wide initiative to address environmental concerns.
1 Research and outline the main aims of Agenda 21.
2 Locally, Local Agenda 21 is concerned with improving quality of life for local residents, as shown in Figure 22.11. In pairs, use the criteria for Oldham to decide whether you feel that the local residents of Homebush in Sydney had their quality of life improved by the 2000 Olympics.
3 For a local urban area of your choice, working as a group, write a development plan towards a sustainable future for the town or city:
   • Consider five features you would like to improve, e.g. transport, housing, employment, open space.
   • List short-term strategies which could be achieved within six months to a year to deal with **three** of these problems.
   • List long-term strategies which could be achieved within five to ten years to deal with **one** of these problems.
   • Decide who will be responsible for managing and funding – paying for! – the change.
   • Outline how you will evaluate the success of your plan and the improvements you have introduced.

The World Bank developed a plan to reduce the gap between rich and poor, the developed and the developing, this was known as the Strategy 21 goals. The World Bank was concerned about the uneven progress of development. The flows of trade and capital within the global economy bring benefits to millions of people, but poverty and suffering still persist for many others. In a truly integrated world, issues like disease, environmental degradation, civil strife, and criminal activity need global solutions.

(adapted from the World Bank web site)

Figure 22.11 World Bank goals for the twenty-first century.

In May 1996 the Development Assistance Committee of the OECD published Shaping the twenty-first century, a policy paper. The paper called for a global partnership to work towards a new development strategy which focused on the following key goals:

- Economic well being: reducing the proportion of people in extreme poverty by half before 2015.
- Social improvements: achieve universal primary education in all countries by 2015. Progress towards gender equality and the empowerment of women by eliminating gender disparities in primary and secondary education by 2005. Reducing by two-thirds the mortality rates for infants and children under 5 and by three-fourths the mortality rates for mothers by 2015. Providing access to reproductive health services for all individuals of appropriate age no later than 2015.
- Environmental sustainability and regeneration: implementing national strategies for sustainable development by 2005 to ensure that by 2015 to reverse the loss of environmental resources globally and nationally.

These goals are expressed in global terms but must be pursued by all countries. Achieving them will also require building capacity for effective, democratic, and accountable governance, protection of human rights, and respect for the rule of law.

Figure 22.12 Shaping the twenty-first century.

You are to prepare for a World Conference on sustainability. Your role, as conference organiser, is to decide on a suitable location for the conference.
- Identify possible cities to host the conference, including those studied in the previous chapters.
- Evaluate each one in terms of its success as a sustainable or environmentally-friendly city. Decide on your criteria for this, for example, how much do you wish to accept the Sydney criteria for the 'green' Olympics?
- Justify your choice of city by explaining the extent to which it is or is working towards sustainability.

Summary
You have learnt that
a) Urban developments are costly in terms of the development of infrastructure needed to support them;
b) Urban development may be judged according to whether or not they are sustainable. 'Sustainability' may be judged on social, economic or environmental criteria;
c) Often, desirable changes are limited by legacies of the past, such as pollution or waste disposal;
d) It is possible to create urban developments that are environmentally friendly in terms of building design, materials, energy and water consumption, and the extent to which they recycle resources
e) At a larger scale, current trends in urbanisation throw doubt about the sustainability of urban areas and the impact they have socially and environmentally.

Ideas for further study
1 Compare the 'green' developments in Sydney with changes that are taking place in urban areas close to you. Which features are environmentally friendly? Which are not?

2 Find out how far your local council has gone in trying to implement 'Agenda 21' principles from the Rio summit. How far has it gone in trying to recycle waste, to use brownfield instead of greenfield sites for urban development, and to promote biodiversity?

References and further reading

BBC 'Australia 2000' series, programme 6 on urban change in Sydney.

Earthscan (1991) 'Caring for the earth; a strategy for sustainable living'

Earthscan (2000) 'State of the World 2000'

Changing urban environments: summary

Enquiry questions	Key ideas and concepts	Chapters and case studies in this book
Introducing urban environments 2.6a What are the characteristics of urban environments? 2.6b How and why do contrasting urban environments develop? 2.6c How do the nature and importance of urban environments vary spatially?	• Definition of urban area, to include housing and population density, nature of employment, high level functions. • Concept of urbanisation – relationship to urban growth. Importance of demographic, migratory and economic processes. Cycle of urbanisation, suburbanisation, counter-urbanisation and re-urbanisation. • There are contrasts in levels of urbanisation at global and continental scales. • Millionaire cities and the emergence of mega-cities of global importance (world cities)	• *Introducing urban environments* • *Introducing rural environments* • *Chapter 17 - Urbanisation in MEDCs – Manchester* • *Chapter 18 - Urbanisation in LEDCs – Mumbai*
Changing urban environments 2.7a What functions occur in urban areas? 2.7b What impact has this had on land use zoning in urban areas? 2.7c What changes are taking place in the city centre and at the rural-urban fringe, and what conflicts have arisen? 2.7d How and why does quality of life vary within urban environments?	• Spatial development of land use within urban areas. Factors influencing this, such as cost and access. • Land use zoning. Nature of CBD, inner and outer urban areas in MEDCs and LEDCs. Reasons for contrasts. Relation to concept of urban models • Centrifugal forces v centripetal forces in city centres. Issues of people, employment and services. • Competition and conflicts at the rural-urban fringe. • Assessment of quality of life. Issues of inequality and social justice. Location and problems of 'poor' areas in both MEDCs and LEDCs - zones of deprivation, poverty and social exclusion.	• *Chapter 19 Urban functions in Manchester and in Mumbai*
The challenge of managing urban environments 2.8a What are the challenges of managing urban environments? 2.8b How do planners and decision-makers attempt to resolve these challenges? 2.8c What is the balance between private and public provision? 2.8d How successful are the strategies of planners and decision-makers in improving the quality of life for all urban dwellers?	***Choose two*** *of the following to illustrate the enquiry questions in 2.8, using examples from countries at **differing states of development*** – Managing housing stock - contrasting LEDC and MEDC issues. Getting the right mix of types and price in MEDCs. Brownfield-greenfield issues. Renewal and gentrification in centre. Issues of homelessness and substandard housing for the urban poor in inner areas of MEDCs and in many areas of LEDCs.	• *Chapter 19 Housing issues in Manchester (gentrification and urban poverty) and Mumbai*
	– Managing movement in cities - problems of providing public transport. Coping with the motor car and management of the daily commute. The unsustainability of the motor car. Creation of cycle ways, walking zones. Redesign of land use to cut down journey to work.	• Chapter 20 Transport issues and proposed solutions in Bangkok and Melbourne.
	– Managing environmental problems in the city, issues of waste, dereliction, air pollution, noise, water supply, issues of environmental health of the inhabitants. Greening the city – role of urban ecosystems.	• Chapter 21 Urban environmental problems and issues in New York City, e.g. open space, waste disposal, health • Chapter 20 - Traffic pollution and safety in Bangkok.
Urban Futures 2.9 What will the future hold for urban areas?	• Regenerating and re-imaging the city in MEDCs, for example flagship projects, heritage tourism, cultural quarters. Attracting new employment, e.g. hi-tech, service, R&D. Assessment of sustainability of projects (economic viability). • Planning urban areas for the future - the creation of the ideal environment	• Chapter 19 – Urban regeneration in Manchester • Chapter 22 - Environmental cities e.g. Sydney's Green Olympics, Agenda 21 in Manchester
Can urban environments be made sustainable?	• The creation of manageable urban areas for the 21st century, including sustainable cities. Features of sustainable cities in terms of energy use, waste disposal, pollution, land use, transport. • Managing the Brown Agenda in LEDC cities.	• Chapter 19 - Urban developments in Mumbai - the Brown Agenda • Chapter 22 - Environmental cities e.g. Sydney's Green Olympics, Agenda 21 in Manchester

Index

320